Techniques in Life Science and Biomedicine for the Non-Expert

Series Editor
Alexander E. Kalyuzhny, University of Minnesota
Minneapolis, MN, USA

The goal of this series is to provide concise but thorough introductory guides to various scientific techniques, aimed at both the non-expert researcher and novice scientist. Each book will highlight the advantages and limitations of the technique being covered, identify the experiments to which the technique is best suited, and include numerous figures to help better illustrate and explain the technique to the reader. Currently, there is an abundance of books and journals offering various scientific techniques to experts, but these resources, written in technical scientific jargon, can be difficult for the non-expert, whether an experienced scientist from a different discipline or a new researcher, to understand and follow. These techniques, however, may in fact be quite useful to the non-expert due to the interdisciplinary nature of numerous disciplines, and the lack of sufficient comprehensible guides to such techniques can and does slow down research and lead to employing inadequate techniques, resulting in inaccurate data. This series sets out to fill the gap in this much needed scientific resource.

Shalini Mani • Manisha Singh • Anil Kumar

Animal Cell Culture: Principles and Practice

Shalini Mani
Centre for Emerging Diseases, Department
of Biotechnology
Jaypee Institute of Information Technology
Noida, Uttar Pradesh, India

Manisha Singh
Centre for Emerging Diseases, Department
of Biotechnology
Jaypee Institute of Information Technology
Noida, Uttar Pradesh, India

Anil Kumar
Director, Education
Rani Lakshmi Bai Central Agricultural
University
Jhansi, Uttar Pradesh, India

ISSN 2367-1114   ISSN 2367-1122   (electronic)
Techniques in Life Science and Biomedicine for the Non-Expert
ISBN 978-3-031-19487-0   ISBN 978-3-031-19485-6   (eBook)
https://doi.org/10.1007/978-3-031-19485-6

This Springer imprint is published by the registered company Springer Nature Switzerland AG
The registered company address is: Gewerbestrasse 11, 6330 Cham, Switzerland

Preface

The cell culture techniques serve as an imperative and very useful model system in current biological research and are an important tool in understanding varied cellular physiological processes along with screening of toxic/therapeutic compounds for various disorders. This research domain has successfully progressed significantly over the years and proved to be used in a multitude of areas like – drug/vaccine production, functional enzyme, growth factors, toxicity analysis and many more. Equally, the research community has witnessed a lot of advancements and up-gradation through the inclusion of new assays, concepts, qualitative and quantitative techniques, etc., which has added enormous value in this ever-growing area of life sciences. They have proved to be a beneficial tool for healthcare scientists and provide a model system for physicochemical evaluation of selected cells to be studied. Their analysis and behaviour help in determining the exact pathway along with their metabolic rate, and the interaction with drugs proves to be a useful tool for drug screening programmes, clinical trials, and pharmaceutical designing. The *in vitro* cell culture systems help in exploring the principles of Ayur-biology and nutritional biology through development biomarker(s)–based bioassays. The knowledge derived from such experimentation helps not only for screening of herbal medicines, drugs, nutraceuticals, pharmaceuticals and cosmeceuticals but also to understand the mode of action and other information needed to evaluate their efficacy.

Early cell culture research techniques were mainly focused on discovering methods for culturing a diverse array of cells from many species, but today, cell culture methods have added new dimensions to the science research and its authenticity. And with an increasing count of assays, analysis techniques and advanced instrumentation added every now and then to gain in-depth basic knowledge of this area are an utmost requirement now, and we believe that there are many who can benefit from the combined knowledge and experience of basics of cell culture and its application. With this aim, this book provides the in-depth crisp and relevant knowledge on the topic right from distinguishing the cell lines, laboratory design and safety to aspects of cryopreservation, quality control and cell line authentication, to students and researchers new to cell culture handling. The issue of cell line misidentification and cross contamination has been recognised as a significant problem in recent years, and solutions are addressed here. Additionally, a series of detailed protocols is also provided that is routinely used in the cell culture laboratories. Therefore, this book is intended as an in-depth text of cell culture with end-to-end answers to all the queries for beginners as well as the broader research community. This comprehensive research-based guide provides the kind of intensive description and implementation advice that is crucial for getting optimal results in the laboratory.

This book also provides the updated and latest cell culture techniques with more accepted methods of experiments. The well-researched and detailed protocols, mentioned in the different chapters of the book, can be readily used in laboratory practices by the researchers and students. The volume begins with a detailed and informative introduction chapter that includes the basic background of cell culture techniques and further explains standard layout of a cell culture lab, good lab practices in cell culture lab, maintaining sterility in cell culture lab, media preparation, and selection of suitable media. After explaining the basic techniques used in cell culture such as cell counting, toxicity estimation, passaging, maintenance and freezing of cell lines, the book is dedicated to discussing few advanced techniques in cell culture lab such as stem cell culture and 3D cell culture too. The importance and application of these techniques in varied research fields, and a plethora of good practical advice along with the trouble shooting for the expected technical glitches while performing the experimentation are also discussed. Later in the book, we have discussed the application of the various assays or analytical tools discussed in the previous chapters to the readers with the wholesome view and concept development of the related topic. As animal cell culture is mostly represented as experimental model systems in basic and medical sciences, so we have discussed its application in some selected areas of study such as pathological studies, biotechnology, cellular development and differentiation, biomarker identification, and genetic manipulation. We have also discussed the ethical perspectives and the guidelines that should be followed while practising the cell culture techniques in these areas of research. Thus, the book summarises the broad information on mainly basic, fundamental and specialised knowledge which includes historical perspective, utilisation, R&D efforts, present status and the importance being given by policymakers for use of animal cell culture techniques, tools for development of superior biologically important products like vaccine, interferon, monoclonal antibodies even now availability of stem cells for not only diagnostics but also for prophylactic, preventive and therapeutic purposes for healthy society. The book is written keeping in view syllabi of different research institutions, researchers and students as well requirement of the industry. It will serve as instructional material for researchers in biomedical science, veterinary science, process engineering, biochemistry and biotechnology, and reference material for those working in industry and R&D labs.

Noida, Uttar Pradesh, India   Shalini Mani
Noida, Uttar Pradesh, India   Manisha Singh
Jhansi, Uttar Pradesh, India   Anil Kumar

Contents

1   **Overview to Animal Cell Culture** 1
    Shalini Mani
    1.1   Introduction 1
    1.2   Types of Cell Cultures 3
        1.2.1   Primary Cell Culture 4
        1.2.2   Secondary Cell Culture 5
        1.2.3   Cell Line and Cell Strain 5
    1.3   Ethical Considerations in Animal Tissue Culture 6
    1.4   Common Nomenclatures in Animal Cell Culture 7
    References... 9

2   **Cell Culture Laboratory**..................................... 11
    Anvi Jain, Aaru Gulati, Khushi R. Mittal, and Shalini Mani
    2.1   Laboratory Design, Planning and Layout 11
        2.1.1   Ventilation 12
        2.1.2   Requirements.................................... 12
        2.1.3   Services 12
        2.1.4   Layout 13
    2.2   Equipment and Materials for Cell Culture Laboratory 18
        2.2.1   Aseptic Area.................................... 19
        2.2.2   Incubation and Culture........................... 32
        2.2.3   Preparation and Sterilisation 36

3   **Good Laboratory Practices in Animal Cell Culture Laboratory**
    and Biosafety Measures 53
    Kumari Yukta, Mansi Agarwal, Mekhla Pandey, Khushi Mittal,
    Vidushi Srivastava, and Shalini Mani
    3.1   Introduction 53
    3.2   Biosafety Measures in Animal Cell Culture Laboratory 55
        3.2.1   Biosafety Level 1 (BSL-1)........................ 56
        3.2.2   GLPs for Biosafety Level 1 56
        3.2.3   Biosafety Level 2 (BSL-2)........................ 57

    3.2.4   GLPs for Biosafety Level 2   57
    3.2.5   Biosafety Level 3   57
    3.2.6   GLPs for Biosafety Level 3   58
    3.2.7   Biosafety Level 4 (BSL-4).   58
    3.2.8   GLPs for Biosafety Level 4   58
    3.2.9   Safe Laboratory Practices   59
    3.2.10 Common Good Lab Practices (GLPs) in Cell Culture.   59
    3.2.11 Waste Segregation   60
    3.2.12 Types of Waste.   61
    3.2.13 Waste Management   62
  References.   64

4   **Managing Sterility in Animal Cell Culture Laboratory**   65
  Shalini Mani
  4.1   Introduction   65
  4.2   Elements of Aseptic Techniques   66
      4.2.1   Sterile Work Area.   66
      4.2.2   Good Personal Hygiene   66
      4.2.3   Media and Sterile Reagents   67
      4.2.4   Sterile Handling.   67
      4.2.5   Use of Safety Cabinets.   68
      4.2.6   Culture Sterility   68
  4.3   Biological Contamination   70
      4.3.1   Bacteria   70
      4.3.2   Yeasts.   71
      4.3.3   Moulds   71
      4.3.4   Viruses   72
      4.3.5   Mycoplasma.   72
      4.3.6   Cross-contamination   73
  4.4   Aseptic Technique Checklist   73
  References.   74

5   **Media and Buffer Preparation for Cell Culture**   77
  Sakshi Tyagi and Shalini Mani
  5.1   Introduction   77
  5.2   Media.   77
      5.2.1   Physiochemical Properties of Media   78
      5.2.2   Components of Media   79
      5.2.3   Types of Culture Medium   82
  5.3   Methodology to Prepare Cell Culture Media.   84
      5.3.1   Preparation of Incomplete DMEM Media   84
  References.   87

6   **Properties of Cultured Cells and Selection of Culture Media**   89
  Shalini Mani
  6.1   Introduction   89
  6.2   Systems for Growing Cell Culture.   90

6.3   Morphological Differences in Mammalian Cell 90
6.4   Maintaining Cultured Cells 92
6.5   When to Subculture 92
     6.5.1   Cell Density 93
     6.5.2   Exhaustion of Medium. 93
     6.5.3   Subculture Schedule. 93
6.6   Media Recommendations. 94
     6.6.1   Culture Media 94
     6.6.2   Serum. 94
     6.6.3   pH. 95
     6.6.4 CO_2. 96
     6.6.5   Temperature 96
     6.6.6   Different Culture Mediums for Different Cells 96
References. 97

7   **Selection and Maintenance of Cultured Cells** 99
Divya Jindal and Manisha Singh
7.1   Introduction 99
7.2   Origin and Characterisation of Cells 100
     7.2.1   Cell Differentiation 100
7.3   Selection of Appropriate Cell Line 101
     7.3.1   Techniques for Detachment of Cells 104
     7.3.2   Source of Tissue. 105
     7.3.3   Subculture 106
     7.3.4   Growth Conditions and Characteristics 107
     7.3.5   Other Criteria. 108
7.4   Maintenance of Cell Line. 108
7.5   Media 109
7.6   Conclusion 110
References. 110

8   **Inoculation and Passaging of Adherent and Suspension Cells** 115
Pranav Pancham, Divya Jindal, and Manisha Singh
8.1   Introduction 115
8.2   Subculturing in Monolayer Cultures 116
8.3   Troubleshooting Monolayer Cell Subculturing 117
     8.3.1   Dissociation of Cells 117
     8.3.2   Cell Clump Formation Post-dissociation. 118
     8.3.3   Cells Reattachment Issues 118
     8.3.4   Reduced Viability. 118
8.4   Subculturing in Suspension Cell Cultures 119
8.5   Media Consumption by Cell Lines 119
8.6   Types of Cell Detachment Techniques. 120
     8.6.1   Degree of Cell Adhesion 121
     8.6.2   Use of Detached Cells 121
     8.6.3   Process Compatibility 121

8.6.4   Culture Support 121
8.6.5   Process Scale 121
8.6.6   Regulatory Constraints. 122
8.6.7   Temporal Resolution 122
8.6.8   Compatibility with Sterilisation Methods 122
8.6.9   Production Costs 122
8.7   Different Cell Detachment Methods 122
8.8   Role of Cellular Density in the Subculturing Procedures 123
8.9   Determining the Time Interval and Schedule in Subculturing
        Process 124
8.10 Cellular Synchronisation 124
8.10.1 Centrifugal Elutriation 124
8.10.2 Chemical Blockade 125
8.10.3 Arrest in the City of M. 125
8.10.4 Nocodazole 125
8.10.5 Inhibition of CDK1 125
8.10.6 Colchicine 125
8.10.7 Arrest in S (Arrest in G1/S) 126
8.10.8 Double Thymidine Block. 126
8.10.9 Other Methods of Cell Synchronisation 126
8.11 Cellular Interaction and Signalling Mechanism 127
8.12 Attaining Standard Growth Cycles and Split Ratios 127
8.13 Using Antibiotics for Contamination Control 128
8.14 Conclusion 129
References. 129

9   Counting of Cells. 131
      Divya Jindal and Manisha Singh
      9.1   Introduction 131
      9.2   Cell Growth, Cell Propagation, and Cell Viability. 132
      9.3   Earlier Methods of Cell Counting 133
               9.3.1   Recent Techniques in Cell Counting 136
      9.4   Direct Cell Counting 136
      9.5   Indirect Cell Counting 139
      9.6   Errors in Cell Counting 141
      9.7   Conclusion 142
      References. 143

10   Cryopreservation of Cell Lines 147
      Vinayak Agarwal and Manisha Singh
      10.1 Introduction 147
      10.2 Fundamentals of Cryopreservation 148
      10.3 Cryoprotective Agents 149
      10.4 Polymers 150
      10.5 Glycerol 150
      10.6 Dimethyl Sulphoxide (DMSO) 151

10.7   Proteins 151
10.8   Primary Mechanism of Action of Cryoprotectants 152
10.9   The Cell Banker Series 153
10.10   Cryoinjury 154
10.11   Varied Application of Cryopreservation Technique
        on Special Cell Cultures/Cells........................... 155
10.12   Female Germ Cells (Oocytes and Embryos) 155
10.13   Male Germ Cells (Sperm, Testicular Tissue and Semen) 156
10.14   Hepatocytes 156
10.15   Stem Cells 156
10.16   Other Frequently Cryopreserved Cells 157
10.17   Optimum Cooling Rate 157
10.18   Influence of Warming Rate 158
10.19   Limitation of Cryopreservation 158
10.20   Vitrification 159
10.21   Conclusion... 159
References... 160

11   Resuscitation of Frozen Cell Lines 163
     Vinayak Agarwal and Manisha Singh
     11.1   Introduction 163
     11.2   Freezing Cell Lines (Cryopreservation)..................... 164
     11.3   Thawing Cell Lines 164
     11.4   Cell Thawing Techniques................................ 165
             11.4.1   Hand-Warming............................... 165
             11.4.2   Water Bath.................................. 166
             11.4.3   Specialised Devices 166
             11.4.4   Bead Bath................................... 167
     11.5   Material and Reagents Employed for Resuscitation 167
             11.5.1   Materials 168
             11.5.2   Reagents and Solvents......................... 168
     11.6   Methodology for Cell Revival 168
             11.6.1   Equipment Setup 168
             11.6.2   Thawing Protocol............................. 168
     11.7   Critical Parameters for Resuscitation...................... 169
     11.8   Conclusion... 170
     References... 171

12   Isolation and Culturing of Cells from Different Tissues 173
     Sakshi Tyagi and Shalini Mani
     12.1   Introduction 173
     12.2   General Methods to Isolate Cells from Tissues 174
             12.2.1   Mechanical Dissociation 175
             12.2.2   Enzymatic Dissociation........................ 176
             12.2.3   Primary Explant Technique 178
             12.2.4   Chemical Dissociation......................... 179

12.3   Segregation of Viable and Non-viable Cells   180
12.4   Summary   180
References.   180

13   Co-culture Techniques   183
Vinayak Agarwal, Manisha Singh, and Vidushi Srivastava
13.1   Introduction   183
13.2   Co-cultures.   184
13.3   Extracellular Microenvironment   185
13.4   Variables in Co-cultures.   185
        13.4.1   Cell Communication During Co-cultures   185
        13.4.2   Cell Culture Medium.   186
        13.4.3   Cell Population in Co-cultures.   187
        13.4.4   Volume of Co-cultures.   188
13.5   Model Systems in Co-cultures.   188
        13.5.1   Computational Models.   188
        13.5.2   In Vitro Tissue Models.   189
13.6   Potential Challenges with Co-cultures.   191
13.7   Future Perspective   192
13.8   Conclusion   192
References.   192

14   3D Cell Culture Techniques   197
Madhu Rani, Annu Devi, Shashi Prakash Singh, Rashmi Kumari,
and Anil Kumar
14.1   Introduction   197
14.2   3D Cell Culture Versus 2D Cell Culture Systems   199
14.3   Overview of 3D Cell Culture Techniques   200
14.4   Methods of 3D Cell Culture Techniques   200
        14.4.1   Scaffold-Based Techniques   201
        14.4.2   Scaffold-Free Techniques   203
14.5   Stem Cells in 3D Spheroids and Organoids.   205
14.6   Applications of 3D Cell Culture Techniques   207
14.7   Challenges in 3D Culture Techniques   208
14.8   Conclusion   209
14.9   Future Prospects.   210
References.   210

15   Stem Cell Culture Techniques   213
Rashmi Kumari, Madhu Rani, Amrita Nigam, and Anil Kumar
15.1   Introduction   213
15.2   Historical Background.   214
15.3   Properties of Stem Cells.   216
15.4   Sources of Stem Cells   216
        15.4.1   Human Umbilical Cord   217
        15.4.2   Bone Marrow.   217

15.4.3 Adipose Tissue. 217
15.4.4 Amniotic Fluid. 217
15.4.5 Dental Pulp 218
15.5 Classification of Stem Cell Types Based on Origin. 218
15.5.1 Embryonic Stem Cells. 218
15.5.2 Adult Stem Cells 219
15.5.3 Pluripotent Stem Cells. 219
15.6 Classification of Stem Cells Based on Potency 219
15.6.1 Totipotent. 219
15.6.2 Pluripotent 219
15.6.3 Multipotent 220
15.6.4 Oligopotent 220
15.6.5 Unipotent. 220
15.7 Stem Cell Culture Techniques 220
15.7.1 Stem Cell Lines. 221
15.7.2 Standardisations for Specific Uses. 221
15.7.3 Maintenance of Aseptic Condition 221
15.7.4 Safety Features for Stem Cell Culturing 222
15.7.5 Technique of Embryonic Stem Cell Culture
(Fig. 15.3) 223
15.7.6 Thawing of Frozen Stocks. 224
15.7.7 Gelatin Coating of Culture Plates 225
15.7.8 Media Change 225
15.7.9 Passaging. 226
15.7.10 Cryopreservation 227
15.7.11 Precautions and Troubleshooting Tips (Table 15.3) 228
15.8 Limtations and Challenges in Stem Cell Research 228
15.9 Applications of Stem Cells 231
15.10 Conclusion. 233
15.11 Future Prospective 233
References. 233

16 **Identification and Removal of Biological Contamination
in the Media and Cell Suspensions** 235
Vaishnavi Shishodia, Divya Jindal, Sarthak Sinha,
and Manisha Singh
16.1 Introduction 235
16.2 Cell Culture Contaminants. 236
16.2.1 Biological Contaminants 236
16.2.2 Chemical Contaminants. 238
16.2.3 Inorganic Ions 239
16.2.4 Organic Compounds 239
16.3 Role of Antibiotics and Antimycotics 240
16.4 Sources of Contamination 240
16.4.1 Biological Sources. 240
16.4.2 Chemical Sources 241

16.5   Testing and Detection of Different Contaminants in Media   242
16.6   Testing and Detection of Different Contaminants
        in Cell Cultures   242
        16.6.1   Bacteria   242
        16.6.2   Fungi   243
        16.6.3   Virus.   243
        16.6.4   Mycoplasma.   244
        16.6.5   Yeast.   244
16.7   Problem of Cross-Contamination by Other Species   245
        16.7.1   Methods for the Identification and Elimination
                  of Cross-Contamination.   245
16.8   Effects of Contamination on Media and Cell Suspensions   246
16.9   Prevention of Possible Contamination and Safe Handling of Cell
        Cultures – Maintaining Aseptic Conditions.   246
16.10   Conclusion   247
References.   248

17 **Analysis of Cell Growth Kinetics in Suspension and Adherent
    Types of Cell Lines**   251
    Vaishnavi Shishodia, Divya Jindal, Sarthak Sinha,
    and Manisha Singh
    17.1   Introduction   251
    17.2   Kinetic Characterisation of Cell Culture   252
    17.3   Cell Kinetics in Batch Culture.   253
            17.3.1   Lag Phase.   254
            17.3.2   Log Phase.   254
            17.3.3   Time Profile of the Concentration of Cells, Nutrients
                      and Metabolites   255
            17.3.4   Variation of Cell Morphology   255
            17.3.5   Determination of Cell Population Heterogeneity
                      for Kinetic Studies.   256
            17.3.6   Cell Kinetics in Continuous Culture   257
            17.3.7   Advantages of Continuous Culture   257
            17.3.8   Applications of Continuous Culture   258
    17.4   Influences of Physiological Conditions and Rate Equations   258
            17.4.1   Effect of Nutrients   258
            17.4.2   Effect of Temperature   259
            17.4.3   Effect of pH   259
            17.4.4   Dissolved Oxygen and Oxygen Uptake Rate   259
            17.4.5   Dissolved Carbon Dioxide.   260
            17.4.6   Osmolality and Salts   260
            17.4.7   The Rate Law for Cell Growth, Death
                      and Productivity.   260
    17.5   Conclusion   262
    References.   263

18 In Vitro Cytotoxicity Analysis: MTT/XTT, Trypan Blue Exclusion 267
Shalini Mani and Geeta Swargiary
18.1 Introduction 267
18.2 Categories of In Vitro Cytotoxicity and Cell Viability Assays . . . 268
    18.2.1 Dye Exclusion Methods. 269
    18.2.2 Colorimetric Methods 269
    18.2.3 Fluorometric and Luminometric Assays 270
18.3 Commonly Used In Vitro Cytotoxicity Analysis Methods 270
    18.3.1 Trypan Blue Exclusion (TBE). 270
    18.3.2 MTT Assay 273
    18.3.3 XTT Assay. 276
    18.3.4 Sulforhodamine B (SRB) Assay 278
    18.3.5 LDH Assay. 280
References. 283

19 Applications of Animal Cell Culture-Based Assays 285
Pallavi Shah, Anil Kumar, and Rajkumar James Singh
19.1 Introduction 285
19.2 Biomarker Identification 285
    19.2.1 Cell Culture-Based Cancer Biomarker Identification . . . 286
19.3 Genetic Manipulation. 288
    19.3.1 General Scheme for Genetic Engineering of Animal
        Cells in Culture for Protein Production 288
    19.3.2 Applications of Genetic Manipulation of Animal
        Cells in Culture 289
    19.3.3 Advanced Editing Tools. 290
19.4 Pathological Studies. 291
19.5 Pharmaceutical Studies 292
    19.5.1 Drug Screening in Cell Lines 292
19.6 Stem Cell Research 295
19.7 Cellular Development and Differentiation. 299
19.8 Hybridoma Technology 300
References. 301

20 Ethical Issues in Animal Cell Culture 305
Divya Jindal, Vaishanavi, and Manisha Singh
20.1 Introduction 305
20.2 Ethical Concerns in Handling Animal Cell Culturing 306
    20.2.1 Acquisition of Cell Culture 306
    20.2.2 Authentication of Cell Culture. 307
    20.2.3 Characterization of Cell Culture 307
    20.2.4 Isolation of Cell Lines 308
20.3 Ethical Concerns Associated with Chimera Effect in Tissue
    Engineering 308

20.4 Ethics Related to Development and Preservation
       (Cryopreservation) of Cell Cultures. 309
20.5 Ethical Conduct in Cell Culture Handling. 309
       20.5.1 Maintaining the Instability. 310
       20.5.2 Contamination and Non-specific Identification. 310
       20.5.3 Transfer of Cell Line Between Laboratories 311
       20.5.4 Use of Equipment and Media 311
20.6 Dilemmas in Tissue Engineering and Tissue Banking. 311
20.7 Cell Culturing Associated with Intellectual Property and Legal
       Rights. 312
       20.7.1 Laws Concerning Intellectual Property 313
20.8 Public and Scientific Community Opinions on Cell Culturing . . . 314
20.9 Conclusion. 315
References. 315

21 **Common Troubleshooting Methods in Cell Culture Techniques** 317
   Khushi R. Mittal and Shalini Mani
   21.1 Introduction 317
   21.2 Troubleshooting. 318
         21.2.1 Why Do Sometimes Cells Not Stay Viable After
                 Thawing the Stock?. 318
         21.2.2 Why Do Cells Seem to Grow Slowly Sometimes? 319
         21.2.3 What Is the Troubleshooting in Cell Counting? 319
         21.2.4 What Causes Cell Clumping? 319
         21.2.5 What Leads to a Cross-Linked or Misidentified Cell
                 Line? 320
         21.2.6 Why Do Sometimes Cells Do Not Revive After Being
                 Cryopreserved? 320
         21.2.7 What Are the Sources of Contamination and How
                 Should They Be Rectified? 320
         21.2.8 What Causes Unexpected Cell Detachment?. 321
         21.2.9 What Is the Troubleshooting in Cytotoxicity Assay? . . . 321
         21.2.10 What Triggers a Rapid pH Shift in the Medium? 322
         21.2.11 What Leads to Precipitation in the Medium and Either
                 Causes a Change in pH or No Change in pH?. 322
   21.3 Conclusion 322
   References. 323

Index. 325

Chapter 1
Overview to Animal Cell Culture

Shalini Mani

1.1 Introduction

The technique of taking cells from an animal or plant and growing them in an artificially controlled environment is known as cell culture. In many areas of the life sciences, cell culture has become a necessary tool. It lays the groundwork for researching cell proliferation, differentiation and product creation under tightly regulated settings. Cell culture has also allowed scientists to map nearly the entire human genome and examine the intracellular and intercellular signalling mechanisms that control gene expression.

From its origins in developmental biology and pathology, this discipline has evolved into a tool for molecular geneticists, immunologists, surgeons, bioengineers and pharmaceutical manufacturers, while remaining a critical tool for cell biologists, whose input is critical for the technology's continued development. As the new potential for genetic manipulation, whole animal cloning and tissue transplantation emerge, ethical as well as technical problems about gene therapy and tissue replacement are becoming increasingly important.

Despite major advancements in animal cell and tissue culture from the late 1800s, progress in animal tissue culture stopped until the early 1950s due to the lack of a viable cell line. The successful proliferation of cells generated from Mrs Henrietta Lacks' cervical cancer was demonstrated for the first time in the early 1950s. Mrs Henrietta Lacks' cells in culture revolutionised medical and biological research, enabling for major cellular, molecular and therapeutic breakthroughs, including the development of the first efficient polio vaccine (Rodríguez-Hernández et al., 2014; Del Carpio 2014). This culture is now known as HeLa, and there are

S. Mani (✉)
Centre for Emerging Diseases, Department of Biotechnology, Jaypee Institute of Information Technology, Noida, India

© The Author(s), under exclusive license to Springer Nature Switzerland AG 2023
S. Mani et al., *Animal Cell Culture: Principles and Practice*, Techniques in Life Science and Biomedicine for the Non-Expert,
https://doi.org/10.1007/978-3-031-19485-6_1

more than 60,000 research publications, till 2017, on this cell line. Additionally, different scientific studies using HeLa cells have been engaged in multiple Nobel Prize-winning breakthroughs (Del Carpio 2014; Masters Masters, 2002; Schwarz et al., 1985). For the scientific study, animal cell culture is an important technique. Cell culture technology's usefulness in biological science has long been recognised. The isolation of cells from a tissue before creating a culture in an appropriate artificial environment is the first step in animal cell culture. Disaggregation employing enzymatic or mechanical procedures can be used to separate the cells from the tissues. The isolated cells are normally produced from an in vivo environment, but they can also be derived from an existing cell line or cell strain. Cell culture technologies have become an important tool for evaluating the efficacy and toxicity of new medications, vaccines and biopharmaceuticals, as well as for assisted reproductive technology.

Animal cell culture is one of the most essential and versatile procedures used in today's research. The following factors can be investigated using animal cell culture as a model system:

- Drug development and screening
- Carcinogenesis and mutagenesis
- Cellular physiology and biochemistry in their natural state
- Drugs and hazardous chemicals' potential effects on cells

Furthermore, it allows for dependable and repeatable results, making it an important model system in cellular and molecular biology.

Mammalian cell culture necessitates a favourable growing environment. Nutritional and physicochemical requirements are separated in environmental settings. A substrate or medium that offers support and necessary elements such as amino acids, carbohydrates, vitamins, minerals, growth factors, hormones and gases are all required (O_2, CO_2). All of these variables influence physical and chemical variables like pH, osmotic pressure and temperature. The majority of cells in animal tissue culture are anchorage-dependent and require solid or semisolid support in the form of a substrate (adherent or monolayer culture), whereas others can be cultivated directly in the culture media (suspension culture). Under controlled laboratory circumstances, animal, plant and microbial cells are always cultivated in a specified culture media. Microorganisms are more sophisticated than animal cells. It is challenging to estimate the optimal nutritional requirements of animal cells produced in vitro due to their genetic complexity. In comparison to microorganisms, animal cells require more nutrients, and they normally only thrive when connected to properly coated surfaces. Despite these obstacles, various types of animal cells, both undifferentiated and differentiated, can be successfully grown.

Tissue culture is the process of maintaining and propagating cells in vitro under ideal conditions. Animal tissue culture is the process of cultivating animal cells, tissue or organs in a controlled artificial environment. Animal tissue culture was first recognised as an important technique during the creation of the polio vaccine, which used primary monkey kidney cells (the polio vaccine was the first commercial product generated using mammalian cell cultures). Though these primary monkey kidney cells were linked to a number of drawbacks (Van Wezel et al., 1978; Stones 1976; Beale 1981; Van Steenis et al., 1980), including:

1. The possibility of contamination by unknown substances (risk of contamination by various monkey viruses is high).
2. The majority of cells require anchoring and can only be cultivated effectively when attached to a solid or semisolid substrate (obligatorily adherent cell growth).
3. For virus generation, the cells are not well described.
4. Donor animals are in short supply because they are on the point of extinction.

Gay discovered that human tumour cells can produce continuous cell lines in 1951. As previously mentioned, the cell line regarded to be the first human continuous cell line was obtained from a cancer patient, Henrietta Lacks, and HeLa cells are still widely utilised. The most widely used resource in modern laboratories is continuous cell lines and are produced from human malignancies. Aside from advancements in cell culture, several media have been investigated, most of which are based on specific cell nutritional requirements, such as serum-free media, beginning with Ham's fully defined medium in 1965. Later on, hormones and growth factors were added to serum-free media in the 1970s to improve their performance. Thousands of cell lines are currently available, and a variety of media are available for their formation and maintenance.

1.2 Types of Cell Cultures

Animal tissue culture can be classified into two types; cultures that allow cell–cell interactions and stimulate communication or signalling between cells and cultures that do not allow cell–cell interactions and do not encourage communication or signalling between cells.

The first category is consisting of three different types of culture systems: organ cultures, histotypic cultures, and organotypic cultures are the three types of culture systems in the first category. On the other hand, cultures in monolayers or suspensions fall under the second type. Histotypic culture is the cultivating of cells for their re-aggregation to generate tissue-like structure, whereas organ culture is the culture of native tissue that retains most of the in vivo histological properties (Freshney 2005). Individual cell lineages are produced from an organ and then cultivated independently in a 3D matrix to explore interactions and signalling between homologous cells in histotypic cultures. Organ cultures are in vitro cultures of complete embryonic organs or tiny tissue pieces that retain their tissue architecture, i.e. the typical distribution of various cell types in the given organ (Edmondson et al., 2014).Cells from various origins are mixed together in specified proportions and spatial relationships in an organotypic culture to reform a component of an organ, i.e. the recombination of distinct cell types to create a more defined tissue or organ (Edmondson et al., 2014).

1.2.1   Primary Cell Culture

This is the first culture (a freshly isolated cell culture) or culture obtained by enzymatic or mechanical means straight from animal or human tissue (Freshney, 1987). These cells are slow-growing and diverse and have all of the characteristics of the tissue from where they came. The major goal of this culture is to keep cells growing on an appropriate substrate, which can be in the form of glass or plastic containers, in a regulated environment. They have the same karyotype (number and appearance of chromosomes in the nucleus of a eukaryotic cell) as the original tissue because they were taken directly from it. Once subcultured, primary cell cultures can give rise to cell lines that can either perish after a few subcultures (known as finite cell lines) or continue to proliferate indefinitely (known as indefinite cell lines) (these are called continuous cell lines). Normal tissues usually produce finite cell lines, but malignant cells/tissues (which are frequently aneuploid) produce continuous cell lines. Nonetheless, there are also unusual examples of non-tumorigenic continuous cell lines generated from normal tissues, such as MDCK dog kidney, fibroblast 3T3 and others. Mutation is thought to play a role in the evolution of continuous cell lines from primary cultures, altering their characteristics in comparison to finite lines (Jedrzejczak-Silicka, 2017). The possibility of genotypic and phenotypic variation can be increased by serial subculturing of cell lines over time. In contrast to primary hepatocytes, bioinformatic investigations based on proteomic characteristics showed that the Hepa1–6 cell lines lacked mitochondria, indicating a metabolic pathway rearrangements (Jedrzejczak-Silicka, 2017). With the introduction of advanced technology such as 3D culture, the utilisation of primary cells is becoming more common and yielding better results. There are two types of primary cells produced directly from human or animal tissue utilising enzymatic or mechanical procedures (Kim et al., 2020):

Adherent cells, also known as anchorage-dependent cells, are cells that require adhesion for proliferation. In other words, these cells have the ability to adhere to the culture vessel's surface. These cells are frequently obtained from organ tissues, such as the kidney, where the cells are stationary and entrenched in connective tissue.

Suspension cells, also known as anchorage independent cells, do not require any attachment or support in order to thrive. All suspension cells, such as white blood cells and lymphocytes, are extracted from the blood system and suspended in plasma. Cells grown from primary cultures have a short life span for a variety of reasons, i.e. they cannot be kept permanently. The exhaustion of the substrate and nutrients caused by a rise in cell populations in a primary culture can affect cellular activity and lead to the buildup of large amounts of hazardous metabolites in the culture. This could eventually lead to cell growth suppression. When a secondary culture or subculture must be produced to ensure continuing cell development, this stage is known as the confluence stage (contact inhibition).

1.2.2 Secondary Cell Culture

The first passaging of cells, a changeover to a different type of culture system, and the first culture derived from a primary culture are all examples of this (Segeritz & Vallier, 2017). This is commonly done after cells in adherent cultures have used up all of the available substrate or when cells in suspension cultures have exceeded the medium's capacity to sustain further growth and cell proliferation has slowed or stopped entirely. The primary culture must be subcultured in order to maintain appropriate cell density for continuing growth and to induce further proliferation. Secondary cell culture is the term for this procedure.

1.2.3 Cell Line and Cell Strain

A cell line is formed when a primary culture is subcultured or passaged. A continuous cell line is one in which cells continue to develop indefinitely throughout repeated subculturing, whereas finite cell lines incur cell death after several subcultures.

A cell line is a permanently formed cell culture that will multiply indefinitely if a suitable fresh media is continuously provided, whereas cell strains have been acclimated to culture but have a finite division capacity, unlike cell lines (Geraghty et al., 2014). A cell strain can be derived from either a primary culture or a cell line. This is accomplished through the selection or cloning of certain cells with predetermined qualities or characteristics (e.g. specific function or karyotype). In conclusion, the primary culture is the first culture that emerges from the in vivo environment. To create cell lines, this primary culture can be subcultured multiple times. Cell lines are immortalised or transformed cells that have lost control over division as a result of mutations or genetic abnormalities, or as a result of a primary cell being transfected with immortalising genes (Masters, 2002). Because most cell lines come from malignancies, they are tumorigenic (Verma et al., 2020). Cells produced from a primary cell line do not have this problem, although they are difficult to maintain. In most cases, primary cell cultures require a nutritional medium with a high concentration of various amino acids, minerals, and, on rare occasions, hormones or growth agents (McKeehan et al., 1990).

Primary cell cultures can be used effectively for two to four passes, after which the risk of contamination is higher than with cell lines. Primary cell cultures, on the other hand, have their own set of benefits too. The most common advantage of using primary cell culture is mainly due to their higher similarity with the organ system, they are isolated from.

1.3   Ethical Considerations in Animal Tissue Culture

Animal tissue culture procedures frequently use animal or human tissues, necessitating the development of animal research safety and ethics guidelines, commonly known as medical ethics. Animal handling brings up a slew of concerns that aren't present when using animal tissue. To begin research or study a human sample in the form of foetal materials or biopsy samples, the approval of the patient or his or her relatives is required in addition to the consent of local ethics bodies (Festing & Wilkinson, 2007).

A donor consent document in the required format should accompany any samples taken from a human donor. When working with human tissue, keep the following points in mind (Geraghty et al., 2014):

- The patient's or relative's permission to use tissue for research.
- Ownership of specimens, specifically cell lines and derivatives, with the recipient agreeing not to trade or transfer the cell lines and derivatives.
- Genetic modification consent, especially in the case of cell lines.
- For the commercial usage of cell lines, a patent or intellectual property right is required.
- The guidelines should be updated to reflect the most recent advancements in animal tissue culture research.

These recommendations are designed to provide appropriate information to newcomers, as well as to those participating in training and instruction. As using these informations, they may be better aware of the cell culture concerns, and if needed, they can effectively deal with those concerns. The following are the key areas of attention in the guidelines:

- Cell line acquisition
- Cell line authentication
- Cell line identification
- Cell line cryopreservation
- Cell line development
- Cell line instability
- Legal and ethical considerations in the creation of cell lines from human and animal tissues
- Microbial infection of the cell line
- Cell line misidentification
- Equipment selection and maintenance
- Cell line transfer between laboratories

1.4   Common Nomenclatures in Animal Cell Culture

The following are definitions of terms commonly used in animal tissue culture, particularly in the context of cell lines:

Adherent Cells. Cells that have the ability to stick to the culture vessel's surface using the extracellular matrix.

Immortalisation: Immortality is a term used to describe the state of being immortal. Obtaining a state of cell culture in which cells continue to proliferate.

Attachment efficiency: Within a certain period after inoculation, the fraction of cells that really cling to the surface of the culture vessel.

Passaging. Cell transfer is the process of moving cells from one culture vessel to another. Subculturing is a more descriptive name for the process of subdividing cells before transferring them to several cell culture containers.

A passage number indicates the number of times a cell line has been subcultured. When adherent cell cultures reach confluence (when they completely cover the surface of the cell culture tube), some will cease proliferating, and others will die if they are kept in this state for longer periods of time. As a result, adherent cell cultures necessitate recurrent passaging, which necessitates subculturing once the cells have reached confluence. In suspension cultures, where suspended cells utilise their culture media quickly, regular passaging is essential, especially when the cell density gets very high.

While culture maintenance necessitates frequent passaging, the process is more painful for adhering cells since they must be trypsinised. As a result, it is not recommended to passage adherent cell cultures more than once every 48 h.

Split ratio: The divisor of a cell culture's dilution ratio.

Generations number: The number of times a cell population has doubled in size. It's worth noting that passing and generation number are not synonymous.

Population doubling time: This value indicates how many times the cell population has doubled since isolation.

Passage number: The number of subcultures that culture has undergone.

Subculture: It is the process of transferring cells from one culture to another in order to start a new one.

Proliferating cells are subdivided during this process, allowing for the formation of new cell lines. When adherent cell cultures reach the confluent stage (i.e. when they completely cover the surface of the cell culture vessel), they will cease growing and will almost surely die if left there for a long time. As a result, adherent cell cultures should be passaged on a regular basis, meaning that when cells reach confluence, a portion of the cells should be passaged or subcultured to a new cell culture vessel. However, because adherent cells must be trypsinised, it is not recommended to subculture adherent cells on a frequent basis (no more than once per 48 h). Suspension cultures with a high cell density, on the other hand, necessitate frequent passaging because they consume medium quickly. The standard cell growth curve

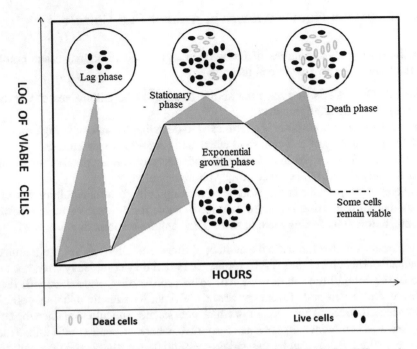

Fig. 1.1 Schematic representation for distribution of live and dead cells during different phases of cell growth in culture conditions

in culture is depicted in the Fig. 1.1, as mentioned- below. Because the cells have not yet accustomed to their new environment, there is less growth during the early lag phase. They grow exponentially as they begin to adapt to their surroundings, which is why this phase is known as the exponential or log phase. All cells are actively growing and consuming media at this moment. If the medium is not changed at this time, growth will come to a halt. As previously stated, the confluent phase occurs when the culture exceeds the medium's capacity. At this point, the culture must be broken down into subcultures. Current research practises necessitating the creation of good models, as good science cannot be accomplished with poor models. Several cell culture procedures, such as stem cell-derived human cells, co-cultures of different cell types, scaffolds and extracellular matrices, tissue architecture, perfusion platforms, organ-on-chip technologies, 3D culture and organ functionality, have been developed in the twenty-first century to overcome the drawbacks of traditional culture procedures and to be more scientifically rigorous (Fang & Eglen, 2017).

Organ-specific methodologies, more broad assessment of cell responses utilising high-content methods and the use of biomarker chemicals can all help to better the biological linkages between such models. A microphysiological model system can be created using these principles. One of the most notable benefits of this type of model system is that it produces results that are more similar to those seen in vivo; yet, managing many factors is a huge difficulty for the animal tissue culture industry.

References

Beale, A. J. (1981). Cell substrate for killed polio vaccine production. *Developments in Biological Standardization, 47*, 19–23.

Del Carpio, A. (2014). The good, the bad, and the HeLa. *Berkley Science Review, 5*.

Edmondson, R., Broglie, J. J., Adcock, A. F., & Yang, L. (2014). Three-dimensional cell culture systems and their applications in drug discovery and cell-based biosensors. *Assay and Drug Development Technologies, 12*(4), 207–218.

Fang, Y., & Eglen, R. M. (2017). Three-dimensional cell cultures in drug discovery and development. *Slas Discovery, 22*(5), 456–472.

Festing, S., & Wilkinson, R. (2007). The ethics of animal research: Talking point on the use of animals in scientific research. *EMBO Reports, 8*(6), 526–530.

Freshney, R. I. (1987). *Animal cell culture: A practical approach*. IRL Press.

Freshney, R. I. (2005). *Culture of animal cells: A manual of basic technique* (5th ed.). Wiley.

Geraghty, R. J., Capes-Davis, A., Davis, J. M., Downward, J., Freshney, R. I., Knezevic, I., Lovell-Badge, R., Masters, J. R., Meredith, J., Stacey, G. N., & Thraves, P. (2014). Guidelines for the use of cell lines in biomedical research. *British Journal of Cancer, 111*(6), 1021–1046.

Jedrzejczak-Silicka, M. (2017). *History of cell culture in new insights into cell culture technology* (S. J. T. Gowder, Ed.). IntechOpen.

Kim, J., Koo, B. K., & Knoblich, J. A. (2020). Human organoids: Model systems for human biology and medicine. *Nature Reviews. Molecular Cell Biology, 21*(10), 571–584.

Masters, J. R. (2002). HeLa cells 50 years on: The good, the bad and the ugly. *Nature Reviews. Cancer, 2*(4), 315–319.

McKeehan, W. L., Barnes, D., Reid, L., Stanbridge, E., Murakami, H., & Sato, G. H. (1990). Frontiers in mammalian cell culture. *In Vitro Cellular & Developmental Biology, 26*(1), 9–23.

Rodríguez-Hernández, C. O., Torres-Garcia, S. E., Olvera-Sandoval, C., Ramirez-Castillo, F. Y., Muro, A. L., & Avelar-Gonzalez, F. J. (2014). Cell culture: History, development and prospects. *International Journal of Current Research Academic Review, 2*, 188–200.

Schwarz, E., Freese, U. K., Gissmann, L., Mayer, W., Roggenbuck, B., Stremlau, A., & Hausen, H. Z. (1985). Structure and transcription of human papillomavirus sequences in cervical carcinoma cells. *Nature, 314*(6006), 111–114.

Segeritz, C. P., & Vallier, L. (2017). Cell culture: Growing cells as model systems in vitro. In *Basic science methods for clinical researchers* (pp. 151–172). Academic.

Stones, P. B. (1976). Production and control of live oral poliovirus vaccine in WI-38 human diploid cells. *Developments in Biological Standardization, 37*, 251–253.

Van Steenis, G., Van Wezel, A. L., de Groot, I. G., & Kruijt, B. C. (1980). Use of captive-bred monkeys for vaccine production. *Developments in Biological Standardization, 45*, 99–105.

Van Wezel, A. L., Van Steenis, G., Hannik, C. A., & Cohen, H. (1978). New approach to the production of concentrated and purified inactivated polio and rabies tissue culture vaccines. *Developments in Biological Standardization, 41*, 159–168.

Verma, A., Verma, M., & Singh, A. (2020). Animal tissue culture principles and applications. In *Animal biotechnology* (pp. 269–293). Academic.

Chapter 2
Cell Culture Laboratory

Anvi Jain, Aaru Gulati, Khushi R. Mittal, and Shalini Mani

2.1 Laboratory Design, Planning and Layout

The primary requisite for the cell culture laboratory that differentiates them from other laboratories is to maintain pathogenic free (i.e. asepsis) condition. The design of a cell culture laboratory majorly depends upon the type of activities performed. Therefore, it is preferred to have a designated work area to minimise the risk of contamination to the cells and to the workers. The design of a lab should have separate work space or either room for medium preparation, washing/sterilising area, transfer area, growth area and areas for cold storage.

In planning for a new infrastructure, multiple factors need to be taken into consideration. The lab should have good water access and disposal services, and also, the new building should incorporate adequate insulation and air conditioning in each room as the need of temperature and airflow requirements is different for each room. The rooms must be constructed with easy-cleaning facilities and with a covering of vinyl coating or any other dustproof surface. While constructing a new building, there is a lot of potential for integrated and creative structure and facilities, which can be built with a comfortable working environment and energy-saving purposes. Thus, for the planning stage following considerations must be addressed:

A. Jain · A. Gulati · K. R. Mittal · S. Mani (✉)
Centre for Emerging Diseases, Department of Biotechnology, Jaypee Institute of Information Technology, Noida, India

© The Author(s), under exclusive license to Springer Nature Switzerland AG 2023
S. Mani et al., *Animal Cell Culture: Principles and Practice*, Techniques in Life Science and Biomedicine for the Non-Expert,
https://doi.org/10.1007/978-3-031-19485-6_2

11

2.1.1 Ventilation

Ventilation and air circulation in cell culture laboratories are essential to the operation. Another key factor required for a cell culture laboratory is pressure balance. Ideally, it is preferred to maintain a positive pressure as compared to the surrounding workspace, in order to prevent the influx of contamination present in the polluted air from outside. To satisfy the needs of pressure, it may be desirable to have a favourable pressure buffer zone in the exterior of the microscope room, preparation room and corridor.

It is advantageous to vent laminar flow hoods to the outside to increase air circulation and prevent excessive heat (300–500 W per hood) from entering the room. Venting hoods to the outside will almost certainly provide the majority of the room's air extraction, leaving only the difficulty of ensuring that airflow via a power plant or an air conditioner would not disrupt the hood's airflow consistency.

2.1.2 Requirements

The culture area must be facilitated with preparation and sterilisation, sterile handling and any other activities. Besides this, centrifugation, microscope, cell counting, incubation and storage at ambient temperature 4 °C, −20 °C, and −196 °C must be considered. Washing of the apparatus or equipment and sterilisation facilities must be positioned (1) next to the aseptic area and (2) on the outer wall in order to allow the heat extraction from the boilers and steam vent of the autoclave. A proper space is required to accommodate different types of equipment such as laminar flow hoods, centrifuge, incubator, microscope, water bath, pipettes, media, reagents, plasticware and glass tubes. The proper spacing of these leads to the proper and systematic working inside the lab.

2.1.3 Services

The major services required in cell culture laboratory involve the availability of hot and cold water, flammable gas (domestic methane, propane, etc.), pressurised air and carbon dioxide. The rooms must be well supplied with filtered air. A sufficient power supply is also needed for the different equipment. An adequate supply of hot and cold water with sinks and proper drainage is preferred in both the preparation and the cell culture areas. A vacuum pipeline can be beneficial for draining culture flasks, but a collection vessel must supplement it and a catch flask, with a waterproof filter placed between the flasks, to avoid fluid, vapour or any other debris from invading the vacuum pipeline and pumps. Apart from this, a slight fall in the floor is recommended for the cleaning purpose as it enables liberal use of water.

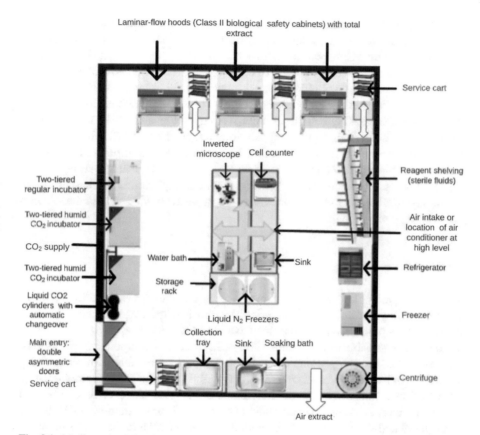

Fig. 2.1   Medium-sized tissue culture laboratory layout

2.1.4   Layout

Sterile handling, incubation, preparation, washup, disinfectants, and preservation are the six key functions that must be accommodated. Build a "sterility gradient" if a single room is used; at one end of the building, the hygienic area for aseptic dealing should be situated, away from the entrance, and washup. The sterilisation facilities must be positioned at the other end, with preparation, preservation and incubation in between. Storage and incubators must be conveniently accessible from the sterile working area, and the preparation area must be near to the washup and sterilisation facilities (Fig. 2.1).

2.1.4.1   Sterile Handling Area

A cell culture laboratory's most important necessity is to establish a sterile handling work area dedicated solely to tissue culture activities. It should be in a quiet corner of the tissue culture lab, away from chemical activity and work on other organisms including bacteria, protozoa and yeast. No congestion or other interruptions that could result in dust or draughts should occur and, if necessary, have an airflow chamber that circulates fresh air across the working environment. If an airflow chamber is not available, the growth or culture process can be performed on a clear bench under a sterile "umbrella" created by a Bunsen burner. A HEPA-filtered air supply is ideal but not always available. If there are no laminar flow hoods available, utilise a separate room or cubicle. A biohazard cabinet level 2 can be useful to treat a poisonous chemical. A plastic-wrapped floored table, ideally with white or neutral grey, should be used in the workplace, in its simplest form, to promote the study of cultures and dissection. Nothing should be kept on the table, and any storage above it should be used solely for sterile tasks (e.g. for holding instruments and pipettes). The table must be either set far from the wall or a plastic sealing strip or mastic sealant should be sealed to the wall.

2.1.4.2   Laminar Flow

The use of laminar flow hoods that spray aseptic air onto the workstation provides greater sterility control at a lesser price than an isolated aseptic space. Separate hoods are recommended because they isolate users and may be passed about, although it is possible to use laminar flow wall or ceiling components in batteries. Choose hoods that suit your area, bench-top or freestanding, and allow enough leg space beneath with aspirators, free area for pumps, etc. Lockable castors must be used on freestanding cabinets so that they can be relocated if desired. The seat height and back angle of the chair should be adjustable to accommodate the height of the hood and should be easy to be pulled up close enough to the front side of the hood to allow for comfortable functioning inside it. For a notebook and materials (that may be needed but are not in immediate need), a mini size cart or trolley should be given beside each hood. Laminar flow hoods have a preferred lateral separation of at least 2 ft in order to facilitate maintenance access and minimise airflow disturbance between the hoods. If the hoods are facing one other, there should be at least 10 ft between the fronts of each hood.

2.1.4.3   Service Bench

It is more practical to place a bench near the sterile handling area for a cell counter, microscope and other essential instruments, dividing or separating the area from the other end of the lab. The service bench should also be able to store pipettes, sterile glassware, plastics, syringes and screw caps.

2.1.4.4   Quarantine and Containment

If enough space is available, designate a separate room as an incubation and/or containment room. This is a totally separated aseptic room with its own incubators, freezers, refrigerators, centrifuge, supplies and laminar flow hood disposal. This room should be isolated from the rest of the suite and must be under negative pressure on the rest of the sterile area. It is possible to treat newly imported biopsies or cell lines here before they are seen to be free from contamination. If local rules permit, the same space may be used as a level 2 containment room at different times. It would also include a pathogen hood or biohazard cabinet with a separate extract and pathogen trap when used at a higher level of containment.

2.1.4.5   Incubation

Besides an airflow cabinet and easily washed benching, the cell culture laboratory will require an incubator or hot room environment to keep the cells at 30–40 °C. The type of incubation required is influenced by temperature, gas phase, size and closeness to the work area.

It's quite likely that normal, non-gassed incubators or a hot space would be enough or that CO_2 and a humid environment would be needed. Significant quantities of sealed flasks or large-volume flasks should be maintained in a hot room, while open plates and dishes should be incubated in a humid CO_2 incubator. A thermostatically controlled hot room or separate incubators should be used for incubation.

Incubator   Incubation space would be 0.2 m³ (200 L, 6 ft³), and shelf space would be 0.5 m² (6 ft²). One or even more humid incubators with controlled CO_2 are almost certainly required.

Hot Room   It may be possible to transform a space inside the laboratory area or a neighbouring room or walk-in cabinets, into a hot room if the space is readily available and usable. To stop cold spots appearing on the walls, the room should be insulated. Line the area with plastic laminate-veneer sheet, separated from the wall by around 5 cm (2 in.) of mineral wool, or fibreglass if insulation is necessary. If wall-mounted shelving will be used, set the position of the studs or straps holding the lining panel to indicate attachment points. To prevent the shelves from sagging, utilise demountable shelving and room shelves that support 500–600 mm (21 in.) apart. It is better to use freestanding shelving units as they are removable, so it is easy to clean the space and the rack. Enable 200–300 mm across racks/shelves, with larger (450 mm) shelves at the bottom and smaller (250–300 mm) shelves above from the eye level. Air can flow due to perforated shelving installed on flexible brackets. There should be no irregularities or bumps in the shelving and must be completely horizontal. Wooden furniture must be avoided as much as is achievable, since it traps the heat and can shelter pests. In any area of the hot room, a small bench, ideally sturdy plastic laminate or stainless steel, should be installed. An inverted microscope, the beaker you intend to inspect and a book should all be able to fit on the table. If you intend to do cell synchrony tests or other sterile procedures at 37 °C, you can however also make room for a compact laminar flow unit with a filter size of 300 × 300 or 450 × 450 mm placed on a wall or on a stand over a portion of the bench. The fan motor must be suited for use in the tropics and must not be kept working continually. If it runs continuously, the room will become warmer, and the motor will burn. Some will want to use the space for incubation of shaker shelves, tubes and other non-tissue culture incubations once a hot room is made available, so the amount of bench space offered can allow for this possibility. Certain microorganisms, such as bacteria or yeast, should be prohibited. Incandescent lighting is favoured to fluorescent lighting, which may degrade the medium. In addition, certain fluorescent tubes have trouble lighting up in a hot room. The flow of air in the room, accuracy and sensitivity of the control gear, the evolution of heat by other apparatus (stirrers, etc.), the nature of the insulation and the position of the thermostat sensor in the room all affect the temperature of the hot room at any point and at any time.

2.1.4.6 Preparation Area

Media Preparation In smaller laboratories, extensive media preparation can be skipped if a trusted material of culture media is available. Smaller labs may choose to buy ready-made media over preparing their own media, even though a large company (about 50 workers performing tissue culture experiments) finds it cost-effective. These labs will then just have to make reagents including salt solutions and ethylenediaminetetraacetic acid (EDTA); store these reagents and water in sterilised bottles and other small products. While the preparation area should stay quiet and clean, in this situation sterilised handling is not needed since every product will be sterilised. If trustworthy commercialised media are hard to come by, the preparation room must be massive to hold a pH metre, coarse and balance, and if space allows then an osmometer too. Stirring and dissolving solutions, as well as bottling and packing different products, would necessitate additional bench space, as will additional atmospheric and refrigerated shelf space. If necessary, an additional horizontal laminar flow hood for filtering and bottling sterile liquids should indeed be installed in the sterile region. For sterility quality management, incubator space must be set aside. At the non-sterile end of the preparation area, heat-stable equipment and solutions may be dry-heated sterilised autoclaved. Both streams finally converge on the storage areas.

Washup The heat and humidity generated by sterilisation and washup facilities are better placed outside the tissue culture lab, as heat and humidity can be difficult to dissipate without raising airflow beyond appropriate standards. Where necessary, autoclaves, distillation devices and ovens should be stored in a separate space with a strong extraction fan. The washup area must have enough space for soaking glassware as well as space for an automatic washing machine (if needed). There ought to be enough bench space to sort pipettes, handle glassware baskets and package and seal sterilisation packs. You'll also require space for a pipette dryer and washer. If you're constructing a lab from the ground up, you can get sinks built in to whatever size you choose. The best products to use for skinks are polypropylene or stainless steel, the former for hypochlorite disinfectants and the latter for radioisotopes. Sinks ought to be deep enough (450 mm, 18 in.) to allow one to manually rinse and wash the larger objects without having to stoop very far. From floor to rim, they should be around 900 mm (3 ft). When bent over the sink, an elevated edge across the top of the sink will contain spillage to protect the operator from getting soaked. The raised edge should wrap around the back of the taps. A single cold water tap, a combination hot and cold mixer, a cold tap for a hose connection for a rinsing system and a nonmetallic or stainless steel tap for deionised water from a tank above the sink would be needed for each washing sink. A centralised deionised water supply must be prevented since the piping will collect bacteria and dirt. It's impossible to scrub when it's coated with algae. Remember to leave ample parking space for carts or trolleys which are commonly used to gather dirty glassware and redistribute fresh sterile stock.

2.1.4.7 Storage

Storage is vital to keep non-sterile and sterile items clearly labelled and isolated. Hence, it is necessary to provide storage. Following storage areas must be easily accessible from the sterile working area:

- Sterile liquids maintained at room temperature (like water, salt solutions, etc.), at $-20\,°C$ or $-70\,°C$ (glutamine, trypsin, serum, etc.), at $4\,°C$ (media).
- Non-sterile and sterile glassware.
- Disposable sterile plastics (e.g. syringes, petri dishes, centrifuge vials and tubes, etc.).
- Non-sterile and sterile items (e.g. stoppers, screw caps, etc.).
- Non-sterile and sterile apparatus like filters.
- Disposable items (e.g. plastic bags, gloves, etc.).
- Storage should also be made for following items.
- Liquid nitrogen to replenish freezers; liquid nitrogen can be stored in one of two ways:
  - Underneath the bench in Dewar's (25–50 L)
  - On a trolley in a huge storage vessel (100–150 L) or in storage tanks (500–1000 L) permanently sited in their own space with proper ventilation or, ideally, outside in stable, weatherproof housing

Since contaminants can build up in liquid nitrogen storage vessels, they should be stored in safe places. This can be avoided by using a perfused wall freezer. The space where the nitrogen is stored and dispensed must have adequate ventilation, preferably with a warning to signal when the oxygen tension falls below safe levels.

Carbon dioxide is stored in separate cylinders and transferred to the laboratory as required (the cylinders should be mounted on the wall or stacked on a bench in a rack).

A CO_2 supply piped to workstations, or a CO_2 supply piped from a pressurised tank of CO_2 that is replenished on a daily basis. (As a general rule, 2–3 people will only need a few cylinders, 10–15 people will likely benefit from a piped supply from a bank of cylinders and more than 15 people will benefit from a storage tank.)

Freezers and refrigerators should be placed at the non-sterile end of the lab, since the doors and compressor fans produce dust and draughts, as well as harbouring fungal spores. Refrigerators and freezers often necessitate cleaning and defrosting on a regular basis, which produces a level and type of operation that can be kept isolated from the sterile operating environment.

The most important aspect of storage areas is easy access for both removal and replenishment of stocks, with older stocks being held in the front. Double-sided modules are convenient since one side can be restocked while the other is in operation.

As a rough estimate, each individual would need 200 L (8 ft^3) of 4 °C storage and 100 L (4 ft^3) of 20 °C storage. For less participants, the amount per person increases. A 250 L (10 ft^3) refrigerator and a 150 L (6 ft^3) freezer may be needed for one user. Of course, these numbers only account for warehouse space; entry and working space in walk-in cold rooms and deep freezer rooms must also be factored in. Separate 20 °C freezers are generally preferable to a walk-in 20 °C space. They're easier to clean and maintain, and they provide more redundancy in the event that one machine crashes.

2.2 Equipment and Materials for Cell Culture Laboratory

A number of equipment are essential in cell culture. The main function of different types of equipment is to facilitate accurate as well as efficient working and also allows performing a large range of assays and analysis. Moreover, it is required to prioritise the need of different equipment into different categories: (1) essential, you can't usually conduct tissue culture without these equipment; (2) beneficial, by the help of these equipment, culture would be achieved more effectively, more quickly or with less effort; and (3) useful addition, items that would enhance working conditions allow for more sophisticated analyses or make your workplace more appealing in general as shown in Table 2.1. Therefore, the following sections briefly list the equipment that is frequently required in the cell culture laboratories.

Table 2.1 Tissue culture equipment

Requirement	Equipment	Purpose
Essential requirement	Laminar flow hood	Maintaining sterile environment
	Incubator	Maintaining optimal temperature, humidity for incubation
	Centrifuge	Used to separate substances from a solution
	Inverted microscope	Viewing and studying cell culture
	Haemocytometer	Determining the concentration of cells in aqueous solution
	Freezer	For storage of media, reagents and serum in $-20\ °C$
	Water bath	Used for incubating sample in water
	Magnetic stirrer	Used to agitate the liquid for maintaining uniform suspension of culture
	Refrigerator	Used for cool storage of media and reagents
	Liquid N_2 freezer	Used for preservation of cell lines and biological material
	CO_2 incubator (humid)	Used to control humidity and CO_2 concentration
Nonessential, but beneficial	PH meter	Used for measuring pH of a aqueous solution
	Autoclave	Sterilisation of glassware, equipment, liquids and solution
	Cell counter	Used for automatic cell counting
	Peristaltic pump	Used to pump out sterile fluids from flasks, filter-sterilisation
	Pipettor(s)	Dispensing measured volumes of liquid accurately
	Roller racks	Use for roller bottle culture
	Drying oven(s)	Used for drying of glassware and plastics
	Fluorescence microscope	Visualisation of specimens with the help of fluorochromes
Additional requirements	Confocal microscope	Used to analyse fluorescence in thick specimens
	Osmometer	Measuring the osmolality in various biological samples
	Colony counter	Counting colonies of bacteria or other microorganisms on a plate
	Conductivity meter	Used to measure conductivity in a solution
	Microtitration plate reader	Used to analyse the chromogenic endpoints of different assays

2.2.1   Aseptic Area

2.2.1.1   Laminar Flow Hood

It is important for laboratories to provide aseptic working conditions to carry out procedures, and, therefore, laminar flow hoods or biological safety cabinets are highly recommended. Laminar flow hoods (Fig. 2.2) provide a sterile working environment to prevent contamination to the cell culture. A horizontal flow hood is less expensive and offers the good sterile protection for your cultures, but it's only good

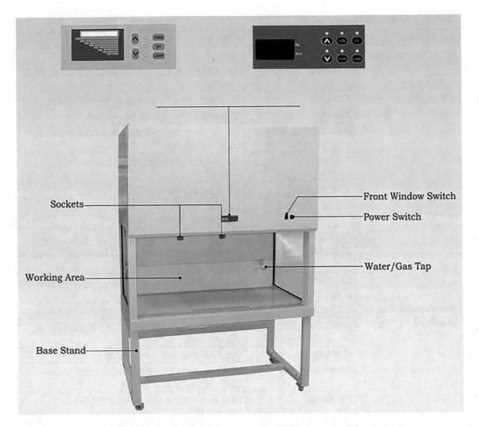

Fig. **2.2** Laminar flow hood. (https://www.indiamart.com/proddetail/laminar-air-flow-cabinet-16764797088.html)

for medium preparation (without antibiotics) and other nontoxic aseptic reagents as well as for culturing non-primate cells. To meet varying research and clinical needs, three types of biohazard cabinets have been developed: Classes I, II and III. Class I biohazard cabinets provide a substantial level of protection to laboratory staff and to the surroundings when used with good microbiological techniques, but they do not shield cultures from contamination. While for handling potentially hazardous materials substances (carcinogenic, cell lines and other toxic substances), Class II or Class III biohazard cabinets are used.

The air is filtered and contaminant-free before being circulated into the cabinet by using HEPA (high-performance particle air filter) filter. Many hoods have UV germicidal lamp which is used to sterilise the material inside while not in use. The hood is fitted with an aspirator pump that extracts and pumps liquids directly into the disinfectant. The following points should be considered while buying and installing new hood:

1. Size – a working area of 1200 mm (4 ft) wide × 600 mm (2 ft) deep is generally accurate.
2. Noise level – it should be at a quiet place, as noisy hoods are exhausting.
3. Functional efficiency – will the integrity of the hood's work space be undermined by airflow from other cabinets, room ventilation or separate air-conditioning units? Will polluted air or aerosols escape out as a result of the turbulence? This condition would necessitate a minimum of 3000 mm (10 ft) of face-to-face separation and 500 mm (2 ft) of lateral separation.
4. Interior – there should be easy access for cleaning and washing both inside the working area and under the work area in case of spillage.
5. Comfort – there should be a comfortable sitting arrangement as some cabinets have awkward ducting under the work surface, leaving no space for your feet, lights or other accessories above your head that touch your head or screens that obstruct your view.
6. Screen – in order to make cleaning and handling bulky culture apparatus simpler, the front screen should be able to be lifted, lowered or fully removed.

2.2.1.2   Service Carts

It is basically a movable cart which is used to place important items near laminar flow hoods. These carts are useful for filling the gap between adjacent hoods and can be quickly removed for hood maintenance. They can also be used to transport goods to and from the hoods, as well as for getting basic supplies restocked by service personnel.

2.2.1.3   Pipette Aids, Automatic Pipetting and Dispensing

Pipettes   Pipettes must be of the blowout type, with a large tip for easy delivery and a graded tip, with the scale's highest point at the top rather than the tip. Disposable plastic pipettes (Fig. 2.3a) or reusable glass (Fig. 2.3b) or disposable plastic Pasteur pipettes (Pastettes) (Fig. 2.3c) may be utilised. Often labs use disposable plastic pipettes, which have the benefit of being presterilised and prepackaged and do not pose the same safety risks as chipped or cracked glass pipettes. They also don't need to be cleaned or plugged which is time-consuming. On the drawback, they're very pricey and if they're single-packed take a long time to use. Glass pipettes can be more cost-efficient in a large laboratory, but the ease, safety of handling and greater durability of plastic disposables are preferred.

Pipette Cans   Pipettes are kept clean in the hood with usage of pipette cans (Fig. 2.4) that can be locked while not in use. Sterilisation of glass pipettes is typically done in aluminium- or nickel-plated steel cans. Square-sectioned cans with dimensions of 75 × 75 mm and a diameter of 300–400 mm are superior to circular cans because they stack more quickly and do not bounce around on the work surface.

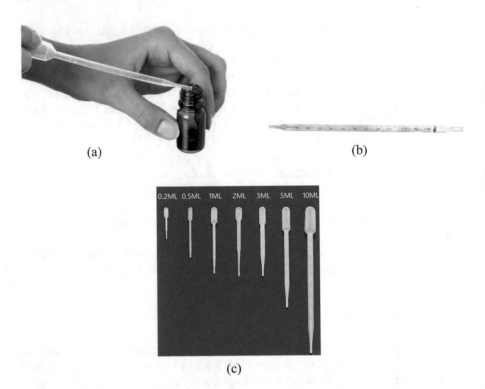

Fig. 2.3 Pipettes: (**a**) 1 ml plastic disposable plastic pipette, (**b**) reusable glass pipette, (**c**) disposable Pasteur pipette. (https://www.aromareadyproducts.com/1-ml-plastic-disposable-pipettes. html, https://www.coleparmer.in/p/pyrex-7079-glass-serological-pipettes/2018, https://www. indiamart.com/proddetail/disposable-pasteur-pipette-21102414988.html)

Pipette Cylinders/Pipette Hoods They must be made of polypropylene (Fig. 2.5) and spread around the lab (one per workstation), with enough in reserve to enable full cylinders to remain in disinfectant for 2 h prior disposal (plastic) or washing (glass).

Pipette Controllers While a rubber bulb and perhaps other proprietary pipetting devices are easy to use and inexpensive, a motorised pipette controller (Fig. 2.6), which can be purchased with a separate or built-in pump and is either rechargeable or mains, vastly improves precision, pace and reproducibility. The feel and weight of the pipette controller during continuous use are the most important factors to consider when selecting one. Pipette controllers typically have a filter at the pipette insert to prevent pollutants from being transferred. Such filters are disposable, although others can be reused after being sterilised.

Pipettors These instruments evolved from Eppendorf micropipettes, but since the operating range has now expanded to 5 mL or more, the name "micropipette" is no longer sufficient, and the instrument is far more generally referred to as a pipettor

Fig. 2.4 Pipette can.
(https://www.daigger.com/
pipette-canisters-i-
bkl1360s)

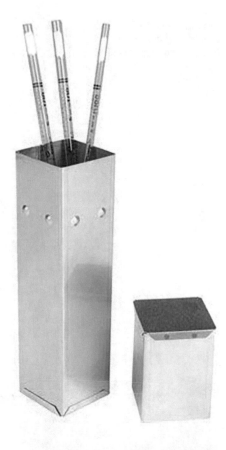

Fig. 2.5 Polypropylene
pipette cylinder. (https://
www.camlab.co.uk/
polypropylene-pipette-
cylinder-jar-autoclavable-
at-121-c-for-20-minutes)

Fig. 2.6 Motorised pipette controller. (https://www.indiamart.com/proddetail/pipette-controller-11803188033.html)

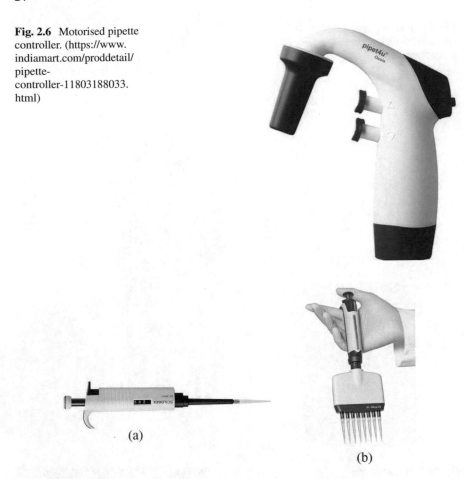

(a)

(b)

Fig. 2.7 Pipettors: (**a**) variable-volume pipetting device, (**b**) Multipoint pipettor with manifold for 8 plastic tips; 4 and 12 are also available https://www.scilogex.com/micropette-single-channel-fixed-pipettors.html, https://www.mt.com/in/en/home/products/pipettes/manual-pipettes/multichannel-pipettes.html

(Fig. 2.7a). Pipettors with several tips (Fig. 2.7b) are available to be used with microtitration plates. Tips can be purchased loose and sterilised in the lab, or they can be purchased sterile and placed in racks that are ready to use. Pipettors can be useful for laboratory work that requires accuracy but does not require further propagation of the cells used.

Syringes Syringes (Fig. 2.8) are often used for filtration and extraction of reagents (drugs, radioisotopes or antibiotics) from sealed vials when used in combination with needles and syringe filter adapters.

Large-Volume Dispensing As the amount of medium in culture vessels exceeds 100 mL, a different approach to fluid distribution is needed. A 100 mL pipette, graduated bottle (Fig. 2.9) or media bag can suffice if only a few flasks are involved, but a peristaltic pump is preferred if larger volumes (>500 mL) or a large number of high-volume replicates are needed. Single fluid transfers of very large volumes (10–10,000 L) are usually accomplished by preparing the medium in a sealed pressure flask, autoclaving it and then displacing it into the culture vessel under positive pressure. Pouring may be used to dispense huge volumes, but it can be limited to a single operation with a fixed amount.

Repetitive Dispensing Small-volume repetitive dispensing can be accomplished by continuous piston movement in a syringe (Fig. 2.10a) or a repeated syringe operation with a two-way valve attached to a reservoir, as shown by the Cornwall syringe (Fig. 2.10b). Avoiding the drying period after autoclaving and flushing the syringe out with a serum-free medium or a salt solution before and after use will help to prevent the valves from sticking. A peristaltic pump (Fig. 2.11a) can also be used for routine serial deliveries, with the added benefit of being able to unlock the pump with a foot switch, freeing up your hands. To prevent contaminating the tubing at the reservoir and supply ends, care must be taken when installing such devices. In general, they're only useful if you're dealing with a huge number of flasks. Only the delivery tube is autoclaved, allowing for high precision and reproducibility over a range of 10–100 mL. In the event of unintended exposure or a change of cell type or reagent, a large number of delivery tubes may be sterilised and kept in storage, allowing for a rapid changeover.

Fig. 2.8 Syringes (https://labsociety.com/lab-equipment/syringes/)

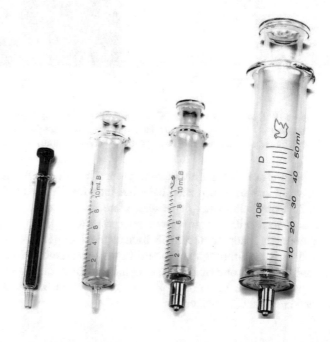

Fig. 2.9 Graduated bottle
dispenser. (https://www.
amazon.com/Graduated-
Dispenser-Dispense-Glass-
Injection-Transparent/dp/
B0836PYJ3P)

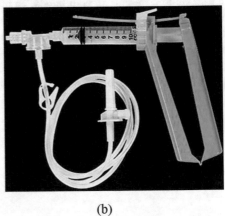

(a)   (b)

Fig. 2.10 Syringe dispensers: (**a**) stepping dispenser, (**b**) 10 ml repeating syringe dispenser
(Cornwall syringe). (https://www.labnetinternational.com/products/labpette-r-repeating-pipette,
https://www.wolfmed.com/fld-disp-syringe-10ml-10-cs.html)

Automation Several attempts to automate cell culture have been made, but very
few devices or instruments have the functionality required for general laboratory
use. Many automatic dispensers, plate readers and other accessories have been
introduced with the advent of microtitration plates. Seeding from one plate to
another is made simpler with transfer devices (Fig. 2.12). In the pharmaceutical
industry, high-throughput screens based on microtitration systems are widely used
and represent one of the most significant uses of automation. Whether a simple
manual system or a complex automated system is selected, different criteria are
used to direct the decision.

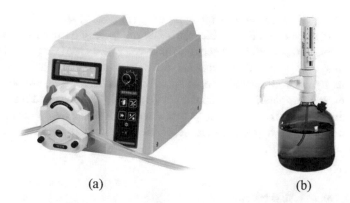

(a) (b)

Fig. 2.11 Automatic dispensers: (**a**) peristaltic dispensing pump, (**b**) Zippette bottle-top dispenser. (https://www.drifton.eu/shop/10-dispensing-tube-pumps/1343-lp-bt100-1f-peristaltic-dispensing-pump/, https://in.vwr.com/store/product/6951003/bottle-top-dispenser-zippette-pro)

Fig. 2.12 Transfer device (Corning Costar Transtar). (https://www.fishersci.se/shop/products/corning-costar-transtar-96-96-well-liquid-transfer-syste m-accessories-reservoir-liners-3/10311441)

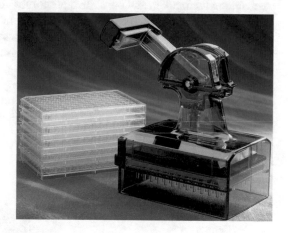

1. Ease of use and cost in terms of time saved.
2. Accuracy, efficiency and reproducibility are essential.
3. Sterilisation ease and impact on accuracy and repeatability.
4. Protection in terms of mechanical, electrical, chemical, biological and radiological systems.

2.2.1.4 Peristaltic Pump

A peristaltic pump can be used to extract spent medium or other reagents from a culture flask, and the effluent can be collected directly into disinfectant in a vented tube, with little chance of aerosol discharge into the atmosphere if the vent has a cotton plug or micropore filter. To avoid waste from splashing back into the vent, the inlet line should be at least 5 cm (2 in.) longer than the outlet line.

Instead of a peristaltic pump, a vacuum pump (Fig. 2.13) similar to that for sterile filtration may be used. If possible, the same pump may be used for both functions; however, a trap would be needed to prevent waste from entering the pump. The effluent should be stored in a tank, and after the completion of work, a disinfectant such as hypochlorite should be added. To avoid fluid or aerosol from being carried over, a hydrophobic micropore filter (e.g. Pall Gelman) and a second trap should be installed in the line to the pump. Also, it is advisable to never draw air through a pump from a hypochlorite reservoir because the free chlorine can corrode the pump and be harmful.

2.2.1.5 Inverted Microscope

Inverted microscopes (Fig. 2.14a) are widely used by the laboratories for observing the cells in culture. There are many simple and affordable inverted microscopes available in the market, and one with a phototube for digital recording or viewing is recommended, which can be connected to the camera/monitor (Fig. 2.14b) for taking pictures of the specimen. The majority of inverted microscopes used for cell culture have magnification objective lenses with magnifications of 4×, 10× and 20× and often even higher magnifications.

The Nikon labelling ring is a useful accessory for inverted microscopes. This component is incorporated in the nose piece as an alternative to the objective and can be used to label the bottom of the dish where a colony or patch of cell can be found.

Fig. 2.13 Liquid suction system/vacuum pump receiver. (https://www. directindustry.com/prod/ ibs-integra-biosciences/ product-39224-290127. html)

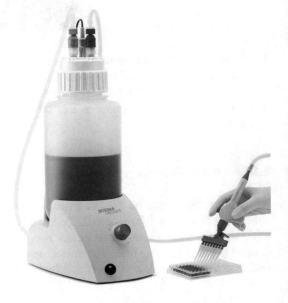

2.2.1.6   Centrifuge

Centrifugation is used to increase the concentration of cells or to remove a reagent or media from cell suspensions on a regular basis. In order to avoid cell disruption, a low-speed bench-top clinical centrifuge (Fig. 2.15a) with proportionally controlled braking must be used. For satisfactory cell sedimentation, 80–100 g (gravitational force) is sufficient. Higher speeds can cause damage to the pellet or cells, as well as promote agglutination. If large-scale suspension cultures are planned, a large-capacity refrigerated centrifuge (Fig. 2.15b), such as 4 × 1 L or 6 × 1 L, will be needed.

2.2.1.7   Cell Counter

A cell counter (Fig. 2.16a, b) is extremely useful and is needed when more than two or three cell lines are carried and also for determination of precise quantitative growth kinetics. Cultured cell counting can be done in a number of different indirect and direct methods.

Haemocytometer Slide

The direct method uses an etched graticule slide along with the thick coverslip (Fig 2.17). It is the easy and cheapest way in assessing the cell viability by using dye exclusion method.

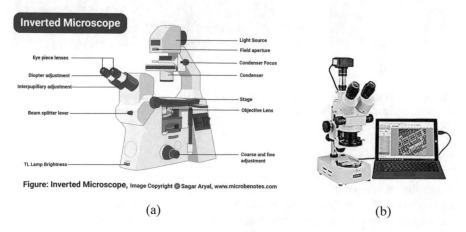

(a)

(b)

Fig. 2.14 Inverted microscope: (**a**) a simple inverted microscope, (**b**) an inverted microscope linked to a CCD camera and monitor. (https://microbenotes.com/inverted-microscope/, https://www.microscopemaster.com/omax-90x-usb3-18mp-digital-trinocular-zoom-stereo-microscope.html)

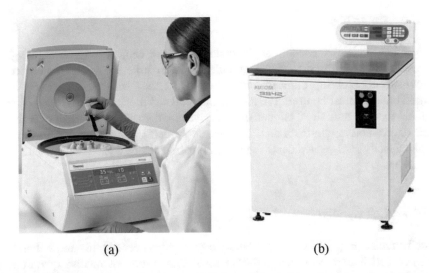

(a)   (b)

Fig. 2.15 Centrifuge: (**a**) a bench-top clinical centrifuge, (**b**) a large-capacity refrigerated centrifuge. (https://www.businesswire.com/news/home/20160204005001/en/New-Benchtop-Centrifuge-Incorporates-Hybrid-Rotor-for-Application-Flexibility, https://www.centrifuge.jp/products/model-9942/)

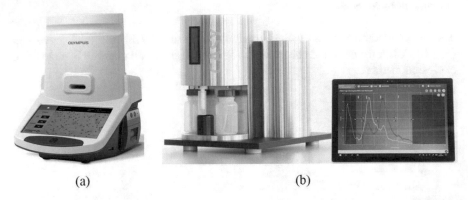

(a)   (b)

Fig. 2.16 Cell counter: (**a**) automated cell counter, (**b**) CASY electronic cell counter. (https://www.selectscience.net/products/cell-counter-model-r1/?prodID=204303, https://www.bioke.com/webshop/ols/5651738.html)

Electronic Cell Counter

A number of companies now sell models that range in complexity from basic particle counting to automated cell counting and size analysis (Fig. 2.16b). Bench-top flow cytometers are an alternative to electronic particle counting and can also provide additional parameters, such as viability via diacetyl fluorescein (DAF) uptake, apoptotic score and DNA material, if required.

Cell Sizing

Most midrange and high-range cell counters (Scharfe, Beckman Coulter) provide cell size analysis as well as the ability to download data to a PC, either directly or through a network.

2.2.1.8 CCD Camera and Monitor

In the lab, digital cameras and sensors have proven to be very useful. It is advised to choose a camera with a high resolution but not a high sensitivity, as the normal sensitivity is normally adequate and high sensitivity will result in over illumination. For phase contrast observation in living populations, black and white typically provide greater resolution and is very satisfactory. For record shots, a still digital camera would suffice; however, a charge-coupled device (CCD) camera (Fig. 2.14b) can facilitate real-time viewing and can be used for time lapse recording.

2.2.1.9 Dissecting Microscope

A dissecting microscope (Fig. 2.18) will be required to dissect tiny pieces of tissue and for counting and picking small colonies in agar plates. It can also be used to count monolayer colonies.

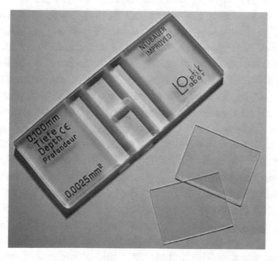

Fig. 2.17 Plastic haemocytometer. (https://www.indiamart.com/proddetail/plastic-hemocytometer-1222249697.html)

Fig. 2.18 Dissecting microscope (https://www.indiamart.com/proddetail/dissecting-microscope-18745403433.html)

2.2.2 Incubation and Culture

2.2.2.1 Incubator

If a hot room is unavailable, an alternative dry incubator may be needed. Even in a hot room, having another incubator next to the hood for trypsinisation is often helpful. The incubator should be huge enough for each person, varying from 50 to 200 L (1.5–6 ft³), and should have temperature control to within 0.2 °C, a safety thermostat that shuts down the incubator if it overheats or, better still, controls it if the first thermostat fails and forced air ventilation. The incubator (Fig. 2.19a) should be corrosion-resistant and easy to disinfect. Since temperature regulation is usually easier in a smaller incubator and if one half fails or needs to be washed, the other can still be used; a double-chamber or two incubators stacked one over the other, separately controlled, are preferable to one big incubator. Furthermore, one may be used for daily access while the other for limited access. Several incubators have a heated water jacket that distributes heat uniformly in the cabinet, preventing cold spots from forming. In the case of a furnace malfunction or a power outage, these incubators can maintain their temperature for longer. New high-efficiency insulation and diffuse surface heater components, on the other hand, have virtually removed the need for a water jacket and made moving the incubator even easier. Incubator shelving is generally perforated to allow for air circulation. In monolayer cultures, however, the perforations may cause inconsistencies in cell distribution, with differences in cell density following the pattern of spacing on the shelves. The differences may be due to areas that cool down more rapidly when the door is opened, or they could be caused by convection currents produced over points of contact relative to holes in the shelf. In experiments where uniform density is necessary, flasks and dishes should be put on a ceramic tile or metal tray, even if no problem occurs during routine maintenance.

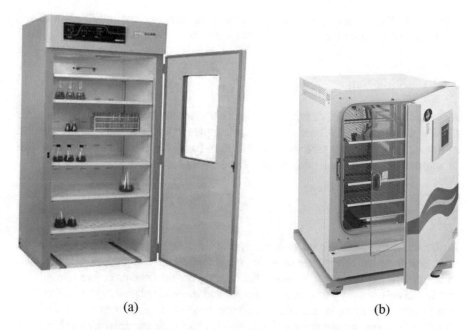

(a)   (b)

Fig. 2.19 (a) Standard incubator, (b) CO_2 incubator. (https://labincubators.net/products/large-capacity-microbiological-incubators, https://www.indiamart.com/proddetail/co2-incubator-19050 641548.html)

2.2.2.2   Humid CO_2 Incubator

While cultures may be incubated in sealed flasks in a normal dry incubator or a hot space, some vessels, such as Petri dishes or multi-well plates, need a regulated environment with high humidity and elevated CO_2 stress. Placing the cultures in a plastic box or chamber (Bellco, MP Biomedicals) is the cheapest way to regulate the gas phase: seal the jar after gassing it with the appropriate CO_2 mixture. Have an open dish of water if the container is not fully filled with dishes to raise the humidity inside the chamber.

CO_2 incubators (Fig. 2.19b) are more costly, but their ease of use and better control of CO_2 stress and temperature (anaerobic jars and desiccators take longer to warm up) make them worthwhile. A controlled environment is accomplished by using a humidifying tray and regulating CO_2 stress with a CO_2-monitoring system, which draws air from the incubator into a sample chamber, tests CO_2 concentration and injects pure CO_2 into the incubator. To keep both the CO_2 level and the temperature uniform, air is circulated around the incubator by using a fan. In addition, since the walls of dry, heated wall incubators appear to stay dry even at high relative humidity, they encourage less fungal contamination on the walls.

The size of the incubator needed will be determined by the number of people that will be using it as well as the types of cultures that will be used. Incubators, particularly humidified ones, require frequent cleaning, so the interior can disassemble easily without leaving inaccessible crevices or corners.

2.2.2.3 Water Bath

A water bath (Fig. 2.20) is a type of laboratory apparatus that consists of a container filled with hot water. It is used to incubate samples in water for an extended period of time at a constant temperature. A cell culture facility requires a 37 °C water bath to warm up the media and other reagents used by cells.

2.2.2.4 Roller Racks

Roller racks are widely used to scale up the monolayer culture. The selection of apparatus depends on the type of scale that includes size and amount of the bottles to be rolled. The scale can be determined using the number of cells required, the maximum cell density that can be achieved, and the surface area of the bottles. The bottles with a diameter of 125 mm (5 in.) and lengths ranging from 150 to 500 mm (6–20 in.) are commonly desired. The overall yield is determined by the length of the bottle, but it is constrained by the size of the rack; the height of the rack defines the number of tiers (i.e. rows) of bottles. The larger racks are less expensive than the small ones, but the small ones allow you with several benefits: (1) allows you to continue operating if one rack needs maintenance, (2) allows you to create your racks steadily and (3) easier to spot in a hot room.

2.2.2.5 Magnetic Stirrer

A magnetic stirrer (Fig. 2.21) is a device commonly used in laboratories which consists of a rotating magnet or a stationary electromagnet that creates rapid stirring action. This helps in quick mixing of solution and chemicals. Any stirrer can be used to dissolve chemicals easily, but the stirrer for enzymatic tissue disaggregation or suspension culture requires a special stirrer with conditions as follows:

Fig. 2.20 Water bath. (https://www. industrybying.com/ water-bath-shakers-bio-technics-LA. SH.WA.744331/)

1. The motor must not heat the culture, i.e. use either the rotating-field type or belt type of drive from an exterior motor.
2. The speed should be limited to 50rpm.
3. The low rpm torque should be able to stir up to 10 L of fluid.
4. The stirrer location should be independently controlled.

2.2.2.6 Temperature Recorder

One can track cell freezing, frozen storage, sterilisation ovens and incubators with the help of recording thermometer (Fig. 2.22) ranging from below −70 °C to about +200 °C, with a single instrument equipped with a resistance thermometer or thermocouple with a long Teflon-coated lead. Temperature management in ovens, incubators and hot rooms should be checked on a daily basis for consistency and stability.

Temperature management in ovens, hot rooms and incubators should be checked on a daily basis for stability and consistency. Recording thermometers should be permanently installed in the hot room, autoclave and sterilising oven with dated records maintained to monitor for suspicious activity on a regular basis, particularly in the event of a problem.

2.2.2.7 Culture Vessels

The choice of culture vessels is influenced by a number of factors including:

• The sampling regime (i.e. are the samples to be obtained concurrently or at intervals over time?).

Fig. 2.21 Magnetic stirrer. (https://www.indiamart. com/proddetail/magnetic-stirrer-4196517588.html)

- If the cell is grown in monolayer or suspension.
- The needed yield (number of cells).

It's essential to specifically mark non-sterile and sterile plastics, as well as non-tissue culture and tissue culture classes, and store them separately. If an appropriate washup and sterilisation service is available, glass bottles with flat sides can be used instead of plastic bottles. Petri dishes (Fig. 2.23a) are significantly less costly than flasks (Fig. 2.23b), but they are more susceptible to spillage and contamination. They are worth considering, at least for use in experiments if not for normal cell line propagation, depending on the pattern of work and the sterility of the environment. Petri dishes are incredibly beneficial for colony-forming assays that require colonies to be stained, counted or separated at the end of the experiment.

2.2.3   *Preparation and Sterilisation*

2.2.3.1   Washup

Sinks or Soaking Baths

Sinks or soaking baths ought to be deep enough just to completely immerse all of the glassware in detergent when soaking, but not so deep that the weight of the glass breaks smaller items at the rim. A sink with a width of 400 mm (15 in.), a length of 600 mm (24 in.) and a depth of 300 mm (12 in.) is perfect.

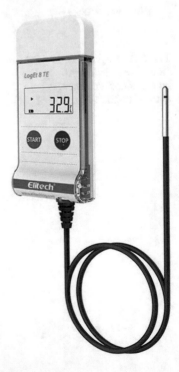

Fig 2.22 Laboratory temperature recorder. (http://www.elitechlog. com/ laboratory-temperature-monitoring/)

Glassware Washing Machine

When the number of glassware becomes overwhelming, an automatic washing machine might be worth considering (Fig. 2.24).

Pipette Washer

Glass pipettes can be cleaned in a regular siphon-type washer (Fig. 2.25) with ease. To prevent uncomfortable raising of the pipettes, the washer should be mounted just above floor level, rather than on the bench, and should be attached to both the deionised and standard cold water sources, so that the final few rinses can be completed in deionised water. A quick changeover valve should be inserted into the deionised water feed line if at all necessary.

Pipette Dryer

Pipettes may be directly moved to an electric dryer (Fig. 2.26) if a stainless steel basket is used in the washer. Pipettes may also be dried on a rack or in a conventional drying oven.

Drying Oven

This should have a wide volume, be fan-driven, and be capable of reaching 100 °C. (Fig 2.27). The sterilising oven can also serve as a drying oven in operation.

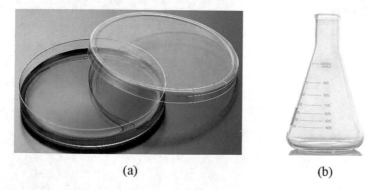

(a)   (b)

Fig. 2.23 Culture vessels, (a) Petri dish (b) conical Erlenmeyer flask. (https://jet.en.alibaba.com/product/60764448373-218294738/9cm_petri_dishes_sterile_petri_dis h_90mm.html, https://www.indiamart.com/proddetail/erlenmeyer-flask-9207089648.html)

2.2.3.2 Preparation of Media and Reagents

Water Purifier

Rinsing glassware, diluting concentrates and dissolving powdered media all require purified water. The first of these criteria is normally fulfilled by deionised or reverse osmosis water, but the second and third necessitate ultrapure water (UPW), which necessitates a three- or four-stage method (Fig. 2.28). To prevent infection with algae or other microorganisms, filtered water should not be preserved but should be recycled through the apparatus on a daily basis. Before reuse, any reservoirs or tubing in the system should be screened for algae on a daily basis, washed with hypochlorite and detergent (e.g. Chloros or Clorox) and thoroughly rinsed in filtered water.

Balances

It might be much cheaper to prepare a few reagents in the laboratory itself for which an electronic balance with an automatic tare (Fig. 2.29) will be required. Depending on the size of the service, it should be efficient in weighing objects ranging from 10 mg to 100 g or even 1 kg. It could be more cost-effective to obtain two balances, one coarse and one fine, so the outlay will be comparable and the convenience and precision will be enhanced.

Fig. 2.24 Laboratory glassware washing machine (https://www.indiamart.com/proddetail/lancer-910-lx-laboratory-free-standing-glassware-washer-dryers-18738894573.html)

Fig. 2.25 Digital pipette washer. (https://www.indiamart.com/proddetail/automatic-pipette-washer-18828975855.html)

Fig. 2.26 Electric pipette dryer. (https://www.thomassci.com/Laboratory-Supplies/Pipet-Controllers/_/ELECTRIC-PIPET-DRYERS)

Hot plate Magnetic Stirrer

A magnetic stirrer with a hot plate (Fig. 2.30), in addition to the ambient temperature stirrers used for suspension cultures and trypsinisation, may be useful for speeding the dissolution of certain reagents.

Fig. 2.27 Laboratory drying oven. (https://www.stanhope-seta.co.uk/product/laboratory-oven/)

Fig. 2.28 Laboratory reverse osmosis and ion exchange water purification system. (https://www.indiamart.com/proddetail/laboratory-water-purification-system-20188644288.html)

pH Meter

For the preparation of media and special reagents, a simple pH metre (Fig. 2.31) is necessary. A pH metre will be needed in the preparation of stock solutions, for routine quality control tests during the preparation of media and reagents and also when phenol red cannot be used.

Conductivity Meter

It is important to conduct quality control procedures when preparing solutions in the laboratory to avoid errors. A conductivity metre (Fig. 2.32) may be used to check ionic concentration against a known standard.

Osmometer

An osmometer (Fig. 2.33) is a valuable tool for adjusting new formulations, checking solutions as they are being prepared, serving as a second line of quality control and compensating for the addition of reagents to the medium.

Osmometers normally operate by lowering a medium's freezing point or raising its vapour pressure. Since you might choose to test an important or limited reagent on occasion, choose one with a small sample volume (less than 1 mL), and the precision ($+10$ mosmol/kg) might be less critical than the reagent's importance or scarcity.

Bottling: Automatic Dispensers

When preparing reagents and media in bulk, aliquoting into containers for sterilisation and use is normally needed. For quantities up to around 50 mL, bottle-top dispensers (Fig. 2.11b) are suitable; beyond that, gravity dispensing from a tank, such as a graduated bottle or plastic bag, is appropriate if precision is not critical. A peristaltic pump is preferable if the amount dispensed is more important. The precision and volume of the dispenser are determined by the tube diameter and dispensing time. A long dispensing time with a narrow distribution tube is more precise, but a wide bore tube is quicker.

Fig. 2.29 Electronic balance. (https://shpuchun. en.made-in-china.com/ product/XytmhvBMkbcV/ China-Ja1002-10mg- Digital-Analytical- Balance-Electronic- Weighing-Precision-Scale- for-Laboratories.html)

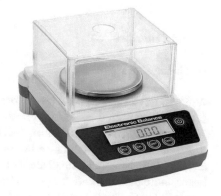

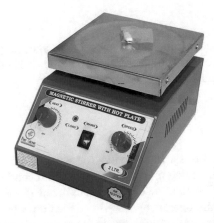

Fig. 2.30 Magnetic stirrer with hot plate. (https://www.amazon.in/Bio-Gene-Magnetic-Stirrer-Plate/dp/B07X95Y44C/ref=asc_df_B07X95Y44C/?tag=googleshopdes-21&linkCode=df0&hvad id=397008728335&hvpos=&hvnetw=g&hvrand=15973143602899877898& hvpone=&hvptwo=&hvqmt=&hvdev=c&hvdvcmdl=&hvlocint=&hvlocphy=1007820&hvtargid =pla-837281400171&psc=1&ext_vrnc=hi)

Fig. 2.31 Laboratory pH
meter. (https://www.
indiamart.com/proddetail/
laboratory-ph-
meter-15063679812.html)

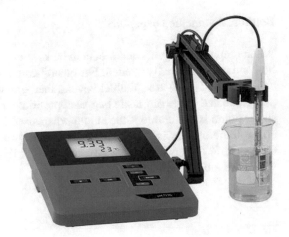

2.2.3.3 Sterilisation

Sterilising Oven

While most sterilisation can be performed in an autoclave, sterilising glassware by dry heat is preferred because it prevents chemical contamination from steam condensate and corrosion of pipette cans. To ensure even heating during the load, such sterilisation would necessitate a high-temperature (160–180 °C) fan-powered oven (Fig. 2.34). When only a small amount of glassware is used, it is more cost-effective, faster, more uniform and cheaper to use two small ovens rather than one large one. You will also be better protected if breakdown takes place.

Fig. 2.32 Bench-top
conductivity meter.
(https://www.coleparmer.
in/i/
oakton-con-550-benchtop-
conductivity-meter-
kit/3541935)

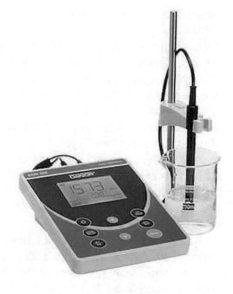

Fig. 2.33 Laboratory
osmometer. (https://www.
medicalexpo.com/prod/
advanced-instruments-inc/
product-80774-671771.
html)

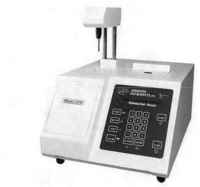

Sterilisation Filters

For sterile filtration, reusable equipment is available, with sizes varying based on the nature of the procedure. Positive pressure filtration, which includes a pump upstream from a pressure tank, is usually used with large volumes of 10 L or more, whereas smaller volumes may be done by a smaller reservoir and a peristaltic pump downstream. However, disposable filters ranging in size from 25 mm syringe adapters to 47 mm in panel, bottle-top adapters or filter flasks are still used in most laboratories. It's also a smart thing to have a limited stock of bigger sizes on hand.

Steam Steriliser (Autoclave)

An autoclave (Fig. 2.35b) is a laboratory apparatus that uses steam under controlled pressure. This system is used to sterilise a number of solutions, contaminated materials, media, glassware and discarded cultures on a daily basis. One of the simplest and cheapest sterilisers known is a domestic pressure cooker that produces 100 kPa (1 atm, 15 lb/in.[2]) above ambient pressure. To sterilise the materials for the class, a small- or medium-sized oven or autoclave would be considered sufficient. Before or after sterilisation, a "wet" cycle (water, salt solutions, etc.) is conducted without the chamber being evacuated. Dry products (instruments, swabs, screw caps and so on) enable the chamber to be evacuated or the air replaced by downward displacement before sterilisation to allow efficient access of hot steam. Moreover, a medium-sized unit can heat up and cool down faster, and it will be more cost-effective for small loads.

Small autoclaves (Fig. 2.35a) have their own steam generator (calorifier), whereas larger autoclaves (Fig. 2.35c) may include a self-contained steam generator, a separate steam generator or the ability to use a steam line. If high-pressure steam is available on demand, it would be the most cost-effective and straightforward way to heat and pressurise the autoclave.

Fig. 2.34 Hot air sterilising oven. (https:// www.scimmit.com/ product-detail/ hot-air-sterilizing-oven/)

2.2.3.4   Storage

Consumables

Tubes, culture vessels and other consumables may be stacked on shelves above adjacent benches or on carts near the hoods. Stable liquids, such as water and PBS, need shelf space as well.

Refrigerators and Freezers

A domestic refrigerator (preferably without an auto defrost freezer) is an adequate and economical instrument for storing reagents and media at 2–8 °C in small cell culture laboratories. It is advisable to have a cold room if the number of people is more than three or four and the laboratory is large. But the walls of the cold room should be cleaned regularly, and old stock should be eliminated in order to minimise the contamination.

Many cell culture reagents need to be stored at −5 °C to −20 °C, and, hence, ultradeep freezer (i.e. −80 °C freezer) is not needed for storage. A domestic freezer is a more cost-effective solution than a laboratory freezer. Although most reagents can tolerate temperature fluctuations in an autodefrost (i.e. self-thawing) freezer, antibiotics and enzymes can't be stored in autodefrost freezer. Most of the essential components of serum are small proteins, polypeptides and simpler organic and inorganic compounds that may be resistant to cryogenic damage, particularly when stored in volumes greater than 100 mL.

Cryostorage Containers

There are mainly two types of liquid nitrogen storage, i.e. liquid and vapour phase, which are available as wide-necked or narrow-necked storage containers. Vapour phase systems are needed for storing biohazardous materials because they reduce the risk of explosion with cryostorage tubes, while liquid phase systems have longer static holding times and are therefore more cost-effective.

For any small scale laboratory, a 35 L freezer with both a narrow neck and storage in canes and drawers in a rack system should accommodate around 500–1000 ampoules for a small laboratory. Larger freezers will accommodate more than 10,000 ampoules and will include models with liquid nitrogen perfused walls, reducing nitrogen consumption and ensuring safe storage without liquid nitrogen in the storage chamber. Narrow-necked containers (Fig. 2.36a) evaporate nitrogen at a slower rate and are more cost-effective, but wide-necked containers (Fig. 2.36b) are easier to access and have a greater storage space.

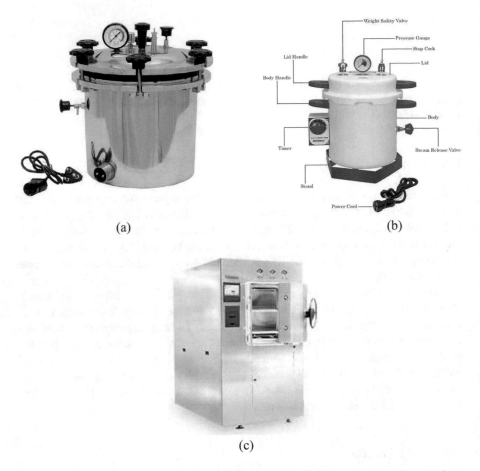

Fig. 2.35 Autoclave: (**a**) small laboratory autoclave, (**b**) labelled laboratory autoclave, (**c**) large laboratory autoclave

Controlled Rate Freezer

While cells can be frozen by simply putting them in an insulated box and freezing them at −70 °C, some cells may need different cooling rates or complex programmed cooling curves. A programmable freezer (e.g. CryoMed, Planer) (Fig. 2.37) allows for variable cooling levels by regulating the rate of liquid nitrogen injection into the freezing chamber, which is controlled by a preset programme.

2.2.3.5   Supplementary Laboratory Equipment

Computers and Networks

Whether or not a computer or terminal is placed in the tissue culture laboratory, keeping track records of cell line maintenance, primary culture and tests makes recovery and study easier in the future. The easiest way to keep track of cell lines is to use a computer database that can also be used to keep track of the nitrogen freezer's inventory. Storage management of plastics, reagents and media can also be improved in larger laboratories. Individual computers can be backed up centrally on a daily basis, and information entered at one place can be recovered elsewhere, which is a major benefit of networking. Photographs taken in a tissue culture laboratory, for example, may be downloaded to a central server and recovered in an office or writing area.

Upright Microscope

For autoradiography, mycoplasma identification and chromosome analysis, an upright microscope (Fig. 2.38) may be needed in addition to an inverted microscope. For mycoplasma testing through fluorescence and fluorescent antibody observation, choose a high-quality laboratory microscope with standard bright field optics up to 100× objective magnification, phase contrast up to at least 40× and ideally 100× objective magnification and fluorescence optics with epi-illumination and 40× and 100× objectives. For photographic archives of permanent preparations, a digital or CCD camera should be installed.

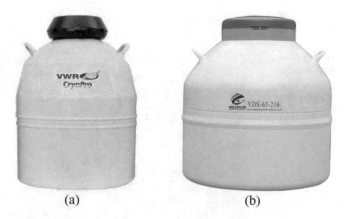

(a)   (b)

Fig. 2.36 Cryostorage containers: (**a**) narrow-necked cryostorage container, (**b**) wide-necked cryostorage container. (https://in.vwr.com/store/product/17612604/cryogenic-storage-tanks-cc-series-cryopro, https://www.alibaba.com/product-detail/Wide-neck-laboratory-series-cryogenic-vessel_1600114 040118.html)

Low-Temperature Freezer

Many tissue culture reagents are stored at 4–20 °C, although certain products from cultures, reagents and drugs may need a temperature of 70–90 °C, at which stage most of the water is frozen and most chemical and radiolytic reactions are seriously hindered. A freezer set to 70–90 °C can also be used to freeze cells within an insulated container. While a chest freezer is more effective at maintaining a low temperature with minimal power consumption, vertical cabinets take up less floor space and are easier to handle. If you go for a vertical cabinet, ensure it has separate compartments. Low-temperature freezers (Fig. 2.39) produce a lot of heat that has to be dissipated in order for them to run properly (or at all). These freezers ought to be located in a well-ventilated or air-conditioned space where the ambient temperature does not exceed 23 °C. If this is not practical, invest in a freezer intended for tropical use; else, you will have regular maintenance issues, and the freezer will have a shorter operating life.

Confocal Microscope

When examined with a confocal microscope (Fig. 2.40), cytological studies of fluorescently tagged cells also benefit from increased resolution. This technique helps the microscope to observe an "optical segment" through the specimen in one focal plane, preventing intrusion from neighbouring cells or organelles that are not in the same focal plane. The data is digitally stored and can be interpreted in a variety of ways, including creating a vertical segment across the sample, which is especially helpful when observing three-dimensional communities like spheroids or filter wells.

Fig. 2.37 CryoMed controlled rate freezer. (https://www.thermofisher. com/order/catalog/product/ TSCM17PA#/TSCM17PA)

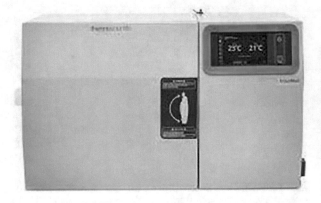

Fig. 2.38 An upright microscope. (https://www.indiamart.com/proddetail/upright-microscope-iled-primo-star-19209523188.html)

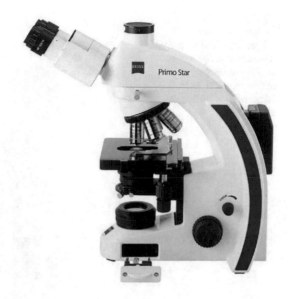

PCR Thermal Cycler

Amplification and identification of complex DNA sequences are used in a range of ancillary techniques in cell line validation, such as mycoplasma detection and DNA profiling. If you want to use these methods, they make use of polymerase chain reaction (PCR) and need a PCR thermal cycler (Fig. 2.41).

2.2.3.6   Specialised Equipment

Microinjection Facilities

Micromanipulators may be used to inject directly into a cell for dye injection or nuclear transplantation, for example.

Colony Counter

Monolayer colonies can be counted by eye or using a dissecting microscope with a felt-tip marker to label off the colonies, but if there are a lot of plates to count, an automatic counter (Fig. 2.42a) or a digital colony counter (Fig. 2.42b) will come in handy. The most basic approach employs an electrode tipped marker pen that counts as you land on a colony. They usually use a magnifying glass to better see the colonies. The next step up in complexity and expense is a programmable electronic colony counter, which uses image analysis software to count colonies. These counters are extremely quick, can handle contiguous colonies and can distinguish between colonies of various diameters.

Fig. 2.39 Low-temperature freezer. (https://profilab24.com/en/laboratory/refrigerators-cooling-technology/carlo-erba-ultra-low-temperature-freezer-lab-ult-338)

Centrifugal Elutriator

The centrifugal elutriator is a centrifuge that has been modified to isolate cells of various sizes. The system is expensive, but it is extremely efficient, particularly for high cell yields.

Flow Cytometer

Flow cytometer (Fig. 2.43) is capable of analysing cell populations according to a vast range of factors that include fluorescence, absorbance and light scatter. A two-dimensional or three-dimensional representation of multiparametric analysis is possible. These devices are often referred to as flow cytometers in the analytical mode,

Fig. 2.40 A confocal microscope. (https://www. zeiss.com/microscopy/int/ products/confocal-microscopes/lsm-900-for-materials-non-contact-surface-topography-in-3d. html)

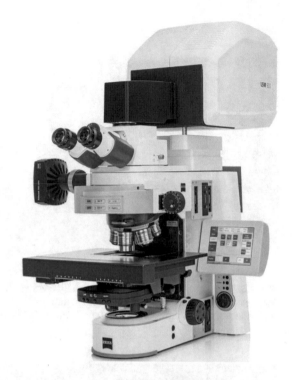

Fig. 2.41 PCR thermal cycler. (https://www. indiamart.com/proddetail/ pcr-thermal-cycler-19171669148.html)

but the signals they provide can also be used in a fluorescence-activated cell sorter to separate individual cell populations with a high degree of resolution. The price is huge ($100,000–200,000), and the best results come from a professional operator. Cheaper bench-top machines (Accuri, Guava) that can calculate cell number and a range of fluorescent parameters are also available.

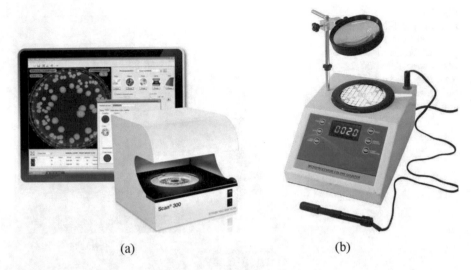

(a)   (b)

Fig. 2.42 Colony counter: (**a**) automated colony counter, (**b**) digital colony counter. (https://www.indiamart.com/proddetail/automatic-colony-counters-14641025948.html, https://www.amazon.in/Instruments-Colony-Counter-Digital-Table/dp/B074J4K9RZ)

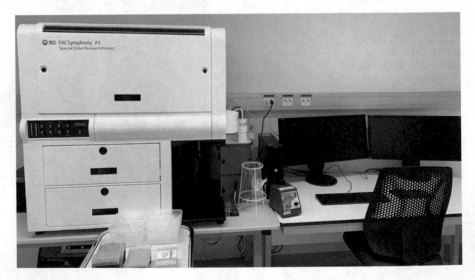

Fig. 2.43 A flow cytometer. (https://wp.unil.ch/fcf/2019/11/bd-symphony-available-at-the-agora/)

Chapter 3
Good Laboratory Practices in Animal Cell Culture Laboratory and Biosafety Measures

Kumari Yukta, Mansi Agarwal, Mekhla Pandey, Khushi Mittal, Vidushi Srivastava, and Shalini Mani

3.1 Introduction

The history of cell culture informs us that while preparing cells for routine usage, several fundamental characteristics of optimal practice must be addressed. The impact of poorly managed or unsatisfactory cell culture procedures becomes increasingly important as cell culture techniques and applications get more complex. The good cell culture practice (GCCP) guidance was created to raise knowledge of a wide range of essential issues in cell culture among employees who are new to using cells in their work, as well as to remind others of the fundamental principles of cell and tissue culture good practices. By encouraging greater international harmonisation, rationalisation and standardisation of laboratory practices, quality control systems, safety procedures, recording and reporting and compliance with laws, regulations and ethical principles, the goal of GCCP is to promote the maintenance of these standards and to reduce uncertainty in the development and application of animal and human cell and tissue culture procedures and products (Hartung et al., 2002). The goal of the GCCP is to foster consensus among all parties involved in the use of cell and tissue culture systems in order to:

1. Establish and maintain best practices in cell and tissue culture.
2. Promote effective quality control systems.
3. Facilitate education and training.
4. Assist journal editors and editorial boards.
5. Assist research funding bodies.

K. Yukta · M. Agarwal · M. Pandey · K. Mittal · V. Srivastava · S. Mani (✉)
Centre for Emerging Diseases, Department of Biotechnology, Jaypee Institute of Information Technology, Noida, India

6. Enable the understanding and application of inferences based on in vitro work.

The GCCP Guidance is based on the six operating concepts listed below (Coecke et al., 2005):

1. Establishing and maintaining an adequate grasp of the in vitro system, as well as the relevant components that may influence it
2. Maintaining the integrity, validity and repeatability of any work undertaken by ensuring the quality of all materials and processes, as well as their usage and application
3. Records of the information needed to track the materials and procedures used, to allow the work to be repeated and to allow the intended audience to comprehend and evaluate the work
4. Establishing and maintaining suitable safeguards to protect people and the environment from any potential dangers
5. Adherence to applicable rules and regulations, as well as ethical values
6. Providing the appropriate and satisfactory education and training for all personnel, to encourage high-quality work and safety

A cell culture laboratory training program should contain the following important culture techniques, procedures and regulations (Geraghty et al., 2014).

Basic Laboratory Procedures
1. Fundamental laboratory methods.
2. Understanding the nature and purpose of standard operating procedures (SOPs).
3. Microscopy.
4. Centrifugation.
5. Operation of the autoclave.
6. Laminar airflow or microbiological safety cabinets, incubators, and cryostorage facilities are used and maintained.
7. Maintenance of equipment.
8. Design and safety in the laboratory.
9. In vitro work risk assessment and risk management.
10. Controlling the quality.
11. Disposal of waste.
12. Optimised procedures for cleaning, disinfection and fumigation.

Basic Culture Procedures
1. Disinfection and sterilisation, as well as sterile technique and aseptic manipulation
2. Culture media preparation, storage and monitoring
3. Techniques for isolating cells and tissues in culture
4. Cell counts and viability testing
5. Subculturing
6. Tests for sterility or bioburden
7. Testing for mycoplasma
8. Cell and tissue cryopreservation, storage and recovery

Advanced and Special Culture Procedures

1. Cell characterisation and authentication are examples of sterile procedures and aseptic manipulation
2. Methods of cell isolation and purification
3. Banking of cells and tissues
4. Differentiation induction
5. Techniques of complex culture (e.g. co-culture, culture on filter inserts, perfusion cultures)
6. Stable cell line selection and transfection
7. Use of bioreactors

Documentation and Record-Keeping

General information and policies of the organisation responsible for the laboratory (operational issues, safety, quality standards)

1. Laboratory data, equipment records, storage records
2. Records of occupational health and training
3. Records of safety

    Records, manuals and information related to quality assurance

Laws and Regulations

All laboratory personnel should be educated on the following institutional, national and international procedures, guidelines, regulations and legislation that apply to their work:

1. Organisation/institute rules and policies
2. Responsibility allocation
3. Microorganism containment
4. Regulations on the use of animals and animal cells and tissues
5. Regulations on the use of human cells and tissues

3.2 Biosafety Measures in Animal Cell Culture Laboratory

A cell culture laboratory has a number of specific hazards associated with handling and manipulating human or animal cells and tissues, as well as toxic, corrosive or mutagenic solvents and reagents, in addition to the safety risks common to most everyday work places such as electrical and fire hazards (Chosewood & Wilson, 2009). Accidental punctures with syringe needles or other contaminated sharps, spills and splashes on skin and mucous membranes, ingestion through mouth pipetting and inhalation exposures to infected aerosols are the most common of these dangers.

The primary goal of any biosafety protocol is to prevent or eliminate potential biological agent exposure in laboratory employees and the surrounding environment. The utmost vital element of safety in a cell culture laboratory is the firm adherence to typical microbiological practices and techniques.

Biosafety Levels The document of Biosafety in Microbiological and Biomedical Research Facilities, written by the Centers for Disease Control (CDC) and the National Institutes of Health (NIH) and published by the US Department of Health and Human Services, contains guidelines and proposals for biosafety in the United States (Chosewood & Wilson, 2009). The document establishes four levels of containment, known as biosafety levels 1 through 4, and outlines the microbiological techniques, safety equipment and facility safeguards for each level of risk associated with handling a specific agent. Biosafety in Biomedical Laboratories defines the levels (the BMBL). The BMBL's biosafety level designations define specific practices, as well as safety and facility standards. There are several conducts in equipment, practices and lab features that can augment the biosafety and can encompass biocontainment.

3.2.1   Biosafety Level 1 (BSL-1)

BSL-1 is the most basic degree of protection used in most research and clinical laboratories, and it's suitable for compounds that aren't known to cause disease in healthy humans. On an open bench or under a fume hood, work is done. When working in the laboratory, standard microbiological procedures are followed. Infectious agents or toxins that are not known to reliably cause disease in healthy humans are studied in BSL-1 labs. This lab does not need to be segregated from the rest of the building. Nonpathogenic strain of *E. coli*, *Bacillus subtilis*, *Agrobacterium radiobacter*, *Aspergillus niger*, *Lactobacillus acidophilus*, *Micrococcus luteus*, *Neurospora crassa*, *Pseudomonas fluorescens* and *Serratia marcescens* are examples of microbes utilised in BSL-1 labs (Burnett et al., 2009).

3.2.2   GLPs for Biosafety Level 1

1. No mouth pipetting is permitted.
2. Wait until you need to utilise a sharp object before uncovering or unwrapping it.
3. Whenever possible, use disposable scalpels with fixed blades. A scalpel blade without the handle should not be used.
4. Decontamination of all work surfaces on a daily basis, both before and after the job, is completed.
5. Use personal protective equipment such as lab coats, eye protectors, protectors and gloves.
6. Handwashing before and after using the labs.
7. Biohazard warning signs must be posted.
8. Spills in these labs are intended to be decontaminated right away. Infectious materials must be decontaminated before being discarded, which is usually done by autoclaving.

3.2.3 Biosafety Level 2 (BSL-2)

Bacteria, viruses and organisms linked to human diseases are among the biological materials employed in a BSL-2 laboratory. These labs are used to research moderate-risk infectious agents or toxins that can cause harm if breathed, swallowed or skin contact is made. As a result, BSL-2 is acceptable for moderate-risk chemicals that cause disease of various severities when ingested or exposed through the skin or mucous membrane. The majority of cell culture facilities should be BSL-2, although the specific standards vary depending on the cell line utilised and the sort of work done (Herman & Pauwels, 2015).

3.2.4 GLPs for Biosafety Level 2

1. Personal protective equipment (PPE), such as sterile clothing and gloves, should be worn. Depending on the situation, eye protection and face shields can also be worn.
2. The laboratory's doors must be self-closing and locked.
3. There should be a sink and an eyewash facility nearby.
4. All actions involving aerosols or splashes that could cause infection must be carried out in a biological safety cabinet (BSC).
5. Depending on the biological risk assessment, BSL-2 labs must contain equipment that can disinfect laboratory waste, an incinerator, an autoclave and/or another method.
6. Biohazard warning signs must be erected.
7. Outside workers are not permitted to enter the premises while the work is being done, as this may raise the danger of contamination.

3.2.5 Biosafety Level 3

The BSL-3 laboratory is required where studies are being done using infectious agents or poisons, which has the ability to get spread through the air and induce potentially fatal effects. BSL-3 laboratory work is mostly specified with suitable government bodies. Laboratory workers are supervised by doctors and may be immunised against the germ they're working with. Before beginning tasks, a BSL-3 office's plan, functional limits, and methodologies should be validated and archived. At least once a year, offices should be archived and checked again. All open controls of biological material should be done under a BSC (preferably Class II or Class III) or in a sealed regulation framework, and special protective clothing should be worn (Herman & Pauwels, 2015).

3.2.6   GLPs for Biosafety Level 3

1. Biosafety cabinets with properly controlled air flow or sealed enclosures must be provided to avoid infection.
2. These labs must be unidirectional to prevent non-laboratory regions from flowing into the lab.
3. Use two self-closing or interlocking doors that are not accessible from the main building corridors.
4. BSL-3 laboratories have restricted and controlled access at all times, as well as sealed windows and wall surfaces and filtered ventilation systems.
5. Near the exit, there must be a hand-free sink and an eyewash station.
6. Wraparound gowns with a solid front, scrub suits or coveralls are frequently required.
7. Access to such a lab is constantly monitored and restricted.

3.2.7   Biosafety Level 4 (BSL-4)

Exotic agents that offer a high individual risk of life-threatening disease from infectious aerosols and for which no therapy is available should be classified as BSL-4. These agents are only allowed to be used in high-security laboratories. BSL-4 laboratories are uncommon. They exist in a few places in the world, such as the United States. Infectious agents or toxins that pose a high risk of aerosol-transmitted laboratory infections and life-threatening sickness for which no vaccination or cure is available are studied in BSL-4 laboratories. These laboratories must be operated with caution and require extensive training. In terms of critical impediments, all work must be conducted in a Class III BSC or in collaboration with the lab specialist in a Class I or II BSC, wearing a full-body, air-provided positive pressing factor suit (Burnett et al., 2009).

3.2.8   GLPs for Biosafety Level 4

1. Before entering, researchers must change their clothes and shower before leaving.
2. The laboratories meet all BSL-3 requirements and are located in safe, isolated zones inside a larger facility or in a separate, dedicated building for this purpose exclusively.
3. Dedicated supply and exhaust ventilation equipment, as well as vacuum and decontamination systems, are also required at this level. Special installation may be required depending on the material and needs.

4. Laboratory staff must wear full-body, air-supplied suits, the most advanced type of personal protective equipment, or a Class III biological cabinet must be present.
5. The laboratory is built to keep toxins from spreading to other areas.

   Other agents may have an antigenic association that is very similar to BSL-4 material; however, there is insufficient data to assign a classification to these agents.
6. These labs are equipped with high-tech features such as specialised supply and exhaust air, vacuum lines, and decontamination zones.

3.2.9 Safe Laboratory Practices

The following suggestions are intended to serve as guidelines for safe laboratory practices only and should not be interpreted as a comprehensive code of practice. Consult your institution's safety committee, and observe local laboratory safety standards and regulations.

- Always put on the proper personal safety equipment. When gloves get contaminated, replace them, and dispose of them with other contaminated laboratory waste.
- Before leaving the laboratory, wash your hands after working with potentially harmful items.
- In the laboratory, do not eat, drink, smoke, handle contact lenses, apply cosmetics or store food for human consumption.
- Adhere to the institution's sharps policies (needles, scalpels, pipettes and broken glassware, for example).
- Minimise the production of aerosols and/or splashes as much as possible.
- Decontaminate all work surfaces with an adequate disinfectant before and after your research, as well as promptly after any spill or splash of potentially contagious material. Even if laboratory equipment is not contaminated, it should be cleaned on a regular basis.
- Decontaminate all potentially infectious materials before discarding them, and report any incidences involving infectious materials exposure to the proper persons (e.g. laboratory supervisor, safety officer).

3.2.10 Common Good Lab Practices (GLPs) in Cell Culture

In addition to specific GCCP, good lab practices (GLPs) are methods used in any laboratory to guarantee that experiments are conducted safely and without injury (Coecke et al., 2005). Microbes are often seen in bioscience laboratories, and some of them can be dangerous to people. Furthermore, experimenting necessitates the use of a variety of substances and laboratory instruments, all of which require expert handling. It is critical to be aware of GLPs in order to avoid accidents and to ensure a prompt response in the event of one.

General GLPs

1. When working in the laboratory, always wear your lab coat.
2. Always know where fire extinguishers, first aid kits, the phone and escape routes are located and how to utilise them.
3. Use a waterproof Band-Aid to cover any cuts or injuries.
4. Do not bring any food or cosmetics inside the lab.
5. All chemical processes that could result in fumes should be carried out in a fume hood with proper headgear.
6. When using test tubes, always point them away from yourself and your neighbours. Keep the test tube at least 15 cm away from your body at all times.
7. Before leaving the lab, always wash your used laboratory glassware.
8. Ensure that laboratory trash is properly disposed of.
9. Before leaving the laboratory, always wash your hands.

GLPs when handling chemicals

1. Become familiar with the various compounds in your laboratory, as well as the safety precautions that apply to them.
2. When working with chemicals, always wear gloves. Before leaving the lab, wash/discard your gloves.
3. When working with chemicals, avoid touching your eyes or face. Depending on the chemical employed, this might produce irritation and redness, as well as other serious side effects.
4. Always remember to replace the lid on the container after applying chemicals.
5. Don't put any leftover chemicals back into the container.
6. Avoid touching chemicals with your bare hands.
7. Never use your mouth to pipette.

GLPs when using lab instruments

1. Turn off the UV light 10 min before using it.
2. Always keep a fume hood as closed as feasible when using it.
3. Do not keep chemicals in the fume hood indefinitely.
4. Before beginning the experiment, always cleanse the counter.
5. When conducting experiments, don't talk.

3.2.11 Waste Segregation

The practice of waste segregation and disposal is an extension of GLPs. Various sorts of trash are generated throughout the experimental process in laboratories. As a result, it is our primary job to autoclave all of the contaminated waste. This is crucial for good trash management and waste minimisation. Many laboratories do not differentiate between hazardous waste and waste that is neither hazardous nor regulated as hazardous. When these diverse forms of waste are combined, the entire must be treated as hazardous waste, which raises the cost of disposal for the waste significantly very high (Alawa et al., 2022). Non-hazardous waste can be disposed of in the regular garbage or sewer when it is safe and permitted by law. This can significantly minimise disposal costs. This is the type of waste separation that is both economical and environmentally sound.

Biohazardous waste is any waste that contains infectious material or that, due to its biological nature, may be detrimental to humans, animals, plants or the environment. Infectious microbiological waste (including contaminated disposable culture dishes and disposable instruments used to transfer, inoculate and mix cultures), pathological waste, sharps and hazardous products of recombinant DNA biotechnology and genetic modification fall into this category (Burnett et al., 2009). The waste products of the cell culture labs are frequently contaminated with:

1. Human, animal or plant pathogens
2. Recombinant nucleic acids (e.g. rDNA)
3. Human/primate blood, blood merchandise, tissues, cultures, cells or different probably infectious material (OPIM)

Before leaving the laboratory, these contaminated wastes should be inactivated (e.g. autoclaved or bleached). Non-inactivated trash should be stored in the generating laboratory, and we will not leave it unattended. The biohazard image should be labelled on waste that is biohazardous to people. Infectious waste should be lined in an uninterrupted manner and inactivated at 1-day intervals.

It is critical for waste minimisation and successful waste management. Because it increases public health protection, waste should be separated according to specific treatment and disposal standards (World Health Organization, 2017). Producers of trash should be responsible for segregation. We can separate waste from culture labs in a variety of methods. Any waste that could cause laceration or puncture injuries must be disposed of as "SHARPS." Sharps must be kept separate from other types of garbage. Non-sharp waste may be mixed with metal sharps and shattered glass, but not with non-sharp waste. Biological waste should not be mixed with chemical waste or other laboratory rubbish, and waste that is to be incinerated should not be mixed with glass or plastics.

3.2.12 Types of Waste

Chemical waste, infectious (biohazard) waste and pathological (large tissue) waste are the three main forms of waste generated in animal cell culture laboratories. This section covers information on various sorts of garbage, as well as their management in cell culture laboratory.

1. General waste (e.g. room waste, which is similar to residential garbage) can be included in the municipal waste stream. Tissue paper, gloves and other items are examples.
2. Highly infectious garbage should be collected on an individual basis wherever possible and autoclaved as soon as practicable. Fluid blood or bulk body fluids, human produced albumin, infectious live or attenuated vaccinations and laboratory wastes are all examples.

   - In addition to infectious waste, small volumes of chemical or pharmaceutical waste are collected. Paraffin/xylene, Wright's stain and Ictotest tablets are among examples. Wastes containing a high concentration of important metals should be collected on a case-by-case basis (e.g. lead thermometers, batteries).
   - If these square measures are headed for combustion, low-level hot infectious waste is also gathered in yellow baggage or containers. Human tissues from pathology and histology labs, as well as corpses from research animal waste, are examples.
   - If sharps are encapsulated, it will be easier to collect them directly within the gold drums or barrels that were used for encapsulation, reducing handling risks.
   - For easier cleanup, double packaging, such as a bag within a holder or instrumentation, is usually advised for dangerous waste and extremely dangerous waste containers (World Health Organization, 2017) (Table 3.1).

3.2.13   Waste Management

In this section, we'll learn about ways for managing and disposing of laboratory waste that poses a chemical hazard, as well as multi-hazardous wastes that pose a combination of chemical, thermal and biological risks. The most basic method for managing laboratory waste is to promote safety while minimising environmental effect and to keep these goals in mind from the start. Before leaving the laboratory, all infectious materials should be disinfected. All infectious items should be destroyed, with the exception of sharps, glasses and plastics (even after disinfection). Sharps must be kept separate from other trash and stored in puncture-resistant containers; all metallic sharps, regardless of their intended purpose, are biohazardous and must be encapsulated before disposal (World Health Organization, 2005).

Disinfect the liquid waste before dumping it into the sewer system. As per good laboratory practice, all biological waste should also be treated before disposal. Biohazardous waste must be appropriately identified and treated, and records must be kept. Personnel who may come into touch with bio-hazardous materials need to be properly trained (World Health Organization, 2005).

The initial responsibility for putting this waste management system in place falls to skilled laboratory staff. These individuals are ideally positioned to comprehend the chemical and physical properties of the materials that are used or created. To

Table 3.1 Summary of different types of waste and division and identification of different containers for collection of respective waste

Type of waste	Colour of container and markings	Type of container
Tissue culture contaminated waste	Autoclave bags	Solid plastic bins
Highly infectious waste	Yellow marked "HIGHLY INFECTIOUS"	Strong, leak-proof plastic bag or container capable of being autoclaved
Other infectious waste (pathological and anatomical waste)	Yellow	Leak-proof plastic bag or container
Sharps	Yellow, marked "SHARPS"	Puncture proof container
Chemical and pharmaceutical waste	Brown	Plastic bag or container
Radioactive waste	–	Lead box, labelled with radioactive symbol
General healthcare waste	Black	Plastic bag
Uncontaminated broken glass	Red bins	Large glass recycling bins, a sharp bin
Centrifuge tubes Fraction collection tubes Disposable flasks Petri dishes Microtitre plates Weigh boats Uncontaminated gloves and gloves contaminated with residual chemicals	Non-hazardous lab. consumables	Offensive waste bags, "LX" waste bins

reduce waste management's environmental impact, there are four tiers: pollution interference and supply reduction; use or distribution of unwanted, surplus materials; treatment, reclamation and employment of waste materials; and treatment, reclamation and employment of waste materials and disposal through burning, treatment or land burial (World Health Organization, 2005).

1. The principles of inexperienced chemistry pollution interference and supply reduction are incorporated into the first rung of this strategy hierarchy. Clearly, the most straightforward way to laboratory waste is to avoid its creation. Lowering the dimensions of laboratory operations, reducing waste production throughout laboratory operations and working on hazardous or less hazardous compounds in chemical procedures are all examples.

2. The second strategic tier entails repurposing undesired materials, dispersing surplus chemicals and reducing risks. Buying only what is required, keeping chemical inventories to avoid duplicate purchases, and reusing excess materials are examples of practices that apply this method. The decision on whether or not a waste is regulated as hazardous is occasionally made by the institution's EHS staff (Environment, Health, and Safety Workers) or by workers of the waste disposal firm.

3. The third strategic stage provides both safety and environmental benefits. If waste cannot be avoided or decreased, the organisation should consider using chemicals that can be safely recovered from waste, as well as the possibility for ill energy from the garbage.
4. Burning, alternate treatment methods and land disposal are included in the fourth and final strategic layer for handling laboratory waste. When making decisions in this tier, consider the waste's environmental fate, as well as its constituents and process by-products, once it leaves the facility or corporation. The purpose, as with alternate tiers, is to reduce the risk to health.

References

Alawa, B., Galodiya, M. N., & Chakma, S. (2022). Source reduction, recycling, disposal, and treatment. In *Hazardous waste management* (pp. 67–88). Elsevier.

Burnett, L. C., Lunn, G., & Coico, R. (2009). Biosafety: Guidelines for working with pathogenic and infectious microorganisms. *Current Protocols in Microbiology, 13*(1), 1A-1.

Chosewood, L. C., & Wilson, D. E. (2009). *Biosafety in microbiological and biomedical laboratories*. US Department of Health and Human Services, Public Health Service, Centers for Disease Control and Prevention, National Institutes of Health.

Coecke, S., Balls, M., Bowe, G., Davis, J., Gstraunthaler, G., Hartung, T., Hay, R., Merten, O. W., Price, A., Schechtman, L., & Stacey, G. (2005). Guidance on good cell culture practice: A report of the second ECVAM task force on good cell culture practice. *Alternatives to Laboratory Animals, 33*(3), 261–287.

Geraghty, R. J., Capes-Davis, A., Davis, J. M., Downward, J., Freshney, R. I., Knezevic, I., Lovell-Badge, R., Masters, J. R., Meredith, J., Stacey, G. N., & Thraves, P. (2014). Guidelines for the use of cell lines in biomedical research. *British Journal of Cancer, 111*(6), 1021–1046.

Hartung, T., Balls, M., Bardouille, C., Blanck, O., Coecke, S., Gstraunthaler, G., & Lewis, D. (2002). ECVAM good cell culture practice task force report 1. *Alternatives to Laboratory Animals, 30*(4), 407–414.

Herman, P., & Pauwels, K. (2015). Biosafety recommendations on the handling of animal cell cultures. In *Animal cell culture* (pp. 689–716). Springer.

World Health Organization. (2005). *Practical guidelines for infection control in health care facilities*. WHO Regional Office for South-East Asia.

World Health Organization. (2017). *Safe management of wastes from health-care activities: A summary*. World Health Organization.

Chapter 4
Managing Sterility in Animal Cell Culture Laboratory

Shalini Mani

4.1 Introduction

The ability to keep cells free of microbes such as bacteria, fungus and viruses is critical for successful cell culture. Non-sterile supplies, medium and reagents, microorganism-laden airborne particles, filthy incubators, dirty equipment, and dirty work surfaces may act as a source of biological contamination in cell culture-based experiments. A cell culture laboratory's most important requirement is to keep an aseptic work environment dedicated solely to cell culture work. A specialised cell culture area within a larger laboratory can still be used for sterile handling, incubation and storage of cell cultures, reagents and medium, even if a separate tissue culture room is preferred. A cell culture hood (i.e. biosafety cabinet) is the simplest and most cost-effective approach to provide aseptic conditions.

When working with cell cultures, it is critical to use sterile techniques. To protect both the cultivated cells and the laboratory worker from infection, the aseptic technique entails a variety of measures. The laboratory worker must understand that all cells handled in the lab are potentially contagious and must be treated with care. When necessary, protective clothing like gloves, lab coats or aprons and eyewear should be worn (Knutsen, 1991). Sharp tools that can puncture the skin, such as needles, scissors, scalpel blades and glass, should be handled with caution. To avoid the possibility of broken or splintered glass, utilise sterile disposable plastic materials (Rooney, 2001).

S. Mani (✉)
Centre for Emerging Diseases, Department of Biotechnology, Jaypee Institute of Information Technology, Noida, India

© The Author(s), under exclusive license to Springer Nature Switzerland AG 2023
S. Mani et al., *Animal Cell Culture: Principles and Practice*, Techniques in Life Science and Biomedicine for the Non-Expert,
https://doi.org/10.1007/978-3-031-19485-6_4

65

4.2   Elements of Aseptic Techniques

Aseptic method, which is used to create a barrier between the environment's micro-organisms and the sterile cell culture, relies on a set of techniques to decrease the risk of contamination from these sources. A sterile work location, proper personal hygiene, sterile reagents and media and sterile handling are all components of aseptic technique (Sanders, 2012).

4.2.1   Sterile Work Area

Using a cell culture hood is the simplest and most cost-effective way to reduce contamination from airborne particles and aerosols (e.g. dust, spores, shed skin, sneezing).

- The cell culture hood should be properly set up and located in an area dedicated to cell culture that is free of drafts from doors, windows and other equipment and has no through traffic.
- The work surface should be clear and cluttered, with only the objects needed for a specific activity; it should not be utilised as a storage space.
- The work surface should be fully disinfected before and after usage, and the surrounding surfaces and equipment should be cleaned on a regular basis.
- Wipe the work surface with 70% ethanol before and throughout work, especially after any spillage, for routine cleaning.
- To sterilise the air and exposed work surfaces in the cell culture hood between usage, use UV light.
- In a cell culture hood, blazing with a Bunsen burner is not essential nor suggested.
- Keep the cell culture hoods turned on at all times, only turning them off when they won't be utilised for a long time.

4.2.2   Good Personal Hygiene

Before and after dealing with cell cultures, wash your hands. Wearing personal protective equipment not only protects you from dangerous materials but also decreases the risk of contamination from shed skin, as well as dirt and dust from your clothing.

4.2.3   Media and Sterile Reagents

Although commercial reagents and media are subjected to stringent quality control to assure their sterility, contamination can occur during handling. To avoid contamination, follow the sterile handling instructions below. Use the appropriate sterilisation process to sterilise any reagents, media or solutions prepared in the laboratory (e.g. autoclave, sterile filter). Sterility is required for any materials that come into direct contact with cultures. Disposable sterile dishes, flasks and pipets, for example, are available directly from manufacturers. Before using reusable glassware, it must be carefully cleaned, rinsed and sanitised by autoclaving or dry heat. To ensure sterility, glassware should be cooked at 160 °C for 90 min to 2 h using dry heat. Autoclave materials may be damaged by very high temperatures for 20 min at 120 °C and 15 pressure. All media, reagents and other solutions that come into contact with the cultures must be sterile; media can be bought from the manufacturer as a sterile liquid, autoclaved if not heat-sensitive or filter-sterilised. Supplements can be added aseptically to the media before filtration or aseptically after filtration. Small Gram-negative bacteria should be removed from culture media and solutions, using filters with pore sizes of 0.20–0.22 micron.

4.2.4   Sterile Handling

Contamination can occur at any stage of the cultured cell handling process. Pipetting medium or other solutions for tissue culture should be done with caution. Before inserting the pipet into the bottle, flame the necks of the bottles and flasks, as well as the tips of the pipets. If the pipet tip comes into contact with a non-sterile surface, it should be discarded and a new pipet procured. Tissue culture forceps and scissors can be disinfected quickly by soaking them in 70% alcohol and burning them.

- Always use 70% ethanol to wipe your hands and your work area.
- Before putting the containers, flasks, plates and dishes in the cell culture hood, wipe the outsides with 70% ethanol.
- Avoid pouring media and reagents straight from bottles or flasks; instead, work with liquids with sterile glass or disposable plastic pipettes and a pipettor. Additionally, it is recommended to use each pipette only once to avoid cross-contamination. Unwrap sterile pipettes only when they are about to be used. Keep your pipettes in the same place where you work.
- To prevent germs and airborne pollutants from entering, always cap bottles and flasks after use and tape or place multi-well plates in resealable bags.
- Never expose a sterile flask, bottle, petri dish or other sterile items to the environment until you are ready to use it. As soon as you're finished, return the cover.
- If you need to put a cap or cover down on the work surface, place it with the opening facing down.
- Only use sterilised glasses and other items.

- When doing sterile treatments, be careful not to chat, sing or whistle.
- To avoid contamination, conduct your studies as quickly as feasible.

4.2.5   Use of Safety Cabinets

Although tissue culture work can be done on an open bench if aseptic procedures are rigorously followed, many labs prefer to execute tissue culture work in a dedicated room or low-traffic area. Biological safety cabinets are needed at the very least to safeguard the cultures as well as the laboratory worker. The movement of air in a laminar flow hood protects the work area from dust and contaminants while also acting as a barrier between the worker and the work surface. There are many different types of safety hoods, and the laboratory should choose one based on the sorts of samples being handled and the forms of potential pathogenic exposure (Spindel, 1989). When it comes to routine maintenance inspections on airflow and filter conditioners, manufacturers' guidelines should be followed. The cabinet should be turned on for at least 5 min before beginning work on a daily basis. Every day and after each use, all work surfaces within and outside the hood should be cleaned and disinfected. For cleaning work surfaces, certain safety cabinets have ultraviolet (UV) lights. UV lamps, on the other hand, are no longer suggested because they are often useless (Knutsen, 1991). UV lamps may give the impression of security because their visible blue glow persists long after their germicidal potency has faded.

Over time as the glass tube of UV light gradually loses its capacity to transmit short UV wavelengths. Additionally, dust on the glass tube, distance from the work surface, temperature and air movement can also impair the effectiveness of UV light. Even if the UV output is sufficient, bacteria or mould spores lying beneath the surface of a substance or outside the direct line of the rays will not be destroyed.

Another rule of thumb is that UV cannot kill anything that can be seen. UV lamps are only effective against microorganisms like bacteria, virus and mould spores; they are not effective against insects or other big species. Although some labs employ UV lights in addition to ethanol wipes to sanitise work areas, the current guideline is to wash down work surfaces with ethanol rather than relying on UV lamps. UV lamps should be replaced when they fall below the minimum requirements for protection, which can be measured using a suitable measuring instrument.

4.2.6   Culture Sterility

Contamination of cultures should be examined on a regular basis. When contamination is present, indicators in the tissue culture media change colour: for example, medium containing phenol red turns yellow due to increased acidity. In polluted cultures, cloudiness and turbidity are also visible. Infected cultures are usually discarded after contamination is detected with a microscope. Keeping infected cultures increases the danger of other cultures being contaminated. It is sometimes possible to save a contaminated cell line by treating it with various antibiotics and antimycotics in an attempt to remove the infection. However, such treatment can have a negative impact on cell development and is frequently ineffective in removing contamination from cultures.

Specimens obtained in the laboratory are frequently not sterile, and cultures made from them can get infected with bacteria, fungus or yeast. Microorganisms can limit growth, destroy cell cultures and cause test results to be inconsistent. The pollutants decrease the medium's nutritional significance and may produce harmful compounds for the cells. To counteract potential impurities, antibiotics (penicillin, streptomycin, kanamycin or gentamycin) and fungicides (amphotericin B or mycostatin) can be added to the tissue culture medium (Table 4.1). Sigma offers an antibiotic/antimycotic solution or lyophilised powder containing penicillin, streptomycin and amphotericin B. The most widely used antibiotic addition is a mixture of penicillin and streptomycin; kanamycin and gentamycin are used alone.

The most often used fungicides are mycostatin and amphotericin B. (Rooney, 2001). The final concentrations of the most regularly used antibiotics and antimycotics are listed in Table 4.1. Combining antibiotics in tissue culture media can be challenging since some antibiotics are incompatible with one another, and one may limit the activity of another. Furthermore, combination antibiotics may be more cytotoxic than separate antibiotics at lower dosages. In addition, long-term antibiotic use can lead to antibiotic resistance in cell lines. As a result, some laboratories add antibiotics and/or fungicides to the medium when starting a culture but remove them from the media used in subsequent subcultures. Prior to use, every tissue culture medium, whether commercially prepared or made in the lab, should be checked for sterility. A small aliquot of each lot of medium is incubated for 48 h at 37 °C and checked for signs of contamination such as turbidity (infected medium will be hazy) and colour change (infected medium will turn yellow if phenol red is used as an indicator). Any tainted material should be thrown away.

Though the use of antibiotics is recommended in case of bacterial contamination, however, it should not be regularly used in cell culture (Giuliano et al., 2019). Because their continued use encourages the development of antibiotic-resistant strains and allows low-level contamination to persist, which can turn into full-scale contamination once the antibiotic is removed from the media, and because they can

Table 4.1 Working concentrations of antibiotics and antimycotics for mammalian cell culture

Antibiotic/antimycotic	Final concentration in culture medium
Penicillin	50–100 U/ml
Streptomycin sulphate	50–100 ug/ml
Kanamycin	100 ug/ml
Gentamycin	50 ug/ml
Mycostatin	20 ug/ml
Amphotericin B	0.25 ug/ml

conceal mycoplasma infections and other cryptic contaminants. Furthermore, some antibiotics may have a cross-reaction with cells, interfering with the biological processes being studied. Hence, antibiotics should only be administered as a last resort and for a short period of time, and they should be eliminated as soon as possible from the culture. Antibiotic-free cultures should be kept in parallel as a control for cryptic infections if they are used for a long time.

4.3   Biological Contamination

Contamination of cell cultures is by far the most common issue in cell culture laboratories and may potentially cause fatal effects. These contaminants can be divided into two main categories: firstly, chemical contaminants, such as those found in cell culture media, sera and water containing endotoxins, plasticisers and detergents too. Secondly, biological contaminants include moulds, yeasts, viruses and mycoplasma as well. Other cell lines may also contaminate the sample. While it is impossible to completely eliminate contamination, it is possible to reduce it by obtaining a full understanding. It is also possible to lessen its frequency and severity, by having a strong awareness of their sources and using good aseptic methods (Verma et al., 2020). This section gives an overview of the most common forms of biological contamination.

4.3.1   Bacteria

Bacteria are a huge group of unicellular microbes that are found all over the world. They really are. Bacteria are of typically a few micrometres in diameter, and they come in a variety of forms, ranging from spherical to elliptical and from spheres to rods to spirals. Because of their widespread distribution, large size and rapid growth rates, bacteria along with yeasts and moulds are the most typically encountered biological contaminants in the culturing of cells. Visual detection of bacterial contamination is simple. If the culture is contaminated with bacteria and then within a few days of becoming contaminated, if the culture is examined; then the contaminated cultures usually appear foggy (i.e. turbid), and rarely (in case of heavy contamination) it also appears as a thin layer on the surface of culture (Lincoln & Gabridge, 1998). Sudden drops in the pH of the contaminated culture medium are also a common occurrence.

   The bacteria appear as tiny, moving granules between the cells under a low-power microscope, and the outlines of individual bacteria may be resolved under a high-power microscope. An adherent 293 cell culture contaminated with *E. coli* is depicted in the simulated photos below (Fig. 4.1).

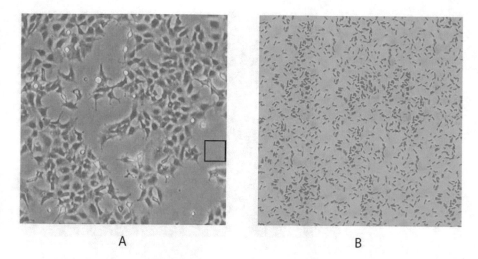

A   B

Fig. 4.1 Virtual phase contrast images of adherent 293 cells infected with *E. coli* that have been simulated. Under low-power microscopy, the gaps between adhering cells display small, shimmering granules, but the individual bacteria are difficult to discern (panel A). Individual *E. coli* cells, which are normally rod-shaped and measure around 2 micron long and 0.5 micron in diameter, can be seen when the area enclosed by the black square is magnified further. The dark square in panel A has 100 micron on each side

4.3.2 Yeasts

Yeasts are unicellular eukaryotic microorganisms belonging to the kingdom fungi. They exist with sizes ranging from a few micrometres (usually) to 40 micrometres (rarely). Similar to bacterial contamination, the cultures contaminated with yeasts become turbid, especially if the contamination is advanced. The pH of a culture, infected by yeasts, changes very little until the contamination gets severe, at which point the pH normally rises (Raju et al., 2014). Yeast appears as discrete ovoid or spherical particles under microscopy, which may bud off smaller particles. Figure 4.2 displays an adherent 293 cell culture infected with yeast 24 h after plating.

4.3.3 Moulds

Moulds are eukaryotic microorganisms that form as multicellular filaments called hyphae and belong to the kingdom fungi. A colony or mycelium is a connected network of these multicellular filaments that contain genetically identical nuclei. The pH of the culture remains steady in the early stages of contamination and then rapidly climbs as the culture gets more severely contaminated and turbid, similar to yeast contamination (Raju et al., 2014).

Fig. 4.2 Phase contrast image showing 293 adherent cells contaminated with yeasts. These yeast cells look like an ovoid particle and during replication bud off smaller particles

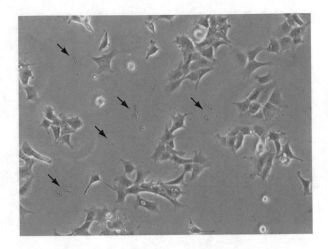

Mycelia appear as slender, wisp-like filaments and sometimes as thicker aggregates of spores under microscope. Many mould species' spores may persist in their latent state in extremely harsh and unfriendly habitats and only gets activated when they come into contact with optimal growth conditions.

4.3.4 Viruses

Viruses are microscopic infectious pathogens that replicate by hijacking the machinery of the host cell. Because of their small size, they are difficult to detect in culture and to eliminate from chemicals used in cell culture labs. Because most viruses have strict host requirements, they usually have no negative impact on cell cultures from species other than their host (Merten, 2002). Using virally infected cell cultures, on the other hand, can pose a major health risk to laboratory employees, especially if human or primate cells are cultivated. Electron microscopy, immunostaining using a panel of antibodies, ELISA assays, and PCR with appropriate viral primers can all be used to detect viral infection in cell cultures (Barone et al., 2020).

4.3.5 Mycoplasma

Mycoplasma are basic bacteria that do not have a cell wall and are the smallest self-replicating organisms. Mycoplasma are difficult to detect because of their small size (usually less than one millimetre); unless they reach exceptionally high concentrations and cause the cell culture to degrade, there are generally no apparent symptoms of infection. Some slow-growing mycoplasma can survive in culture without causing cell death, but they can change the host cells' behaviour and metabolism.

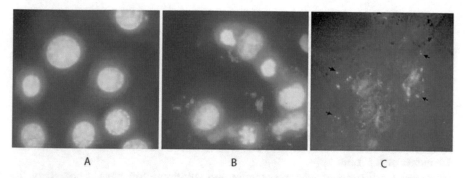

A   B   C

Fig. 4.3 Images showing the mycoplasma-free cultured cells (panel A) and cells infected with mycoplasma (panels B and C)

Chronic mycoplasma infections can cause lower cell growth, decreased saturation density and agglutination in suspension cultures; nevertheless, the only definite approach to detect mycoplasma contamination is to regularly test the cultures with fluorescent staining (e.g. Hoechst 33258), ELISA, PCR,

immunostaining, autoradiography or microbiological assays (Drexler & Uphoff, 2002). Figure 4.3 represents the mycoplasma infections in cultured cells.

4.3.6 Cross-contamination

Though it is not a routine problem as microbial contamination, however, the widespread cross-contamination of many cell lines with HeLa and other fast-growing cell lines is a well-established problem with serious penalties (Nardone, 2007). Cross-contamination can be avoided by obtaining cell lines from reputable cell banks, reviewing the features of the cell lines on a regular basis and using proper aseptic technique. Cross-contamination in your cell cultures can be confirmed by DNA fingerprinting, karyotype analysis and isotype analysis too (Nardone, 2008).

4.4 Aseptic Technique Checklist

The following checklist offers a brief list of suggestions and measures to guide through and achieve a solid aseptic technique.

Work Area
Is the cell culture hood appropriately set up?
Is the cell culture hood in an area free from drafts and through traffic?
Is the work surface organised, and does it contain all the items essential for your experiment?

Did you wipe the work surface with 70% ethanol prior to work?
Are you regularly cleaning and sterilising your incubators, refrigerators, freezers and other laboratory equipment?

Personal Hygiene

Did you rinse your hands?
Are you wearing personal protective equipment?
If you have long hair, is it tied in the back?
Are you using a pipettor to work with liquids?

Reagents and Media

Have you sterilised any reagents, media and solutions you have prepared in the laboratory using the appropriate procedure?
Did you wipe the outside of the bottles, flasks and plates with 70% ethanol before placing them on your work surface?
Are all your bottles, flasks and other containers capped when not in use?
Are all your plates stored in sterile resealable bags?
Does any of your reagents look cloudy? Contaminated? Do they contain floating particles?
Have foul smell? Unusual colour? If yes, did you decontaminated and discarded them?

Handling

Are you working slowly and deliberately, mindful of aseptic technique?
Did you wipe the surfaces of all the items including pipettor, bottles and flasks with 70% ethanol before placing them in the cell culture hood?
Are you placing the caps or covers face down on the work area?
Are you using sterile glass pipettes or sterile disposable plastic pipettes to manipulate all? liquids?
Are you using a sterile pipette only once to avoid cross-contamination?
Are you careful not to touch the pipette tip to anything non-sterile, including the outside edge of the bottle threads?
Did you mop up any spillage immediately and wiped the area with 70% ethanol?

References

Barone, P. W., Wiebe, M. E., Leung, J. C., Hussein, I., Keumurian, F. J., Bouressa, J., Brussel, A., Chen, D., Chong, M., Dehghani, H., & Gerentes, L. (2020). Viral contamination in biologic manufacture and implications for emerging therapies. *Nature Biotechnology, 38*(5), 563–572.

Drexler, H. G., & Uphoff, C. C. (2002). Mycoplasma contamination of cell cultures: Incidence, sources, effects, detection, elimination, prevention. *Cytotechnology, 39*(2), 75–90.

Giuliano, C., Patel, C. R., & Kale-Pradhan, P. B. (2019). A guide to bacterial culture identification and results interpretation. *Pharmacology & Therapeutics, 44*(4), 192.

Knutsen, T. (1991). *The ACT cytogenetic laboratory manual* (2nd edn, pp. 563–587) (M. J. Barch, Ed.). Raven Press.

Lincoln, C. K., & Gabridge, M. G. (1998). Cell culture contamination: Sources, consequences, prevention, and elimination. *Methods in Cell Biology, 57*, 49–65.

Merten, O. W. (2002). Virus contaminations of cell cultures – A biotechnological view. *Cytotechnology, 39*(2), 91–116.

Nardone, R. M. (2007). Eradication of cross-contaminated cell lines: A call for action. *Cell Biology and Toxicology, 23*(6), 367–372.

Nardone, R. M. (2008). Curbing rampant cross-contamination and misidentification of cell lines. *Biotechniques, 45*(3), 221–227.

Raju, S., Mathad, S., & Chandrashekar. (2014). *Animal biotechnology.*

Rooney, D. E. (Ed.). (2001). *Human cytogenetics: constitutional analysis: A practical approach.* Oxford University Press.

Sanders, E. R. (2012). Aseptic laboratory techniques: Volume transfers with serological pipettes and micropipettors. *JoVE, 63*, e2754.

Spindel, W. (1989). *Biosafety in the laboratory: Prudent practices for the handling and disposal of infectious materials.* National Research Council.

Verma, A., Verma, M., & Singh, A. (2020). Animal tissue culture principles and applications. In *Animal biotechnology* (pp. 269–293). Academic.

Chapter 5
Media and Buffer Preparation for Cell Culture

Sakshi Tyagi and Shalini Mani

5.1 Introduction

Cell culture is among the most commonly used methods in the field of life sciences. It is the common term used for the exclusion of cells, tissues or organs from a plant or an animal and their consequent placement into a non-natural environment favourable to their survival and/or proliferation. The basic conditions required for the optimal growth of cells include regulated temperature, a substrate for cell attachment and suitable growth medium and incubator that controls pH and osmolality. The utmost crucial and essential stage in cell culture is selection of suitable growth medium. Cell culture media generally constitutes a relevant basis of energy and compounds, which maintain the cell cycle. A distinctive culture medium is composed of a balance of amino acids, inorganic salts, glucose and serum as a basis of growth factors, hormones and attachment factors (Oyeleye et al., 2016).

5.2 Media

In cell culture, media is the most vital element of the culture environment, as it delivers the essential hormones, nutrients and growth factors required for the development of cell. Apart from maintaining the cell growth, media also regulates the osmotic pressure and the pH of the culture.

In initial cell culture experiments, cells uptake their required nutrients, growth factors from the natural media, attained from extracts of tissues and body fluids like

S. Tyagi · S. Mani (✉)
Centre for Emerging Diseases, Department of Biotechnology, Jaypee Institute of Information Technology, Noida, India

lymph, chick embryo extract, serum, etc. It was the necessity for standardisation, quality of media and escalating demand that directed towards the evolution of defined media. In general, the culture environment is controlled relating to the carbon dioxide (CO_2), temperature, pH, osmotic pressure, vital metabolites (like carbohydrates, vitamins, amino acids, peptides and proteins), extracellular matrix, inorganic ions and hormones (Verma et al., 2020).

5.2.1   Physiochemical Properties of Media

In order to support better growth and proliferation of the cultured cells, the culture media is requisite to have definite physicochemical properties (pH, CO_2, buffering, osmolarity, viscosity, temperature, etc.). Some of the important physiochemical properties have been discussed below:

pH   Mostly cells need pH conditions in the alkaline range (7.2–7.4), and proximate regulator of pH is necessary for optimum culture conditions. However, these optimum conditions may vary. Fibroblasts require alkaline conditions (pH: 7.4–7.7), while continuous transformed cell lines prefer more acidic conditions (pH: 7.0–7.4) (Shang et al., 2021).

CO_2   The growth medium regulates the culture's pH and buffers the cells in culture contrary to any variations in the pH. In general, this buffering is attained by involving an organic (like HEPES) or CO_2-bicarbonate-based buffer, since the media's pH relies on the gentle equilibrium of dissolved CO_2 and bicarbonate (HCO_{3-}). That means any variation in the atmospheric CO_2 can change the pH of the medium. In case of usage of media buffered with a CO_2 (HCO_{3-})-based buffer, it is important to promote the usage of exogenic CO_2 particularly if the cells are cultured in open dishes or transformed cell lines are cultured at high concentrations (Freshney, 2005).

Buffering System   Maintenance of pH is principally significant instantly followed by cell seeding during the establishment of cell culture and is generally attained by one of two buffering systems: (1) natural buffering system and (2) chemical buffering system. In the case of a natural buffering system, CO_3/HCO_3 content of the culture medium is balanced by the gaseous CO_2 whereas in a chemical buffering system it is done with the help of zwitter ion known as HEPES. Cultures with natural bicarbonate/CO_2 buffering systems are required to be regulated in an atmosphere of 5–10% CO_2 in air normally provided by CO_2 incubator. Bicarbonate/CO_2 is cost-effective and harmless and also delivers other chemical profits to the cells. HEPES has excellent buffering capacity in the alkaline pH range (7.2–7.4) but is comparatively pricier and could be harmful to few cell types at the concentration more than ~100 nMolar. A controlled gaseous atmosphere is not a requisite in case of HEPES buffered cultures. Phenol red is present as the component of most of the commercially available culture media as an indicator of pH so that the pH status of the media is continually specified by the colour. In case of any colour change (yellow, acid, or purple, alkali), the culture medium should be replaced immediately) (Michl et al., 2019).

Temperature The optimum temperature for cell culture is reliant on (1) host's body temperature from which the cells were isolated, (2) anatomical difference in temperature (e.g. the skin's temperature may be lower than the temperature of the testis) and (3) the integration of a safety aspect to permit for negligible faults in controlling the incubator. As overheating could be a severe problem in comparison to under heating for cell cultures, hence, generally the incubator's temperature is maintained somewhat subordinate than the optimum temperature. Mostly, human and mammalian cell lines are sustained at 37 °C for their optimum growth (Oyeleye et al., 2016).

Osmolality Most of the cultured cells have an equitably varied tolerance for osmotic pressure. The optimal level of osmolality for human cells in cell culture is about 290 mosmol/kg in reference to osmolality of human plasma. However, this level may vary for other species like in case of mice it is 310 mosmol/kg (Freshney, 2005).

5.2.2   Components of Media

Major components of media include inorganic salts, carbohydrates, amino acids, vitamins, fatty acid and lipids, organic supplements, hormones and growth factors, antibiotics, etc.

Inorganic Salts The additions of inorganic salts in media accomplish numerous functions. The salts are primarily those of Na^+, Mg^{2+}, Cl^-, K^+, Ca^{2+}, SO_4^{2-}, PO_4^{3-} and HCO_{3-} are the main constituents that contribute to the osmolality of the medium. Principally these salts aid in holding the osmotic balance of cells and help in maintaining membrane potential by delivery of Na^+, K^+ and Ca^{2+}. All the above-mentioned factors are required by the matrix of the cell as enzyme cofactors and for the attachment of the cell (McLimans, 1972).

Carbohydrates In the form of sugars, carbohydrates are the major energy source. Glucose and galactose are commonly used sugars, though few media also contain fructose or maltose. Media comprising the larger concentration of sugars are capable of promoting the development of an extensive range of cell types. Pyruvate is involved in the formation of some media, as a substitute source of energy. However, glucose is primarily used in cell culture and can pass through the plasma membrane via facilitated diffusion and carry proteins. In cell culture composition, the amount of glucose is present in the range of 1 g/L (5.5 mM) to 10 g/L (55 mM). Numerous standard media are accompanied with 5.5 mM D-glucose (approximately) which comes near to normal blood sugar levels in vivo. These media comprise Ames' Medium, Minimum Essential Medium Eagle (EMEM), Swim's S-77 Medium, Fischer's Medium, Basal Medium Eagle (BME), etc. (McLimans, 1972).

Amino Acids Amino acids are the basic component of proteins. "Essential" amino acids should be incorporated into the culture medium as cells are not capable of synthesising these on their own. Generally, the essential amino acids include cysteine and tyrosine, but some nonessential amino acids may be required. Glutamine is also needed by most cell lines. Though glucose is present in most defined media, it has been indicated that glutamine is used as an energy carbon source by cultured cells in choice to glucose (McLimans, 1972).

Vitamins In cell culture, serum is a vital source of vitamins. Though, most media are also supplemented with vitamins making them constantly more appropriate for an extensive range of cell lines. Vitamins are pioneers for several cofactors. Most of the vitamins, particularly B group vitamins, are vital for cell growth and proliferation. The existence of B12 is necessary for some cell lines. The vitamins usually used in media include thiamine, riboflavin and biotin (Masters, 2000).

Proteins and Peptides These are mostly essential in serum-free media. Albumin, fibronectin, transferrin and fetuin are the most commonly used proteins and peptides. These are generally used in place of those that usually exist through the inclusion of serum to the media (Masters, 2000).

Fatty Acids and Lipids These are significant components in case of serum-free media since they are generally extant in serum, e.g., steroids and cholesterol vital for specific cells (McLimans, 1972).

Organic Supplements A range of other compounds, that includes proteins, nucleosides, peptides, pyruvate, citric acid cycle intermediates and lipids, seems in complex media. However, these components are considered to be essential when there is reduced serum concentration. These compounds support cloning and in upkeeping particular specific cells, even in the existence of serum (McLimans, 1972).

Hormones and Growth Factors Hormones and growth factors are not stated in the formulation of most common media, though they are normally supplemented to serum-free media. Growth factors, chemokines and cytokines are considered as the cell's chemical messengers. They are tiny protein molecules that are naturally released by cells. These molecules bring a physiological effect (proliferation, growth or differentiation) on the target cells. Each stem cell type needed media complemented with definite growth factors with the purpose of maintaining the health, proliferation and differentiation of culture. For instance, fibroblast growth factor-basic (bFGF) is the chief growth factor intricated in sustaining pluripotency for human embryonic stem (ES) and induced pluripotent stem (iPS) cell cultures, whereas leukaemia inhibitory factor (LIF) is the key controller for mouse ES and iPS cells (Masters, 2000)

Antibiotics and Antimycotics Unless well-maintained sterile conditions (like usage of laminar hoods), it is essential to include antibiotics and antimycotics into the media. A vast variety of appropriate arrangements are accessible from comparatively definite antibiotics such as penicillin/streptomycin solutions, to wider spectrum antibacterial/antimycotic agents that include amphotericin B or kanamycin. The selected antibiotics should be nontoxic in nature to the cultured cells. The selection may rely on the kind of contamination encountered in the specific laboratory. Penicillin-streptomycin solution can be added at 0.5–1 mL of solution per 100 mL of cell culture medium for a final concentration of 50–100 IU/mL penicillin and 50–100 µg/mL streptomycin. Gentamicin sulphate, another antibiotic, is used at 50–100 µg/mL. The antimycotic amphotericin B is used at 2.5 µg/mL. These concentrations apply to media that contain serum. In case of serum-free media, this concentration can be reduced to at least 50% (Jacoby & Pasten, 1979).

Serum Among the biological fluids, serum is the most significant one that has been proven to be success for cell culturing. There is usually an addition of 5–20% of serum to media for optimum cell growth. Serum plays a pivotal role as the foundation of growth and adhesion factors. It also provides lipids, hormones and minerals for the culture of cells. Additionally, serum is also involved in regulating penetrability of cell membrane and behaves as a transporter for enzymes, lipids, trace elements and micronutrients into the cell. Most commonly used sera in tissue culture are foetal bovine, bovine calf, adult horse and human serum. Among above-mentioned sera foetal bovine (FBS) and calf (CS) are the most extensively used. Though, usage of serum in media has certain limitations such as complications with standardisation, high value, inconsistency, specificity and undesired effects like activation or repression of development and/or cellular function on particular cell cultures (Bolin et al., 1994).

Note (1) Always make aliquots in smaller working amount (like in 15 ml falcons) from stock (frozen FBS) to avoid contamination and to reduce freeze/thaw cycles. Store at −20 °C (as shown in Fig. 5.1). (2) Thaw frozen FBS in water bath for at least half an hour before use.

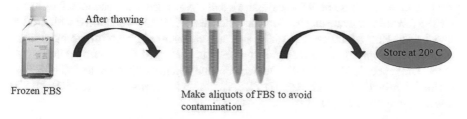

Frozen FBS

After thawing

Make aliquots of FBS to avoid contamination

Store at 20° C

Fig. 5.1 Preparation of aliquots from frozen FBS

Table 5.1 Types of media with their applications

S. no.	Type of media	Examples	Applications
1.	Balanced salt solutions	PBS, Hanks' BSS, Earle's salts D-PBS HBSS EBSS	Form the base of several complex media
2.	Basal media	MEM	Used for primary and diploid culture
		DMEM	Alteration of MEM comprising elevated level of amino acids and vitamins. Used for a varied range of cell types such as hybridomas
		GMEM	Glasgow's altered MEM was developed for BHK-21 cells
3.	Complex media	Iscove's DMEM	Further supplemented alteration of DMEM which offers high-density growth
		RPMI 1640	Formerly acquired for human leukaemic cells It supports a varied range of mammalian cells like hybridomas
		Leibovitz L-15	Intended for CO_2 free environments
4.	Serum-free media	CHO HEK293	Used for serum-free applications

5.2.3 Types of Culture Medium

Depending upon the requirement of different cell lines, culture media can be divided into four types as discussed below and also summarised in the Table 5.1.

Balanced Salt Solutions (BSS) BSS comprises of inorganic salts and may involve HCO_3. and, in few cases, glucose. The most primary media are BSS, such as phosphate-buffered saline (PBS). PBS can be utilised for rinsing of cells and for small incubation period in suspension. BSS is used as the base of the formulation of many complete media, and it is commercially available in the form of Eagle's Minimum Essential Medium (MEM) with Hank's salts (Hanks & Wallace, 1949) or Eagle's MEM with Earle's salts (Earle, 1943), representing which BSS formulation was utilised. Hank's salts (e.g. Hank's BSS (HBSS)) would infer the usage of closed flasks with a gas phase of air, while Earle's salts (e.g. Earle's Balanced Salt Solution (EBSS)) would indicate an elevated concentration of HCO_3. well suited with growth in 5% CO_2. BSS formulae are usually altered – for example, by eliminating glucose or phenol red from Hanks's BSS or by parting out Ca^{2+} or Mg^{2+} ions from Dulbecco's PBS (Dulbecco & Vogt, 1954). Dulbecco's phosphate-buffered saline (D-PBS) is formed deprived of Ca^{2+} or Mg^{2+} (D-PBSA), which are formed distinctly (D-PBSB) and supplemented prior to use if obligatory.

Basal Media   Basal media composed of amino acids, inorganic salts, vitamins and a carbon source like glucose. However, these basal media formulations need to be further complemented with serum. Majority of the cells grow suitably well in this medium. Basic media like MEM, Dulbecco's Modified Eagle Medium (DMEM) and Glasgow Modified Eagle Medium (GMEM) are made up of salts, vitamins, amino acids, glucose and other nutrients. A basal medium is complemented by inclusion of L-glutamine, antibiotics (mostly penicillin and streptomycin sulphate) and generally serum in order to prepare a "complete medium." MEM was evolved from Eagle's Basal Medium (BME) via expanding the range and concentration of the ingredients. Among all media, Eagle's MEM is the most commonly used one. DMEM was evolved for mouse fibroblasts for studying the transformation and propagation of virus. It has double the concentration of amino acids present in MEM, has fourfold vitamin concentrations, utilises double the HCO_3 and CO_2 concentrations to attain improved buffering (Stanners et al., 1971) and has added vitamins and amino acids, as well as lipoic acid and nucleosides. It has been widely used for different cell types such as primary fibroblasts, smooth muscle cells, neurons and glial cells and also for cell lines including HeLa, 293, Cos-7 and PC-12. GMEM is modified MEM, comprising twofold concentration of vitamins and amino acids as compared to original Basal Medium Eagle and is utilised without serum. It is used to explore the genetic factors that strained cell competence. It is utilised for kidney cell lines like BHK-21 (Yao & Asayama, 2017).

Complex Media   Complex media include Iscove's DMEM, RPMI 1640 and Leibovitz L-15. Iscove's modified DMEM (Developed by Iscove and Melchers). It is boosted with numerous vitamins and nonessential amino acids that are absent in DMEM (biotin and cyanocobalamin) and supplementary pyruvate, selenite and HEPES. Transferrin, soybean lipids and bovine serum albumin are used as serum alternatives. With its large concentration of vitamins and amino acids, Iscove's DMEM is complementary for fast proliferating, high-density cell cultures that include COS-7, Jurkat and macrophage cell. RPMI 1640 is Roswell Park Memorial Institute, and it is the modification of RPMI 1630 or McCoy's 5A having the supremacy to culture blood's lymphocytes (Ozturk & Hu, 2005). Basically, it is phosphate-rich and comes under the category of serum-free media. A characteristic pH of this media is 8.0 and utilises a bicarbonate buffering system. The bone marrow cells, B-lymphocytes, hybridoma cells and T-lymphocytes are generally cultured in this type of media. The RPMI 1640 typically constitutes glucose, essential and nonessential amino acids, phenol red and vitamins. Leibovitz's L-15 was developed by Leibovitz in 1963. In this media, the buffering capacity is facilitated by phosphates and free amino acids (basic in nature) in place of sodium bicarbonate ($NaHCO_3$), so that pH of the culture is controlled in ambient air without using CO_2 incubator. As an alternative of glucose, pyruvate (and galactose) is supplemented at an increased concentration in order to regulate pH drops because of the production of lactic acid through the glucose metabolism and to stimulate the CO_2 release from the respiratory chain. This medium promotes the growth of HEP-2 monkey kidney cells and primary explants of embryonic and adult human tissue (Yao & Asayama, 2017).

Serum-Free Media Serum-free media are media developed to grow a definite cell type or execute a definite application in the absenteeism of serum. One of the benefits of usage of serum-free media is the capability to make the media selective for definite cell types by selecting the suitable combination of growth factors (Broedel & Papiak, 2003). This type of media formulations exists for various primary cell culture and cell lines that include recombinant protein generating lines of Chinese hamster ovary (CHO), many hybridoma cell lines, the insect lines Sf9 and Sf21 (*Spodoptera frugiperda*) and cell lines that serve as hosts for viral production (like 293, VERO, MDCK, MDBK) and others.

5.3   Methodology to Prepare Cell Culture Media

All the above discussed media are important and have wide applications. However, DMEM and RPMI160 are the two media which are commonly used for the growth of different cell lines such as primary fibroblasts, glial cells, neuronal cells and neoplastic leukocytes. The stepwise method for preparation of these two media is discussed below in detail.

5.3.1   Preparation of Incomplete DMEM Media

DMEM is one of the most commonly used variations of Eagle's medium. DMEM encompasses quadruple concentration of amino acids and vitamins. In addition, the preparation also contains glycine, serine and ferric nitrate.

Composition of DMEM (as per Manufacturer Guidelines, from Himedia)
DMEM is a basic media, and it is without proteins or growth factors. Hence, it needs to be supplemented to be a "complete" media. It is generally supplemented with 5–10% FBS. Media in powder form is prepared without $NaHCO_3$ because it tends to gas off in the powdered form. DMEM was formerly used to culture mouse embryonic stem cells. It also has wide applications in the formation of viral plaque, in primary mouse and chicken cells and in studies related to contact inhibition. It can also be utilised to culture hybridomas (Inoue et al., 2019). Composition of DMEM media is discussed below in Table 5.2.

Materials Required DMEM, autoclaved water, antibiotics, reagent bottles, filter assembly, gloves, pipettes, tips, $NaHCO_3$, filter assembly, membrane filter, pH strip, Petri plate, syringe filter

Protocol for Preparation Is as Follows
1. Dissolve 19.5 g in 900 ml autoclaved water with continuous, mild stirring until the powder is entirely dissolved. Avoid heating.

2. Add 3.7 g of sodium bicarbonate powder for 1 l of media and keep stirring until mixed.
3. Regulate the pH to 0.2–0.3 pH units below the desired pH (7.4–7.6) by means of 1N HCl or 1N NaOH as the pH tends to upsurge during filtration.
4. Make up the final volume to 1000 ml with autoclaved water.
5. Sterilise the medium instantly by filtering through a sterile filter assembly.
6. Aseptically add sterile supplements like serum, antibiotics, etc. as per the requirement, and distribute the desired amount of sterile media into sterile containers.
7. Store the prepared media at 2–8 °C till use.

Note Keep few ml of media in a Petri plate for sterility check at least for 24 h at 37°in CO_2 incubator. If there is any colour change or turbidity observed after 2–3 days, there could be chance of contamination in media.

Preparation of Incomplete RPMI Media (As Per Manufacturer Guidelines, from Himedia)

RPMI media are a media series defined by Moore et al for the in vitro culture of cells like human normal and cancerous cells. RPMI 1640 is the most frequently used media in the series. An alteration of McCoy's 5A medium, the medium was precisely designed to promote the growth of human lymphoblastoid cells in suspension culture.

Composition of RPMI 1640

RPMI 1640 is based on a bicarbonate buffering system and varies from the commonly used media's in cell culture in its distinct pH formulation (pH 8). RPMI 1640

Table 5.2 Composition of DMEM media (as per manufacturer guidelines, from Himedia)

Inorganic salts (mg/L)	Amino acids(mg/L)	Vitamins(mg/L)	Others(mg/L)
Calcium chloride dihydrate (265), Ferric nitrate nonahydrate (0.1), magnesium sulphate anhydrous (97.72), potassium chloride (400), sodium chloride (6400), sodium dihydrogen phosphate anhydrous (109)	Glycine (30), L-arginine hydrochloride (84), L-cystine dihydrochloride (62.570), L-glutamine (584), L-histidine hydrochloride monohydrate (42), L-isoleucine (105), L-leucine (105), L-lysine hydrochloride (146), L-methionine (30), L-phenylalanine (66), L-serine (42), L-threonine (95), L-tryptophan (16), L-tyrosine disodium salt (103.790), L-valine (94)	Choline chloride (4), D-Ca-pantothenate (4), folic acid (4), nicotinamide (4), pyridoxal 5 phosphate (4), riboflavin (0.4), Thiamine hydrochloride (4), i-inositol (7.2)	D-glucose (4500), HEPES buffer (5958), phenol red sodium salt (15.9), sodium pyruvate (110)

Source: https://www.himedialabs.com/TD/AT151.pdf

promotes the growth of a vast range of suspension cells and grown as monolayers. It is also frequently used to suspend peripheral blood mononuclear cells (Schulte Schrepping et al., 2020). Composition of RPMI media is discussed below in Table 5.3.

Materials Required RPMI 1640, autoclaved water, antibiotics, reagent bottles, filter assembly, gloves, pipettes, tips, NaHCO3, filter assembly, membrane filter, pH strip.

Protocol for Preparation Is as Follows
1. Dissolve 8.4 g in 900 ml autoclaved water with continuous and moderate stirring till the powder is mixed wholly. Do not heat the water.
2. It may be required to lower the pH to 4.0 with 1N HCl to totally dissolve this product. After it has mixed completely, the pH can be elevated to 7.2 with 1N NaOH prior to the addition of sodium bicarbonate.
3. Add 2.0 g of sodium bicarbonate powder for 1 l of media, and stir till its complete dissolution.
4. Regulate the pH to 0.2–0.3 pH units beneath the desired pH with 1N HCl or 1N NaOH since the pH tends to rise during filtration.
5. Make up the final volume to 1000 ml with autoclaved water.
6. Sterilise the medium instantly with the help of sterile filter assembly.
7. Add sterile supplements such as serum, antibiotics, etc. as per the requirement in an aseptic manner, and distribute the desired amount of sterile media into sterile containers.
8. Store the prepared media at 2–8 °C till use.

Table 5.3 Composition of RPMI media (as per manufacturer guidelines, from Himedia)

Inorganic salts (mg/L)	Amino acids (mg/L)	Vitamins (mg/L)
Calcium nitrate tetrahydrate (100), glutathione reduced (1), magnesium sulphate anhydrous (48.84), phenol red sodium salt (5.3), potassium chloride (400), sodium chloride (6000), sodium phosphate dibasic anhydrous (800)	Glycine (10), L-arginine hydrochloride (241), L-asparagine (50), L-aspartic acid (20), L-cystine dihydrochloride (65.2), L-glutamic acid (20), L-glutamine (300), L-histidine hydrochloride monohydrate (20.96), L-hydroxyproline (20), L-isoleucine (50), L-leucine (50), L-lysine hydrochloride (40), L-methionine (15), L-phenylalanine (15), L-proline (20), L-serine (30), L-threonine (20), L-tryptophan (5), L-tyrosine sodium salt (28.830), L-valine (20)	Choline chloride (3), D-biotin (0.2), D-Ca-pantothenate (0.25), folic acid (1), niacinamide (1), pyridoxine hydrochloride (1), riboflavin (0.2), thiamine hydrochloride (1), vitamin B12 (0.005), i-inositol (35), p-amino benzoic acid (PABA) (1)

Source: https://www.himedialabs.com/TD/AT150.pdf

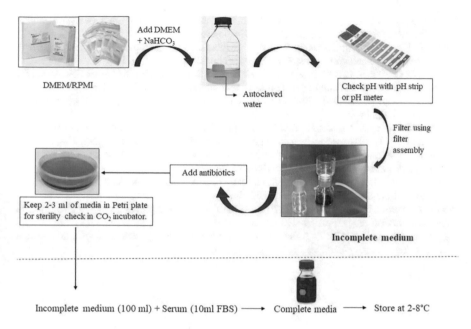

Fig. 5.2 Schematic representation of preparation of DMEM/RPMI media

Preparation of Complete Media For complete media, add 10% FBS to the incomplete media, for instance, for 100 ml of incomplete media (DMEM/RPMI), add 10 ml of FBS. Finally, store the complete media at 2–8 °C till further use (as shown in Fig. 5.2).

References

Bolin, S. R., et al. (1994). *Journal of Virological Methods, 48*, 211.

Broedel, S. E., Jr., & Papiak, S. M. (2003). *The case for serum-free media*. BioProcess International.

Dulbecco, R., & Vogt, M. (1954). Plaque formation and isolation of pure lines with poliomyelitis viruses. *The Journal of Experimental Medicine, 99*, 167–182.

Earle, W. R. (1943). Production of malignancy *in vitro*. IV. The mouse fibroblast cultures and changes seen in the living cells. *Journal of the National Cancer Institute, 4*, 165–212.

Freshney, I. (2005). *Culture of animal cells: A manual of basic technique* (5th ed.). Wiley.

Hanks, J. H., & Wallace, R. E. (1949). Relation of oxygen and temperature in the preservation of tissues by refrigeration. *Proceedings of the Society for Experimental Biology and Medicine, 71*(2), 196–200.

Inoue, K., Ishizawa, M., & Kubota, T. (2019). Monoclonal anti-dsDNA antibody 2C10 escorts DNA to intracellular DNA sensors in normal mononuclear cells and stimulates secretion of multiple cytokines implicated in lupus pathogenesis. *Clinical and Experimental Immunology*.

Jacoby, W. B., & Pasten, I. H. (1979). Chapter 7. In: *Methods in enzymology: Cell culture* (Vol. 58). Academic.

Animal cell culture: Practical approach, 3, by John R. W. Masters. (2000), Oxford University Press.

McLimans, W. F. (1972). Chapter 5. In: G. H. Rothblat, & V. J. Cristofalo (Eds.), *Growth, nutrition and metabolism of cells in culture* (Vol. 1). Academic.

Michl, J., Park, K. C., & Swietach, P. (2019). Evidence-based guidelines for controlling pH in mammalian live-cell culture systems. *Communications Biology, 2*(1), 1–2.

Oyeleye, O. O., Ola, S. I., & Omitogun, O. G. (2016). Basics of animal cell culture: Foundation for modern science. *Biotechnology and Molecular Biology Reviews, 11*(2), 6–16.

Ozturk, S., & Hu, W. S. (Eds.). (2005). *Cell culture technology for pharmaceutical and cell-based therapies*. CRC Press.

Schulte Schrepping, J., Reusch, N., Paclik, D., Baßler, K., Schlickeiser, S., Zhang, B., et al. (2020). Severe COVID-19 is marked by a dysregulated myeloid cell compartment. *Cell, 182*, 1419.

Shang, N., Bhullar, K. S., & Wu, J. (2021). Methodologies for bioactivity assay: Cell study. In *Biologically active peptides* (pp. 155–189). Academic.

Stanners, C. P., Eliceiri, G. L., & Green, H. (1971). Two types of ribosome in mouse–hamster hybrid cells. *Nature: New Biology, 230*, 52–54.

Verma, A., Verma, M., & Singh, A. (2020). Animal tissue culture principles and applications. In *Animal biotechnology* (pp. 269–293). Academic.

Yao, T., & Asayama, Y. (2017). Animal-cell culture media: History, characteristics, and current issues. *Reproductive Medicine Biology, 16*(2), 99–117.

Chapter 6
Properties of Cultured Cells and Selection of Culture Media

Shalini Mani

6.1 Introduction

The cells growing in cultured conditions are characteristically different from normal cells. Mostly, these cultured cells appear as fibroblast-like (spindle-shaped) or epithelial-like (polygonal) in shape. The cells are isolated and cultured from a large number of different tissue types. Hence, these cells are preferably defined and named with respect to their origin of tissues. For instance, neuronal cells, myoblasts and lymphocytes are given these nomenclatures, based upon their origin. Fibroblasts are the cells that grow in the places between other cells and secrete the proteins of the extracellular matrix like collagen. These cells do not associate strongly with one another or with the cell of different nature, though fibroblasts have to ability to easily adhere themselves to a substratum. The epithelial cells are primarily involved in covering the body surface and bounding cavities, for example, the gut or kidney tubules. These cells are laterally bound with each other with the help of tight junction. By binding in this manner, these cells form a sheet-like structure that has a different composition in their apical and basolateral surfaces. These two surfaces are kept at a distance by the tight junctions. The cells, which are taken from an animal and grown in culture conditions, are termed primary cells, till the time they are further subcultured. If the primary cells are successfully established in culture conditions, then it will divide and primary cells, if successfully established in culture, will multiply and will need consistent subculturing. There are numerous stages to this process of transformation. The nature of different environmental factors significantly affects the cells in culture conditions. Some examples are as follows:

S. Mani (✉)
Centre for Emerging Diseases, Department of Biotechnology, Jaypee Institute of Information Technology, Noida, India

© The Author(s), under exclusive license to Springer Nature Switzerland AG 2023   89
S. Mani et al., *Animal Cell Culture: Principles and Practice*, Techniques in Life Science and Biomedicine for the Non-Expert,
https://doi.org/10.1007/978-3-031-19485-6_6

- The kind of the substrate and/or phase in which cells are growing. The substrate for monolayer cultures are solid in nature (e.g. plastic); however, it is liquid for suspension cultures.
- The constituents of the culture medium such as nutrients present in the medium and the physicochemical properties of the medium.
- Supplementation of hormones and growth factors to the medium.
- The temperature of culture incubation.
- Thus, we can grow the different cells in the culture conditions, and interestingly the cultured cells have certain distinguishing features from cells growing in in vivo conditions (Punekar, 2018).

1. The cells, which are generally unable to multiply in vivo, can be grown and multiplied in culture conditions.
2. Cell-to-cell communications are very much low in the cultured cells.
3. The three-dimensional architecture of the in vivo cells is not observed in cultured cells.
4. The cultured cells experience the effect of hormonal and nutritional factors in a significantly different manner than that of cell growth in vivo conditions.
5. In general, the cultured cells cannot undergo for differentiation and certain specialised functions.

6.2 Systems for Growing Cell Culture

There are two elementary systems for growing cells in culture conditions. These systems are commonly referred as adherent culture systems (where cells are grown as monolayers on an artificial substrate) or suspension culture (where cells are free-floating in the culture medium). Most of the cells, which are derived from vertebrates (excluding haematopoietic cell lines as well as few others), are anchorage-dependent and toned to be cultured on a suitable substrate (Merten, 2015). These specific surfaces are pre-treated to permit cell adhesion as well as spreading (i.e. tissue culture-treated). Though, several cell lines may also be altered for suspension culture. Cells growing in suspension culture can be sustained in culture flasks, which are not tissue culture-treated. However, as the culture volume to surface area is amplified beyond which optimum gas exchange is stalled (usually $0.2–0.5$ mL/cm^2), the medium needs the process of agitation. This agitation is generally attained with a magnetic stirrer or rotating spinner flasks (Marquis et al., 1989). The difference in the suspension and adherent cultures are summarised in Table 6.1.

6.3 Morphological Differences in Mammalian Cell

On the basis of their morphological differences, most of the mammalian cells, growing in the culture conditions, may be divided into three basic categories:

Table 6.1 Difference in the suspension and adherent culture system

Adherent culture	Suspension culture
Suitable for most of the cell types, including primary cultures	Suitable for cells already altered for suspension cultures and some natural non-adherent cells, such as haematopoietic cells
Need continuous passaging but permits easy visual examination	Very easy to passage but needs daily viability examination to follow growth patterns
Dissociation of cells from the surface is done either enzymatically (trypsin) or mechanically	No need for enzymatic/mechanical dissociation
The growth of cultured cells is restricted by surface area. This may lead to limitations in the production of yields	The growth is mostly limited by consumption of medium. Thus, scaling up the products is easy by managing the medium
Requirement of tissue culture grade flasks	Here the cells may be maintained in even non-tissue culture grade flasks. However, it may require agitation for the gaseous exchange
Mostly used for the cytology, continuous product harvest and other applications based on cell culture	Used for bulk protein production, batch harvesting, etc.

- Fibroblastic (or fibroblast-like) cells: These cells are mostly bipolar/multipolar as well as have elongated shapes. These cells grow by attaching them to a solid substrate (Verma et al., 2020).
- Epithelial-like cells: These cells are polygonal in shape with added even dimensions. These cells mostly grow in the form of discrete patches, by attaching them to a solid substrate (Verma et al., 2020).
- Lymphoblast-like cells: These are spherical in shape and usually grown in suspension form.

In addition to the above-mentioned basic categories, some cells exhibit their morphological features specific to their dedicated function in the host system. A typical example is neuronal cells. These cells exist in various shapes and sizes; however, they can commonly be segregated into two elementary morphological classes (Kanari et al., 2019). For instance, type I neurons have long axons and are thus primarily used to transfer signals over long distances; however, the type II does not carry an axon at all (Skinner, 2013). A typical neuron projects cellular extensions with several branches from the cell body denoted as a dendritic tree. In case of the unipolar/pseudounipolar neuronal cells, the dendrite and axon are observed to emerge from the same process. However, the bipolar neuronal cells have axon and single dendrite, on the opposite ends of the soma (the central part of the cell containing the nucleus). The neuronal cells may be multipolar in nature, which have more than two dendrites (Kanari et al., 2019).

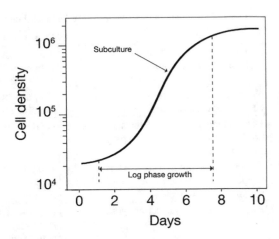

Fig. 6.1 Characteristic growth pattern of cultured cells

6.4   Maintaining Cultured Cells

The cells growing in culture conditions follow a characteristic growth pattern (Fig. 6.1). The semilogarithmic plot displays the cell density versus the time spent in culture. Cells in culture generally multiply ensuing a standard pattern of growth (Ham & McKeehan, 1979). Immediately after the seeding of the cells, the first phase of the culture is called as the lag phase. This is the phase of slow growth, where the cells are adapting to the culture conditions and getting ready for fast growth afterwards. The lag phase is tailed by the log phase (i.e. "logarithmic" phase), a phase where the cells multiply in an exponential manner and consume most of the nutrients present in the growth medium. When all the growth medium is exhausted (i.e. one or more of the nutrients is depleted) or when the cells consume all the available substrates of culture medium, the cells enter the stationary phase (i.e. plateau phase). This is the phase where the cellular proliferation is critically reduced or stops completely. It is important to note that cell culture conditions are significantly different for various cell types. The change in culture conditions compulsory for a specific cell type may cause the deviation from the normal growth and may lead to the appearance of abnormal phenotypes and/or a comprehensive failure of the cell culture (Verma et al., 2020).

6.5   When to Subculture

While growing the cells in culture conditions, it is important to understand the correct time and different criteria for subculturing them (Verma et al., 2020). Here are the following criteria for subculturing the cells, and these different criteria for determining the need for subculture are comparable in both adherent and suspension cultures.

6.5.1 Cell Density

In the case of mammalian cells, adherent cultures should be passaged, before they reach confluency, but care should also be taken that these cells should be in their log phase of the growth cycle. In general, due to the phenomenon of contact inhibition, cells stop growing when they reach the state of confluency, and if they are subcultured and reseeded, they take a comparatively longer time in recovering. Transformed cells can last multiplying even after attaining confluency, but they frequently weaken after around two replications. Likewise, cells growing in the suspension culture system should be passaged when they are in log phase growth, before attaining the confluency (Verma et al., 2020). In the case of suspension culture, these cells start making clumps together in the state of confluence. Also, the medium looks turbid, when the culture flask is swirled (Phelan, 2007).

6.5.2 Exhaustion of Medium

If the pH of the growth medium drops, it usually indicates the accumulation of lactic acid, an acid produced by cellular metabolism. Cells can be poisoned by lactic acid, and the decreased pH can hinder cell growth. The rate of change in pH is determined by the number of cells in the culture, with higher cell concentrations exhaustingmedium faster than those at lower concentrations.

It is therefore recommended to subcultivate when there is a rapid drop in pH (greater than 0.1–0.2 pH units) accompanied by an increase in cell concentration (Phelan, 2007).

6.5.3 Subculture Schedule

The process of passing your cells through a schedule ensures reproducible behaviour, and it lets you monitor their health. From a given seeding density, vary the density of your cultures until you get a consistent growth rate and yield for your cell type. In general, deviations from the established growth patterns indicate either a deteriorated or contaminated culture or a component of your culture system that is malfunctioning (e.g. the temperature is not optimal; the culture medium is too old). An extensive cell culture log should be kept, listing the forms of media used, dissociation procedures followed, split ratios, morphological observations, seeding concentrations, yields and any antibiotics used. The best way to perform experiments and other nonroutine procedures (e.g. changing media types) is in accordance with your subculture schedule. Make sure you don't passage your cells during the lag period or when they have reached confluency and are no longer growing if your experimental schedule does not mesh with the routine subculture schedule.

6.6   Media Recommendations

6.6.1   Culture Media

Providing nutrients, growth factors and hormones for cell growth, as well as regulating pH and osmotic pressure, the culture medium is the most important component of the culture environment. A cell culture medium contains salts, carbohydrates, vitamins, amino acids, metabolic precursors, growth factors, hormones and trace elements. The components required by each cell line are different, and these variations are partly responsible for the wide range of medium formulations (Phelan, 2007). The most common form of carbohydrates supplied is glucose. In some cases, galactose is substituted for glucose in order to reduce the buildup of lactic acid, since galactose has a slower metabolism (Yang & Xiong, 2012). Among the other carbon sources are amino acids (particularly L-glutamine) and pyruvate. The maintenance of many continuous mammalian cell lines can be accomplished on a relatively simple medium like MEM supplemented with serum, and cultures grown in MEM can probably be grown as effectively in DMEM or Medium 199. Be that as it may, when a specialised function is expressed, a more complex medium may be required. Data for selecting the suitable medium for a given cell sort is as a rule accessible in distributed writing and may moreover be referred from the source of the cells or cell banks. Cell culture experiments were initially performed using tissue extracts and body fluids as media, but with the need for standardisation, improved media quality and an increase in demand, defined media were developed. Basal media, reduced-serum media and serum-free media are three basic types of media, which differ in the amount of serum they require.

6.6.2   Serum

The serum is essential for cell culture in basal medium because it contains growth and adhesion agents, hormones, lipids and minerals. Serum also functions as a transporter for lipids, enzymes, micronutrients and trace elements into the cell, as well as regulating cell membrane permeability (Stein, 2007). However, serum in medium has a number of drawbacks, including high cost, difficulty with standardisation, specificity, unpredictability and undesired effects on some cell cultures, such as stimulation or inhibition of growth and/or cellular function (Stein, 2007). Contamination can also represent a severe hazard to successful cell culture experiments if the serum is not received from a reliable source. All Invitrogen and GIBCOR products, including sera, are tested for contamination and guaranteed for quality, safety, consistency and regulatory compliance to address this issue.

Basal Media
The majority of cell lines thrive in basal medium, which contains amino acids, vitamins, inorganic salts and a carbon source like glucose; however, these formulations must be supplemented with serum (Stein, 2007).

Reduced-Serum Media
Using reduced-serum media is another way to lessen the negative effects of serum in cell culture research. Reduced-serum media are nutrient- and animal-derived-factor-enriched basal media formulations that lower the amount of serum required (Rashid & Coombs, 2019).

Serum-Free Media
By replacing the serum with appropriate nutritional and hormonal compositions, serum-free medium (SFM) avoids the problems associated with utilising animal sera (Butler, 2015). Many primary cultures and cell lines, including recombinant protein-producing lines of Chinese hamster ovary (CHO), various hybridoma cell lines, the insect lines Sf9 and Sf21 (*Spodoptera frugiperda*) and cell lines that act as viral hosts (e.g. 293, VERO, MDCK, MDBK) and others, have serum-free media formulations (Butler, 2015; Verma et al., 2020). One of the most significant advantages of employing serum-free media is the flexibility to tailor the medium to specific cell types by selecting the right mix of growth factors (Butler, 2015). Some of the associated advantages of serum-free media are listed below:

- Increased definition.
- More consistent performance.
- Easier purification and downstream processing.
- Precise evaluation of cellular functions.
- Increased productivity.
- Better control over physiological response.
- Enhanced detection of cellular mediators.

6.6.3   pH

At pH 7.4, most typical mammalian cell lines grow effectively, and there is little variation among cell strains. Some altered cell lines, on the other hand, have been shown to grow better in slightly more acidic (pH 7.0–7.4) environments, whereas some normal fibroblast cell lines prefer slightly more basic (pH 7.4–7.7) environments (Yang & Xiong, 2012).

6.6.4 CO_2

The growth medium regulates the pH of the culture and protects the cells against fluctuations in pH. This is usually accomplished by using an organic (e.g. HEPES) or CO_2-bicarbonate-based buffer. Changes in atmospheric CO_2 can affect the pH of the medium, which is dependent on a delicate equilibrium of dissolved carbon dioxide (CO_2) and bicarbonate (HCO3 –). When utilising medium buffered with a CO_2-bicarbonate-based buffer, exogenous CO_2 is required, especially if the cells are grown in open plates or transformed cell lines are cultured at high concentrations (Yang & Xiong, 2012). Despite the fact that most researchers utilise 5–7% CO_2 in the air, most cell culture studies need 4–10% CO_2. However, each medium has a suggested CO_2 pressure and bicarbonate concentration to attain the correct pH and osmolality.

6.6.5   Temperature

The ideal temperature for cell culture is mostly determined by the body temperature of the host from whom the cells were extracted and to a lesser extent by anatomical temperature variations (e.g. the temperature of the skin may be lower than the temperature of skeletal muscle). For cell cultures, overheating is a more critical problem than underheating; as a result, the temperature in the incubator is frequently adjusted slightly lower than the ideal temperature. For optimal growth, most human and animal cell lines are kept around 36 °C to 37 °C (Phelan, 2007; Yang & Xiong, 2012).

6.6.6   Different Culture Mediums for Different Cells

Different common types of culture medium and their preparations protocols are already discussed in Chap. 5, and it is evident that most of the mediums have many components in common, though there is a little variation in the culture medium recommended for different cell lines (Yang & Xiong, 2012). Here we are summarising different culture mediums used for commonly used cell lines (Table 6.2).

Table 6.2 Summary of commonly used cell lines and recommended medium for their growth in cell culture

Cell line	Cell type	Tissue	Recommended medium
293	Fibroblast	Human embryonic kidney	MEM +10% FBS
A549	Epithelial	Human lung carcinoma	F-12 K+ 10%FBS
A9	Fibroblast	Mouse connective tissue	DMEM+10%FBS
BALAB/3 T3	Fibroblast	Mouse embryo	DMEM+10%FBS
Caco-2	Epithelial	Human colon adenocarcinoma	MEM + 20%FBS and NEAA
CHO-K1	Epithelial	Hamster ovary	F-12 K+ 10%FBS
COS-1, COS-3 and COS-7	Fibroblast	Monkey kidney	DMEM+10%FBS
H9	Lymphoblast	Human T-cell lymphoma	RPMI 1640 + 20%FBS
HeLa	Epithelial	Human cervix carcinoma	MEM + 10%FBS and NEAA
HEp-2	Epithelial	Human larynx carcinoma	MEM + 10%FBS
K562	Lymphoblast	Human myelogenous Leukaemia	RPMI 1640 + 10%FBS
L2	Epithelial	Rat lung	F-12 K+ 10%FBS
Jurkat	Lymphoblast	Human lymphoma	RPMI 1640 + 10%FBS
MCF-7	Epithelial	Human breast carcinoma	DMEM+10%FBS

References

Butler, M. (2015). Serum and protein free media. In *Animal cell culture* (pp. 223–236). Springer.

Ham, R. G., & McKeehan, W. L. (1979). [5] Media and growth requirements. *Methods in Enzymology, 58*, 44–93.

Kanari, L., Ramaswamy, S., Shi, Y., Morand, S., Meystre, J., Perin, R., Abdellah, M., Wang, Y., Hess, K., & Markram, H. (2019). Objective morphological classification of neocortical pyramidal cells. *Cerebral Cortex, 29*(4), 1719–1735.

Marquis, C. P., Low, K. S., Barford, J. P., & Harbour, C. (1989). Agitation and aeration effects in suspension mammalian cell cultures. *Cytotechnology, 2*(3), 163–170.

Merten, O. W. (2015). Advances in cell culture: anchorage dependence. *Philosophical Transactions of the Royal Society B: Biological Sciences, 370*(1661), 20140040.

Phelan, M. C. (2007). Techniques for mammalian cell tissue culture. *Current Protocols in Neuroscience, 38*(1), A–3B.

Punekar, N. S. (2018). In vitro versus in vivo: concepts and consequences. In *ENZYMES: Catalysis, kinetics and mechanisms* (pp. 493–519). Springer.

Rashid, M. U., & Coombs, K. M. (2019). Serum-reduced media impacts on cell viability and protein expression in human lung epithelial cells. *Journal of Cellular Physiology, 234*(6), 7718–7724.

Skinner, F. K. (2013). Moving beyond Type I and Type II neuron types. *F1000Research, 2*, 19.

Stein, A. (2007). Decreasing variability in your cell culture. *BioTechniques, 43*(2), 228–229.

Verma, A., Verma, M., & Singh, A. (2020). Animal tissue culture principles and applications. In *Animal biotechnology* (pp. 269–293). Academic.

Yang, Z., & Xiong, H. R. (2012). Culture conditions and types of growth media for mammalian cells. *Biomedical Tissue Culture, 1*, 3–18.

Chapter 7
Selection and Maintenance of Cultured Cells

Divya Jindal and Manisha Singh

7.1 Introduction

Cell culturing and cell proliferation are some of the essential components in cell-based assays, in pre-clinical and biomedical research. This specialised in vitro technique is highly practised in cancer research, recombinant gene formation, vaccine development or stem cell studies (Oyeleye et al., 2016). Moreover, rapid advancements in genomics and proteomics have resulted in the expanded application of cell culture techniques, making it possible to study as well as predict physiological and metabolic effects in higher species (Baust et al., 2017; Philippeos et al., 2012). This technique has offered a standard to demarcate the unmitigated human genome and dissever the intercellular and intracellular signalling pathways that eventually control gene expressions (Masters, n.d.). Further, the growth of cells can be described by cell division or differentiation (Lynn, 2009), and they are grown under specific conditions of temperature and humidity, along with adequately designed nutrition. The cells for cell culturing are mainly isolated from various organs through either mechanical or enzymatic actions, and then subsequently grown in a conditioned environment.

Cell cultures are maintained in two forms, that is primary and secondary cell cultures. In the primary cell culture type, the cells are directly isolated from the host tissue; however, in secondary cell culture, it is sub-cultured (Bayreuther et al., 1991); and based on their origin, these primary cell cultures grow as an adherent or suspension. The adherent cells are anchorage-dependent and hence disseminate as monolayers, while on the other hand, suspension cells do not adhere to the surface of the culture vessel and float in the nutrient medium. When the above-mentioned

D. Jindal · M. Singh (✉)
Department of Biotechnology, Jaypee Institute of Information Technology (JIIT), Noida, Uttar Pradesh, India

© The Author(s), under exclusive license to Springer Nature Switzerland AG 2023
S. Mani et al., *Animal Cell Culture: Principles and Practice*, Techniques in Life Science and Biomedicine for the Non-Expert,
https://doi.org/10.1007/978-3-031-19485-6_7

culture is sub-cultured, it forms the secondary culture or cell line (Malm et al., 2020; Adekanmbi & Falodun, 2016; Uysal et al., 2018). For the growth of cells in vitro, conditions don't impersonate in vivo environments with parameters like pH, carbon dioxide, oxygen, nutrition and temperature. Additionally, these cells need a sterile environment to grow with an optimum incubation period along with the nutrient medium or media that plays an important role in escalating their growth (Verma et al., 2020). With the aid of varied applications, the practice of cell culturing is limited to strict aseptic and environment control without bacterial or fungal growth. For an appropriate cell line, the criteria are governed by their species, functional characteristics, normal or transformed, continuous or finite growth conditions and so on (Uysal et al., 2018; Remotti et al., 1997). These cells are then maintained with optimum nutrients and passaged well to reach their confluence, followed by sub-culturing or passaging (Verma et al., 2020; Jokela & Labarge, 2015).

7.2 Origin and Characterisation of Cells

There are diverse cell types available for culturing, and due to the enriched accessibility of selective nutrient media along with growth hormones for specific cells, it's comparatively easier to culture them. Also, the cell-specific marker proteins have made it viable to identify the lineage from where a lot of cultures are derived. During proliferation, precursor cells may dominate the growth instead of the differentiated cell types, and therefore, the cell line appears to be heterogeneous (e.g. epidermal keratinocytes, stem cells, mature cells, etc.). Further, many cultures encompass a uniform yield of differentiated cells and show a low cell density of 10^4 cells per cm^2, while non-differentiated cells are at higher cell densities of 10^5 cells per cm^2. The higher cell density of cells can retrace the cell cycle after trypsinisation and may decline their cell density. There are multiple lineages found in culture heterogeneity and can only be sustained by using discerning selective conditions which allow them to survive and differentiate (Bhatia et al., 2019), leading to choosing some common phenotype. However, because of the cooperative nature of the growth regulator, the cell population may hold various phenotypes that are only evidenced by cloning. There are many factors like serum, calcium ions (Boyce & Ham, 1983), interaction of matrix and cells (Thomson et al., 1997; Berdichevsky et al., 1992), hormones (Rooney et al., 1994; McCormick & Freshney, 2000) or cell culture densities that play an important role in cell proliferation.

7.2.1 Cell Differentiation

The cell culture medium initiates and sustains the cell growth with its proliferation. Physiological conditions like low cell density, calcium concentration and growth factors promote cell proliferation to a larger extent, while high calcium concentration and growth promoters induce cytostasis and differentiation. Additionally, the serum concentration plays a prominent role in differentiation; its lower concentration elevates differentiation in oligodendrocytes (Raff et al., 1983), whereas its higher concentration disseminates squamous differentiation in bronchial epithelium cells (Lechner et al., 1983). Another factor responsible for cell differentiation is polarity along with the cell shape of the cell, facilitating substrate plasticity.

7.3   Selection of Appropriate Cell Line

1. **Cell Types**

    Selection of cell type will depend upon the process that needs to be performed, as in DNA synthesis or apoptosis, and whether they are transformed or normal. Cells need to be competent; however, surfactant synthesis requires cells that express surfactant proteins (Segeritz & Vallier 2017). There can be some species that require species-specific cultures and conditions with acquired biosafety restrictions and may require DMEM, EBSS, MEM or any other, depending upon their origin and the functions they exhibit. Some may exert properties like retaining drug-metabolising enzymes, synthesis of proteins, neuronal differentiation, formation of domes, synthesis of haemoglobin and several others.

2. **Morphology of Cells in Culture**

    Cells can be distinguished into different categories depending upon their shape and morphology (Table 7.1).

Table 7.1 Different cell types and their morphology

S. no.	Cell types	Morphology	References
1	Epithelial cells	Are polygonal with regular dimensions and propagate attached to the substrate.	McKeehan (1990)
2	Lymphoblast cells	Are spherical, generally propagate in suspension lacking the attachment to the surface.	Werner and Noé (1993)
3	Neuronal cells	Made up of a cell body comprising nucleus and organelles, along with dendrites, branched to process synaptic information from other neuronal cells.	Franze and Guck (2010)
4	Muscle cells	Are characterised for contraction, help in forming outer walls of hollow organs and produce motions. They are formed of myoblasts and connective tissues.	Maestroni (2012)
5	Stem cells	Are undifferentiated or differentiated, consisting of many lipids and glycogen.	Courtot et al. (2014)
6	Fibroblastic cells	Maybe bipolar or multipolar with elongated structures. They are found to propagate by attaching to a substrate.	Borg and Caulfield (1980)
7	Cancer cells	Form suspension of aggregated cells, anchored on culture plates.	Fennell and Jablons (2018)

3. Cell Line Selection

Normally, cell line selection is a monotonous and laborious procedure. This is performed after transfection of cells with the help of dilution cloning to evade non-producing and low-producing cells. Selection of cell lines may take up time, to screen cells for the stability of a specific clone, befalling a big challenge in the biopharmaceutical industry (Carroll & Al-Rubeai, 2004; Browne & Al-Rubeai, 2007). For the selection and screening of cell lines, numerous approaches have been categorised, like flow cytometry, FACS and matrix-based secretion assays, out of which FACS is most commonly utilised for cell production estimation (Gallagher & Kelly, 2017). The following factors play an important role in the selection of a cell line:

  (i) Origin of cell line
  (ii) Cell type (normal or transformed)
  (iii) Cell line type (infinite or finite cell line)
  (iv) Precedents in the growth of cells
  (v) Efficiency of cloning
  (vi) Cell number
  (vii) Availability of cell line and growth factors
(viii) Doubling time

4. Verification of Cell line

To verify the cell line origination and avoid inappropriate identification, cell lines are confirmed with the help of DNA-based techniques, by performing a small tandem repeat analysis and comparing it for cell line and primary culture. This method is being registered in NCBI International Database to avail unique identity of these primary cultures (NCBI Resource Coordinators, 2016). The cell lines can also be verified using Karyotyping, widely utilised to detect abnormalities in chromosomes, and another method is SNP (single nucleotide polymorphisms) profiling (Didion et al., 2014; Liang-Chu et al., 2015) using specific primers of species and allowing the detection of mutation and genetic drift (Pruckler et al., 1995). This is achieved by probing the sequences linked with species-specific cyt c oxidase 1 (CO1), generally termed DNA barcoding (Hebert et al., 2003a, b).

5. Misidentification of Cell Line

Mainly cell lines are wrongly identified due to cross-contamination, majorly supplemented with continuous cell lines as the cells get replaced by one another. Other factors which are responsible for misleading of selection of cell lines are:

  (i) Modification in cellular behaviour and morphological changes
  (ii) Development of two cell lines in a single cell line at the same time
  (iii) Failing in maintenance of cell culture methodology
  (iv) Incorrect labelling of flasks
  (v) Improper sterilisation

6. Cell Culture Types

(a) **Adherent and Suspension Culture:** Cell culture is organised in two different forms when they grow, that is adherent and suspension cell culture. In
adherent culture, the cells are arranged in a monolayer and are anchorage-
dependent as on artificial substrate, while suspension cultures are free-
floating and do not depend on the substrate for their attachment, hence
multi-layered. These cell require periodic passaging and their growth is
traced by continuous cell counting. Because adherent cultures are monolayers, growth is constrained by surface area and the cell yield percentage is
also compromised. In suspension culture, the concentration of cells limits
the growth which permits it to scale up the cell yield (Philippeos et al., 2012;
Alberts et al., 2002).

(b) **Cell Growth:** The quality of cell lines is established by their growth curve,
formed by taking samples at different intervals throughout the growth cycle.
When particular cells are seeded, they initially enter a phase of no growth
called the *lag period,* permitting the cells to rejuvenate from trypsinisation
and supporting them to re-enter the cell cycle. Once the cells are re-entered,
they move to the *log phase*, where the exponential growth of cells is seen, in
which the population of cells is doubled. When cells achieve their confluency, they enter a stationary phase in which the growth becomes static and
the growth fraction is nearly zero. At this point, the cells need to be subcultured before their plateau is reached to escalate the growth fraction with
recovery time (Freshney, 2006) (Fig. 7.1).

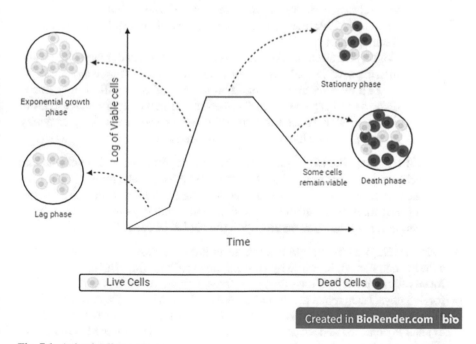

Fig. 7.1 Animal cell growth curve

7.3.1   Techniques for Detachment of Cells

Various methods have been discussed in the literature to assess cell detachment. The commonly used methods involve mechanical, enzymatic detachment and primary explant techniques.

1. **Mechanical Detachment:** It is required to remove and separate soft tissues, assessed by carving or harvesting the specific tissue. This is executed by sieving, syringing or pipetting to separate cells, only used when cell viability in critical yield is not very essential. This method is commonly used because of its quick and easy, economical nature.

2. **Enzymatic Detachment:** This method involves the use of enzymes such as Trypsin, collagenase and many others, allowing hydrolysis of connective tissues and Extracellular matrix (ECM). This method is used due to its property of huge recovery of cells deprived of cell viability.

   (a) **Trypsin Based:** Generally, crude trypsin is used since it consists of other proteases which can remove unwanted proteins. Cells can also bear crude trypsin very well, and their respective effects can also be neutralised by serum or inhibitors. Less toxic pure trypsin can be used for disaggregation of cells if showing specificity in its action. Trypsin is used in two approaches:

      (i) Warm trypsinisation: This is extensively used in the preceding steps, when sliced tissue is washed with a salt solution, followed by their transfer to a warm trypsin solution at 37 °C and stirred well at intervals of 30 min. After incubation, the cells are separated and dispensed into a fresh medium.

      (ii) Cold trypisinisation: Cells are dissected and washed, placed in ice and then exposed to cold trypsin for a day, followed by incubation at 37 °C for half an hour and mixing. The mixing of cells will allow them to disperse and separate from the medium. It is widely used as it doesn't cause cellular damage due to the constant acquaintance of enzymes, resulting in more harvesting of viable cells.

   (b) **Collagenase Based:** Collagenase breaks peptide bonds between amino acids and mainly exists in ECM of connective tissues and muscles. The cells are treated with basal salt solution containing antibiotics, followed by washing and transferring them to a medium consisting of collagenase and incubating for 1–5 days depending upon the choice of cells (Fig. 7.2).

3. **Primary Explant Technique:** In this technique, the tissue is suspended in salt solution and nicked, followed by washing and settling. Since the tissues are uniformly disseminated on the growing surface, a fresh medium is added and kept for 4–5 days. Keep replacing the exhausted medium with the fresh one, untill the preferred outgrowth of cell is achieved. When an ideal growth is achieved, the explant is detached and suspended in a new medium. This method is most suitable for a small number of tissues, as it may affect the viability of cells.

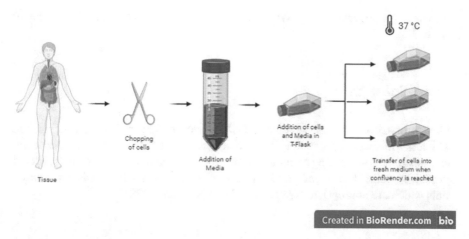

Fig. 7.2 Sub-culturing of Animal cell lines

7.3.2   Source of Tissue

1. **Embryo or Adult:** Cells originated from embryos survive better than from adult tissues. This presumably indicates a lower degree of specialisation and a higher degree of proliferative potential (Vazin & Freed, 2010), which is because adult tissues have less growth potential and much higher non-replicating specialised cells, partaking structured and aggregated ECM, causing transmission and proliferation to be difficult and shorter life span. While embryonic tissues have advantages over adult tissues, for example, 3T3 cell lines, WI-28, fibroblasts or myoblasts are much more tranquil to culture than adult tissues like epithelium and endocrine tissues. Henceforth, selective media have been designed in a way to characterise specific cell types, with serum acting as a growth inhibitor and differentiation promoting (Passier & Mummery, 2003; Czyz et al., 2003).
2. **Normal or Neoplastic:** In general, normal tissues produce cultures with a finite duration, whereas tumour cells will form continuous cell lines. Many continuous cell lines are acquired from normal tissues, being non-tumorigenic, like BHK-21 and 3T3 fibroblasts (Gey & Gey, 1936). Normal cells are derived from undifferentiated stem cells, and their differentiation is complemented by a cessation in cell differentiation, which could be absolute. Although many normal cells can differentiate and likewise de-differentiate, and continue to proliferate and re-differentiate, as in the case of fibrocytes and endothelium tissues. At the same time, some may differentiate and are then impotent to proliferate, like squamous epithelium or neurons.

Neoplastic cells undergo partial differentiation and maintain the ability to divide like B16 melanoma cells and hepatomas. Tumour cells frequently proliferate in the syngeneic host, postulating an easy approach to generating high-yielding cells, with less purity. If the natural host is unavailable, these tumour cells can disseminate in animals, compromising the system's immunology (Glovanella et al., 1976) (Table 7.2).

7.3.3 Subculture

When primary culture is sub-cultured, they acquire lower growth fraction and exhibit tissue-specific properties. It allows the development of culture, termed a cell line, capable of performing cloning, characterisation and uniformity with the loss of specialised cells. The major provision is their high amount of consistent components which are appropriate for prolonged use (Fischer, 1978).

Table 7.2 Commonly used Animal Cell lines

Cell line	Species	Tissues	Cell type	Properties	Medium	References
Normal						
3T3-L1	Mouse	Whole embryo	Fibroblastoid	Contact inhibition is the ability to differentiate between adipose tissues	DIEM	Green and Kehinde (1974)
MRC-5	Human	Embryo lung	Fibroblastic	Contact inhibition	Glucose-DMEM	Jacobs et al. (1970)
MDCK	Dog	Kidney	Epithelial	"Domes" formation	EMEM	Gaush et al. (1996) and Taub and Saier (1979)
Transformed						
HeLa-G2	Human	Hepatoma	Epithelial	Able to retain a few drug-metabolising enzymes	DMEM, 10% FBS	Knowles et al. (1980)
HeLa-S₃	Human	Cervical carcinoma	Epithelial	Fast growth and propagation in suspension	SFM	Resau and Cottrell (1989)
Caco-2	Human	Colorectal cancer	Epithelial	Formation of the strong monolayer with polarised transport	MEM, 20% FBS and NEAA	Fogh et al. (1977)
MCF-7	Human	Breast cancer	Epithelial	Positive oestrogen receptor	MEM, 10% FBS, NEAA and 10µl/ml insulin	Soule et al. (1990)

DMEM Dulbecco's Modified Eagle Medium, *FBS* Fetal bovine serum, *EMEM* Eagle's Minimum Essential Medium, *SFM* Serum-free media, *MEM* Minimum Essential Medium, *NEAA* Non-essential amino acids solution

1. **Finite or Continuous Cell Lines**

When the cells are sub-cultured, they may die (finite cell line) or be transformed into continuous cell lines (infinite cell line). The presence of continuous cell lines is identified by an alteration in cytomorphology with an increased growth rate and cloning efficiency and a decrease in serum and anchorage dependence (Alberts et al., 2002). Normal cultured cells can become continuous cell lines deprived of being malignant, but malignant tumours engender to culture with tumorigenic cells. One of the exclusive importance of these cell lines is their fast growth rates and high cell yield, utilising less amount of serum and easing their maintenance to grow in suspension. The inconvenience of using these cells involves additional chromosomal instability and deficiency of tissue-specific makers.

2. **Sub-culturing of Adherent Cells**

To carry out the subculture of adherent cells, the cells need to be transported into suspension. The degree of adhesion varies with different origins of cells, followed by treatment with proteases; for example, trypsin is used to detach the cells from the flask surface. In some cases, where enzyme treatment becomes harmful to the cell, mechanical detachment methods like cell scrappers can be used and followed by the addition of new media for their growth (Kurashina et al., 2017).

3. **Sub-culturing of Suspension Cells-**

The majority of cultures proliferate as a monolayer, attached to the substrate while some propagate in suspension-like hematopoietic cells. These suspension cultures have been benefited in having dilution only instead of trypsinisation with no constraint for large surface area and alleviated harvesting with leeway to achieve a "steady state". Since the cells are already suspended in a medium, there is no treatment of enzymes for surface detachment, at a faster rate (Liu et al., 2014). These suspension cells are maintained by daily feeding nutrients to achieve confluency. Confluency is reached by diluting the cell culture and expanding them or by diluting the cells to a seeding density suitable for cell lines (Ricardo & Phelan, 2008).

7.3.4   Growth Conditions and Characteristics

Selection of appropriate cell line is insured by the growth rate, efficiency of cloning, ability to grow in suspension or density of their saturation. For instance, to express R-protein (Recombinant protein) in high numbers, a cell line with the ability to grow in suspension and a high growth rate is required. Typical cell growth conditions include:

(a) Temperature: The ideal temperature for culture depends upon the body temperature of the host organism. High temperatures may cause damage to the growing cell. The optimum temperature required for cell culture is 37 °C (Hanks & Wallace, 1949).

(b) Carbon dioxide: The specific medium directs the pH of the cell culture and buffers the cells versus transformation in pH. Buffering of cells is attained by using organic or carbon dioxide bicarbonate-based buffer, as pH depends upon the balance between CO_2 and HCO_3^-. Henceforth, it is required to take exogenous carbon dioxide. However, it is recommended to use 4–10% CO_2 in a cell culture environment, but it has been recommended that each media has its CO_2 and bicarbonate concentration to attain an ideal pH and osmolality (Frahm et al., 2002; Dezengotita et al., 1998).

(c) pH: Many animal cell lines' optimum pH is 7.4, which may differ with transformed cells, with a slight change in acidic and basic environments (Mackenzie et al., 1961).

(d) Relative humidity: It is the amount of water vapour available in the air. The optimum relative humidity of an animal cell line is 95%, which may vary with cell type and function (Tang et al., 2015).

7.3.5 Other Criteria

Selection also depends upon the availability of sufficient stocks, well-characterised cell lines, validatory data, cloning efficiency and the stability of the cell line.

7.4 Maintenance of Cell Line

As soon as the culture is introduced, a primary or sub-culture, there is a need to periodically change the medium or feeding, followed by sub-culturing in the division phase of cells. In non-proliferating cells, the media still required to be changed to stabilize the cells to metabolize, leading to exhaustion and degradation of constituents. Duration of the media replacement depends upon metabolism and rate of growth. For example, HeLa cell lines are required to be sub-cultured in a week, and media is replaced every fourth day due to their rapidly growing nature. Slow proliferating cells, generally transformed ones, require to be sub-cultured in two, three or four weeks, and the medium is replaced within weeks. Cell lines may change their characteristics when maintained with the different standardised regimes. The maintenance needs to be optimised and consistent (Freshney, 2010).

1. **Cell Morphology:** The culture needs to be examined at regular intervals to identify the absence of contamination or any signs of deterioration in cell organelles, between the natural cellular organisation and physiological state of cells from impurities. Such identification infers that the culture needs the replacement of the medium and may show some toxicity or senescence of the cell line. Similarly, cell morphology help in recognising the sign of cross-contamination and misidentification (Freshney, 2010).

2. **Replacement of Medium:** Replacement of medium is conferred by four factors. These are:

   (a) pH drop: Some cells lose viability when pH drops from 7 and in such conditions, the change of fresh medium is required (Eagle, 1973).
   (b) Cell concentration: At high concentration, the culture uses up medium much faster as compared to lower concentrations.
   (c) Cell type: Generally, normal cell division terminates at high cell density due to cell cramming and depletion of growth factors. Transformed cells deteriorate quickly at high cell densities without the medium being changed daily or when sub-cultured (Biscop et al., 2019).
   (d) Morphological deterioration: Regular examination of cell lines checks the progress of deterioration, which becomes irreversible, and is most likely to enter apoptosis (Nelson et al., 2013).
   (e) Volume, depth and surface area: The typical volume-to-surface-area ratio of media is 0.2–0.5 ml/cm^2, based on the oxygen necessity of cells. High oxygen necessity cells need a shallow medium and low necessity cells perform better with a deep medium (Noorafshan et al., 2016).
   (f) Growth factors: They are used to supplement the media, and mostly derived from animal serum, preferably from fetal bovine serum and equine serum.

7.5 Media

1. **Development of Media**

   Initially, the cell is used in a culture in natural-based media composed of body fluids like chick embryo extract, lymph, etc. With the advancement in the proliferation of cells, there comes the requirement of quality media which leads to the initiation of chemical-based media analogue to body fluids (Ham, 1965). Hence, Eagle's Basal Media and Minimal Media has been introduced in the 1950s, supplemented with human and calf serum. Extraction and proliferation of cells require specific lineage, needing selective media. DMEM and RPMI, are usually used, supplemented with serum. Some complex media are also used, like Ham's F12, which contains higher amounts of amino acids and vitamins (Fischer et al., 1948; Yao & Asayama, 2017).

2. **Complete Media:** Complete media indicates a medium that contains all the supplements and components required by a cell to survive, like a serum, hormones, essential and non-essential amino acids, vitamins or growth factors (Pixley, 1985; Eagle, 1955).

   1. Amino acids: Both essential and non-essential amino acids are present in the medium, totally depending on the requirement of the cell to compensate for the cell's incapacity to produce or lose them. Glutamine is essential for most cells, even though some may utilise glutamate as a source of energy and carbon (Xiong & Yang, 2012; Butler & Christie, 1994).

2. Vitamins: Eagle's MEM includes hydrophilic vitamins, widely used in cell maintenance. Biotin is commonly added, confining with serum-free media (Schnellbaecher et al., 2019).
3. Salts: Salts are mainly involved in maintaining the osmolality of the medium by sodium, potassium, magnesium, calcium, chloride, phosphate, and bicarbonates. Divalent cations are mandatory for adherent cells, for example, Cadherin (Yamada & Geiger, 1997).
4. Glucose: Glucose is added to most of the media for energy, entering into glycolysis and forming lactic acid. This lactic acid is accumulated, as seen in transformed cells (Fonseca et al., 2018).
5. Organic supplements: Other components like proteins and peptides are also needed, generally when serum concentration is comparatively low. They generally assist in cloning and maintaining cells. Proteins are the carriers of lipids and minerals, assist in cell attachment, and increase viscosity to reduce shear stress while mixing (Moran et al., 1967).
6. Hormones and growth factors: They have a key role in proliferation when using serum-free media (Lee et al., 1997; Hayashi & Sato, 1976). PDGF (platelet-derived growth factor) is widely used with mitogenic properties, present in serum. Hormones like insulin promote mitogenic activities and assess the uptake of glucose and various amino acids by IGF-I receptors (Fox & Kelley, 1978).
7. Antibiotics- Antibiotics are used to lessen the extent of contamination in culture (Eagle, 1955; Ryu et al., 2017).

7.6   Conclusion

Cell culture is an indispensable technique applied in medicine and diagnostic methods, nearly unbiased with the growth of various organisms by overcoming the implication of transgenic cell culture technology. The chapter describes the various possibilities and methods to perform cell culture to employ translational research problems and essential considerations for performing cell culture techniques. The chapter has also revealed important practices for effectively functioning with cell lines and describes various conditions and media required by the cell to undergo proper cell growth and development.

References

Adekanmbi, A. O., & Falodun, O. I. (2016). Antibiogram of Escherichia coli and Pseudomonas strains isolated from wastewater generated by an Abattoir as it journeys into a receiving river. *Advances in Microbiology, 6*, 303–309.

Alberts, B., Johnson, A, & Lewis J. (2002). Isolating cells and growing them in culture. In *Molecular biology of the cell*. W. W. Norton & Company.

Baust, J. M., Buehring, G. C., Campbell, L., Elmore, E., Harbell, J. W., & Nims, R. W. (2017). *Best practices in cell culture: An overview. In vitro cellular & developmental biology Animal* (pp. 669–672). Tissue Culture Association.

Bayreuther, K., Francz, P. I., Gogol, J., Hapke, C., Maier, M., & Meinrath, H. G. (1991). Differentiation of primary and secondary fibroblasts in cell culture systems. *Mutation Research, 256*, 233–242.

Berdichevsky, F., Gilbert, C., Shearer, M., & Taylor-Papadimitriou, J. (1992). Collagen-induced rapid morphogenesis of human mammary epithelial cells: The role of the alpha 2 beta 1 integrin. *Journal of Cell Science, 102*, 199207.

Bhatia, S., Naved, T., & Sardana, S. (2019). *Introduction to pharmaceutical biotechnology. Characterization of cultured cells*. IOP Publishing.

Biscop, E., Lin, A., Boxem, W. V., Loenhout, J. V., Backer, J. D., & Deben, C. (2019). Influence of cell type and culture medium on determining cancer selectivity of cold atmospheric plasma treatment. *Cancers, 11*, 1287.

Borg, T. K., & Caulfield, J. B. (1980). Morphology of connective tissue in skeletal muscle. *Tissue and Cell, 12*, 197–207.

Boyce, S. T., & Ham, R. G. (1983). Calcium-regulated differentiation of normal human epidermal keratinocytes in chemically defined clonal culture and serum-free serial cultures. *The Journal of Investigative Dermatology., 81*, 33s.

Browne, S. M., & Al-Rubeai, M. (2007). Selection methods for high-producing mammalian cell lines. *Trends in Biotechnology, 25*, 425–432.

Butler, M., & Christie, A. (1994). Adaptation of mammalian cells to non-ammoniagenic media. *Cytotechnology, 15*, 87–94.

Carroll, S., & Al-Rubeai, M. (2004). The selection of high-producing cell lines using flow cytometry and cell sorting. In *Expert opinion on biological therapy* (pp. 1821–1829). Ashley Publications Ltd.

Courtot, A.-M., Magniez, A., Oudrhiri, N., Féraud, O., Bacci, J., Gobbo, E., et al. (2014). Morphological analysis of human induced pluripotent stem cells during induced differentiation and reverse programming. *Biores Open Access, 3*, 206–216.

Czyz, J., Wiese, C., Rolletschek, A., Blyszczuk, P., Cross, M., & Wobus, A. M. (2003). Potential of embryonic and adult stem cells in vitro. *Biological Chemistry, 384*, 1391–1409.

Dezengotita, V. M., Kimura, R., & Miller, W. M. (1998). Effects of CO2 and osmolality on hybridoma cells: growth, metabolism and monoclonal antibody production. *Cytotechnology, 28*, 213–227.

Didion, J. P., Buus, R. J., Naghashfar, Z., Threadgill, D. W., Morse, H. C., 3rd, & de Villena, F. P. (2014). SNP array profiling of mouse cell lines identifies their strains of origin and reveals cross-contamination and widespread aneuploidy. *BMC Genomics, 15*, 1.

Eagle, H. (1955). Nutrition needs of mammalian cells in tissue culture. *Science, 122*, 501–514.

Eagle, H. (1973). The effect of environmental pH on the growth of normal and malignant cells. *Journal of Cellular Physiology, 82*, 1–8.

Fennell, D. A., & Jablons, D. M. (2018). Chapter 13 – Stem cells and lung cancer: In vitro and in vivo studies. In H. I. Pass, D. Ball, & G. V. Scagliotti (Eds.), *IASLC Thoracic Oncology*. Elsevier.

Fischer, L. (1978). The concept of the subculture of the aging reconsidered.

Fischer, A., Astrup, T., et al. (1948). Growth of animal tissue cells in artificial media. In *Proceedings of the society for experimental biology and medicine society for experimental biology and medicine* (pp. 40–46). Blackwell Science.

Fogh, J., Wright, W. C., & Loveless, J. D. (1977). Absence of HeLa cell contamination in 169 cell lines derived from human tumors. *Journal of the National Cancer Institute, 58*, 209–214.

Fonseca, J., Moradi, F., Valente, A. J. F., & Stuart, J. A. (2018). Oxygen and glucose levels in cell culture media determine Resveratrol's effects on growth, hydrogen peroxide production, and mitochondrial dynamics. *Antioxidants, 7*, 157.

Fox, I. H., & Kelley, W. N. (1978). The role of adenosine and 2'-Deoxyadenosine in mammalian cells. *Annual Review of Biochemistry, 47*, 655–686.

Frahm, B., Blank, H. C., Cornand, P., Oelssner, W., Guth, U., Lane, P., et al. (2002). Determination of dissolved $CO(2)$ concentration and $CO(2)$ production rate of mammalian cell suspension culture based on off-gas measurement. *Journal of Biotechnology, 99*, 133–148.

Franze, K., & Guck, J. (2010). The biophysics of neuronal growth. *Reports on Progress in Physics, 73*, 094601.

Freshney, R. (2006). *Basic principles of cell culturs* (pp. 1–22). Wiley.

Freshney, R. I. (2010). *Culture of animal cells: A manual of basic technique and specialized applications*. Wiley.

Gallagher, C., Kelly, P. S. (2017). Selection of high-producing clones using FACS for CHO cell line development. In *Methods in molecular biology* (pp. 143–152). Academic.

Gaush, C. R., Hard, W. L., & Smith, T. F. (1996). Characterization of an established line of canine kidney cells (MDCK). In *Proceedings of the society for experimental biology and medicine society for experimental biology and medicine* (931–935). Blackwell Science.

Gey, G. O., & Gey, M. K. (1936). The Maintenance of Human Normal Cells and Tumor Cells in Continuous Culture: I. Preliminary Report: Cultivation of Mesoblastic Tumors and Normal Tissue and Notes on Methods of Cultivation. *The American Journal of Cancer, 27*, 45.

Glovanella, B. C., Stehlin, J. S., Santamaria, C., Yim, S. O., Morgan, A. C., Williams, L. J., Jr., et al. (1976). Human neoplastic and normal cells in tissue culture. I. Cell lines derived from malignant melanomas and normal melanocytes. *Journal of the National Cancer Institute, 56*, 1131.

Green, H., & Kehinde, O. (1974). Sublines of mouse 3T3 cells that accumulate lipid. *Cell, 1*, 113–116.

Ham, R. G. (1965). Clonal growth of mammalian cells in a chemically defined, synthetic medium. *Proceedings of the National Academy of Sciences of the United States of America, 53*, 288–293.

Hanks, J. H., & Wallace, R. E. (1949). Relation of oxygen and temperature in the preservation of tissues by refrigeration. In *Proceedings of the society for Experimental Biology and Medicine Society* (pp. 196–200). Blackwell Science.

Hayashi, I., & Sato, G. H. (1976). Replacement of serum by hormones permits growth of cells in a defined medium. *Nature, 259*, 132–134.

Hebert PD, Cywinska A, Ball SL, deWaard JR. (2003a) Biological identifications through DNA barcodes. Proceedings of the Biological Sciences 270, 313-321.

Hebert, P. D., Ratnasingham, S., & deWaard, J. R. (2003b). Barcoding animal life: Cytochrome C oxidase subunit 1 divergences among closely related species. *Proceedings of the Biological Sciences, 270*, 96.

Jacobs, J. P., Jones, C. M., & Baille, J. P. (1970). Characteristics of a human diploid cell designated MRC-5. *Nature, 227*, 168–170.

Jokela, T, & Labarge, M. (2015). Culture of animal cells: A manual of basic technique and specialized applications. In *Culture of cancer stem cells* (pp. 583–587).

Knowles, B. B., Howe, C. C., & Aden, D. P. (1980). Human hepatocellular carcinoma cell lines secrete the major plasma proteins and hepatitis B surface antigen. *Science, 209*, 497–499.

Kurashina, Y., et al. (2017). Efficient Subculture Process for Adherent Cells by Selective Collection Using Cultivation Substrate Vibration. *IEEE Transactions on Bio-Medical Engineering, 64*, 580–587.

Lechner, J. F., McClendon, I. A., & LaVeck, M. A. (1983). Shamsuddin AM, Harris CC. Differential control by platelet factors of squamous differentiation in normal and malignant human bronchial epithelial cells. *Cancer Research, 80*, 5915–5921.

Lee, J.-J., Kwon, J.-H., Park, Y. K., Kwon, O., & Yoon, T.-W. (1997). The effects of various hormones and growth factors on the growth of human insulin-producing cell line in serum-free medium. *Experimental & Molecular Medicine, 29*, 209–216.

Liang-Chu, M. M., Yu, M., Haverty, P. M., Koeman, J., Ziegle, J., Lee, M., et al. (2015). Human biosample authentication using the high-throughput, cost-effective SNPtrace(TM) system. *PLoS One, 10*.

Liu, J., Han, X. M., Liang, L., Liu, Q. C., Xu, Y. H., Yang, C. M., et al. (2014). Establishment of a cell suspension culture system of endangered Aquilaria sinensis. *CTA Pharmaceutica Sinica, 49*, 1194–1199.

Lynn, D. E. (2009). Chapter 39 – Cell culture. In *Encyclopedia of insects* (pp. 144–145). Academic.

Mackenzie, C. G., Mackenzie, J. B., & Beck, P. (1961). The effect of pH on growth, protein synthesis, and lipid-rich particles of cultured mammalian cells. *The Journal of Biophysical and Biochemical Cytology, 9*, 141–156.

Maestroni, A. (2012). Cell morphology and function: The specificities of muscle cells. In L. Luzi (Ed.), *Cellular physiology and metabolism of physical exercise* (pp. 9–15). Springer.

Malm, M., Saghaleyni, R., Lundqvist, M., Giudici, M., Chotteau, V., Field, R., et al. (2020). Evolution from adherent to suspension: Systems biology of HEK293 cell line development. *Scientific Reports, 10*.

Masters, J. (n.d.). *Animal cell culture*. Oxford University Press.

McCormick, C., & Freshney, R. I. (2000). Activity of growth factors in the IL-6 group in the differentiation of human lung adenocarcinoma. *British Journal of Cancer, 82*, 881–890.

McKeehan, W. L., Barnes, D., Reid, L., Stanbridge, E., Murakami, H., & Sato, G. H. (1990). Frontiers in mammalian cell culture. In vitro cellular & developmental biology. *Journal of the Tissue Culture Association, 26*, 9–23.

Moran, E. T., Summers, J. D., & Pepper, W. F. (1967). Effect of non-protein nitrogen supplementation of low protein rations on laying hen performance with a note on essential amino acid requirements. *Poultry Science, 46*, 1134–1144.

NCBI Resource Coordinators. (2016). Database resources of the National Center for Biotechnology Information. *Nucleic Acids Research, 46*, D8.

Nelson, L. J., et al. (2013). Profiling the impact of medium formulation on morphology and functionality of primary hepatocytes in vitro. *Scientific Reports, 3*, 2735.

Noorafshan, A., Motamedifar, M., & Karbalay-Doust, S. (2016). Estimation of the cultured cells' volume and surface area: Application of stereological methods on vero cells infected by Rubella virus. *Iranian Journal of Medical Sciences* 37–43.

Oyeleye, O., Ola, S. I., & Omitogun, O. G. (2016). Basics of animal cell culture: Foundation for modern science. *Academic Journals, 11*, 6–16.

Passier, R., & Mummery, C. (2003). Origin and use of embryonic and adult stem cells in differentiation and tissue repair. *Cardiovascular Research, 58*, 324–335.

Philippeos, C., Hughes, R. D., Dhawan, A., & Mitry, R. R. (2012). ntroduction to cell culture. In *Methods in molecular biology* (pp. 1–13). PMC.

Pixley, S. K. R. (1985). *Volume 1: Methods for preparation of media, supplements, and substrata for serum-free animal cell culture*. Alan R. Liss, Inc.

Pruckler, J. M., Pruckler, J. M., & Ades, E. W. (1995). Detection by polymerase chain reaction of all common Mycoplasma in a cell culture facility. *Pathobiology: Journal of Immunopathology, Molecular and Cellular Biology, 63*, 9–11.

Raff, M. C., Miller, R. H., & Noble, M. (1983). A glial progenitor cell that develops in vitro into an astrocyte or an oligodendrocyte depending on culture medium. *Nature, 303*, 390–396.

Remotti, P. C., Löffler, H. J. M., & van Vloten-Doting, L. (1997). Selection of cell-lines and regeneration of plants resistant to fusaric acid from Gladiolus × grandiflorus cv. 'Peter Pears'. *Euphytica, 96*, 237–245.

Resau, J. H., & Cottrell, J. R. (1989). Use of organ explant and cell culture in cancer research. In E. K. Weisburger (Ed.), *Mechanisms of carcinogenesis* (pp. 157–159). Springer.

Ricardo, R., & Phelan, K. (2008). Trypsinizing and subculturing mammalian cells. *Journal of Visualized Experiments, 16*, e755.

Rooney, S. A., Young, S. L., & Mendelson, C. R. (1994). Molecular and cellular processing of lung surfactant. *FASEB Journal: Federation of American Societies for Experimental Biology, 8*, 957–967.

Ryu, A. H., et al. (2017). Use antibiotics in cell culture with caution: genome-wide identification of antibiotic-induced changes in gene expression and regulation. *Scientific Reports, 7*, 1.

Schnellbaecher, A., Binder, D., Bellmaine, S., & Zimmer, A. (2019). Vitamins in cell culture media: Stability and stabilization strategies. *Biotechnology and Bioengineering, 116*, 1537–1555.

Segeritz, C.-P., & Vallier, L. (2017). *Cell culture: Growing cells as model systems in vitro. Basic science methods for clinical researchers* (pp. 151–172). Academic.

Soule, H. D., Maloney, T. M., Wolman, S. R., Peterson, W. D., Jr., Brenz, R., McGrath, C. M., et al. (1990). Isolation and characterization of a spontaneously immortalized human breast epithelial cell line, MCF-10. *Cancer Research, 50*, 6075–6086.

Tang, W., Kuehn, T. H., & Simcik, M. F. (2015). Effects of temperature, humidity and air flow on fungal growth rate on loaded ventilation filters. *Journal of Occupational and Environmental Hygiene, 12*, 525–537.

Taub, M., Saier, M. H. (1979). An established but differentiated kidney epithelial cell line (MDCK). In *Methods in enzymology* (Vol. 58, pp. 552–560). Academic.

Thomson, A. A., Foster, B. A., & Cunha, G. R. (1997). Analysis of growth factor and receptor mRNA levels during development of the rat seminal vesicle and prostate. *Development, 12*, 42431–42439.

Uysal, O., Sevimli, T., Sevimli, M., Gunes, S., & Eker, S. A. (2018). Chapter 17 – Cell and tissue culture: The base of biotechnology. In *Omics technologies and bio-engineering*. Academic.

Vazin, T., & Freed, W. J. (2010). Human embryonic stem cells: Derivation, culture, and differentiation: A review. *Restorative Neurology and Neuroscience, 28*, 589–603.

Verma, A., Verma, M., & Singh, A. (2020). Animal tissue culture principles and applications. *Animal Biotechnology*, 269–293.

Werner, R. G., & Noé, W. (1993). Mammalian cell cultures. Part I: Characterization, morphology and metabolism. *Arzneimittel-Forschung, 43*, 1134–1139.

Xiong, H.-R., & Yang, Z. (2012). Culture conditions and types of growth media for mammalian cells. *Biomedical Tissue Culture*.

Yamada, K. M., & Geiger, B. (1997). Molecular interactions in cell adhesion complexes. *Current Opinion in Cell Biology, 9*, 76–85.

Yao, T., & Asayama, Y. (2017). Animal-cell culture media: History, characteristics, and current issues. *Reproductive Medicine and Biology, 16*, 99–117.

Chapter 8
Inoculation and Passaging of Adherent and Suspension Cells

Pranav Pancham, Divya Jindal, and Manisha Singh

8.1 Introduction

The utilisation of adherent and suspension cell lines plays a crucial role in scientific research; it has various applications in the industrial and academic sectors such as vaccine production, antibody production, cytotoxicity, drug metabolism test and production of therapeutics. Additionally, these culture procedures offer a safe and cost-effective methodology for understanding the cell life cycle. Recent developments in protocols and standard operating procedures have increased the efficiency and quality of research work done on cell lines, hence, making it a significant tool for tackling and understanding biological mysteries. The cell line propagation and cultivation can be performed via two methods – adherent cell culture and suspension cell culture. Application of these techniques can offer an in-depth analysis of growth kinetics of any cell line of choice. Adherent cells also known as anchorage-dependent cell lines require attachment to a particular surface to propagate and reach confluency. Such cells are obtained from organ tissues where the cells are physiologically bound in connective tissues. Similarly, suspension cells are also known as anchorage-independent cell lines that require no physical attachment to any surface and grow in a suspension state in media. These cells are obtained from bodily fluids and the circulatory system.

The proliferation of cell lines possesses various challenges as multiple factors play a crucial role in determining the growth life cycle of select cell lines. Proliferation, nutrition consumption, secondary metabolite secretion and migration properties are considered dynamic and may prove difficult to control unless the

P. Pancham · D. Jindal · M. Singh (✉)
Department of Biotechnology, Jaypee Institute of Information Technology, Noida, Uttar Pradesh, India

© The Author(s), under exclusive license to Springer Nature Switzerland AG 2023
S. Mani et al., *Animal Cell Culture: Principles and Practice*, Techniques in Life Science and Biomedicine for the Non-Expert,
https://doi.org/10.1007/978-3-031-19485-6_8

specific conditions are fulfilled as complex in vivo signalling mechanisms related to cell–cell and cell–matrix interactions are strenuous to replicate and can vary across different cell lines. Before culturing, cell lines are prepared from primary cultures which then are seeded into either adherent or suspension cultures. The majority of cell lines are adherent cell lines and need rigorous effort and constant monitoring to maintain a stable growth activity. These cells grow in restricted surface area which may lead to lower yield capacity. After reaching above a certain level of confluency, these cells are dissociated via the trypsinisation method to extract cells for cell passage. Cell passage is done frequently to keep the cell line healthy. Unlike adherent cell culture, suspension cell line cultures provide higher yield capacity and are comparatively easier to passage with the dilution technique. Moreover, many adherent cell lines can adapt and grow in a suspension state and can be utilised for their advantageous properties. Additionally, Suspension cell lines do not require enzymatic or mechanical trypsinisation techniques to harvest cells which require the use of harsh chemical reagents or extensive force which may damage cell lines. Furthermore, the suspension cell lines require mechanical agitation as compared to adherent cell lines for adequate gas exchange and are favourable in commercial scale-up setup. However, optimum growth can be achieved in both adherent and suspension cell lines through media optimisation and alteration of growth conditions. Further, these cell lines can be stored under ideal storage conditions for future utilisation.

An additional factor that needs close monitoring is cell growth kinetics; in-depth analysis of the growth cycle can provide a clear understanding of cell line growth patterns, which further can help in determining the seeding density, split proportions, morphological characteristics and use of antibiotics in case of contamination. Further, cell synchronisation, passage number and the rate of cell population doubling are some key parameters that can help in steady growth (Bhatia et al., 2019; Dorfman et al., 1980; Kaur & Dufour, 2012; Segeritz & Vallier, 2017).

8.2   Subculturing in Monolayer Cultures

It is frequently required to subculture cultures to maintain them healthy and active. The most prevalent methods of subcultivation involve the use of proteolytic enzymes like trypsin or collagenase to break down intercellular and cell-to-substrate connections. To keep anchorage-dependent cell lines growing in monolayers in exponential development, they must be subcultured at regular intervals. The cells are subcultured when they are at an exponential growth cycle end (approximately confluency of 70–90%). Moreover, the product information sheet includes the subculturing process for each ATCC cell line, as well as split ratios and recommended medium replenishment timeline. Monolayer subcultivation requires the breakage of both inter- and intracellular which includes cell-to-surface connections. A gentle tap with the palm against the side of the culture flask will dislodge some loosely adhered cells. Many require proteolytic enzymes like trypsin/EDTA to break down their

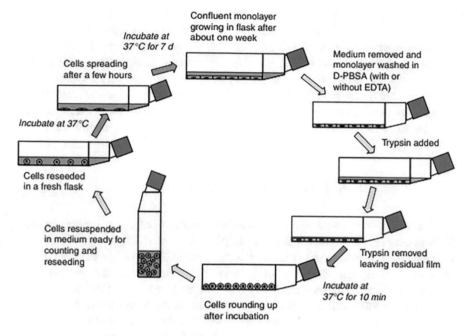

Fig. 8.1 Subculturing in monolayer cultures

protein attachment bonds. For some cell lines, the use of mechanical force such as scraping is recommended to remove the cells in the cultured flask.

The cultured cells with the help of mechanical force or proteolytic enzymes can be dissociated into a suspension state before being diluted to the recommended concentration. Further, this diluted suspension culture is put into a new culture vessel with the fresh medium. After inoculation, the cells in the fresh medium reattach as monolayers and propagate. Since each cell line is different, incubation periods and temperatures, as well as the number of washes and solution compositions, may differ. To avoid harm from the dissociation solution, keep the cultured flask under a microscope at all times during the dissociation process (ATCC, 2022a, b; Freshney, 1983). The detailed method of subculturing in monolayer culture is described in Fig. 8.1.

8.3 Troubleshooting Monolayer Cell Subculturing

8.3.1 Dissociation of Cells

Removal of cells can prove to be a challenge when either the dissociating agents have been rendered inactive by inhibitors in the media (such as serum). Before adding dissociating solution, rinsing the cell monolayer with PBS can provide quick resolution and proper cell dissociation. Further as an alternative, instead of using

PBS, the use of the trypsin–EDTA solution for the initial rinse of the monolayer can provide the desired results if the dissociating solution is weak. Another factor that greatly influences the difficulty in cell removal is high enzyme and EDTA concentrations. To increase the effectiveness of the dissociating solution, the incubation of cells is required at the appropriate temperature. As the cells may have been in a confluent state for far too long, leading to a reduction in the cell-to-cell connections junction space which blocks the dissociation agents to reach the substrate-cell interface. Hence, as a precaution subculture the cells before they reach confluency.

8.3.2   Cell Clump Formation Post-dissociation

Sometimes due to the harsh dissociation technique, due to cell lysis, the cell genomic DNA is exposed. This can occur by either pipetting excessively or the dissociating solution being formulated in a higher concentration above the recommended threshold concentration (i.e. the pH or osmolality of the buffer was incorrect). Further breakage of exposed DNA strands can be facilitated by adding DNAse to the cell suspension. Further cell clumping can be avoided in the future through gentle cell pipetting, reduced incubation time, utilisation of weaker dissociation solution or lower incubation temperature. Furthermore, keep the cell suspension on ice before dilution and dispersion of clumped cells into the fresh media, if there is a time delay between removing the cells from the flask growth surface and seeding a fresh flask. Another factor that plays a crucial role in clump formation is centrifugation at a higher speed and longer duration than required.

8.3.3   Cells Reattachment Issues

The dissociation protocol was prolonged which deprived the cell membrane of essential attachment proteins. The medium lacked adequate serum or adherent factors. Further, utilisation of a protein-coated flask or the addition of attachment factors to the medium such as poly-L-lysine, gelatin, collagen or fibronectin can reduce the probability of cell reattachment issues. Moreover, centrifugation did not inactivate or eliminate the dissociating solution. The addition of serum or enzyme inhibitors such as soybean trypsin inhibitor can facilitate reattachment; later the cells can be centrifuged and resuspended in a fresh medium.

8.3.4   Reduced Viability

Use strong dissociation agents prepared with incorrect pH or osmolality which can be avoided by freshly prepared agents. Before reseeding, avoid keeping cell suspension at a high cell concentration for a long period. Keep the cells refrigerated. There was a problem with the media. Use the recommended formulation protocol, and double-check that it contains all of the necessary ingredients (ATCC, 2022a).

8.4 Subculturing in Suspension Cell Cultures

Primary cultures, either finite or continuous cell lines, rely on anchorage-dependent growth. Other cell lines, particularly cells which are originating from hematopoietic cells, possess the ability to proliferate in suspension cultures. The advantages of cell proliferation in suspension over monolayer propagation are numerous as dilution is all that is required for subculturing. As there is no damage associated with proteolytic enzyme dispersal, there is a reduction in growth lag; another advantage it offers is the usage of less laboratory space. Further, the scale-up capability is enhanced and bioreactors usage is feasible. Generally, cultures are seeded from 2×10^4 to 5×10^5 cells/mL. Cells planted at too low a density will have a growth lag, develop extremely slowly or die off altogether. Excess increase in cell density causes the exhaustion of vital nutrients in the medium which further leads to quick cell death (ATCC, 2022b). The brief difference of monolayer and suspension culture is mentioned in Table 8.1.

8.5 Media Consumption by Cell Lines

Each cell type has a specific requirement which requires an ideal medium for optimum growth. A variety of providers offer chemically specified media in liquid or powdered form. Although it is more expensive than ready-to-use, sterile media offers the advantage over media preparation hassle. The manufacturer's instructions must be followed while rehydrating powdered media with tissue culture-grade

Table 8.1 Monolayer vs. suspension culture

	Monolayer culture	Suspension culture
Culture requirements	Cyclic maintenance is required	Steady-state
	Trypsin passage	Dilution
	Limited by the surface area ratio	Volume (gas exchange)
Growth properties	Contact inhibition	Homogeneous suspension
	Cell interaction	
	Diffusion boundary	
Useful for	Cytology	Bulk production
	Mitotic shake-off	Batch harvesting
	In situ extractions	
	Continuous product harvesting	
Applicability	Most cell types, including primaries	Only transformed cells

water. Filter-sterilise the medium before transferring it to sterile vials. In a 4 °C refrigerator, the prepared medium can be kept for up to 1 month. Laboratories that need a big amount of medium can make their own using standard recipes. This is the most cost-effective method, but it is a time-taking process. Animal cells can be cultivated in either a wholly natural medium or a synthetic also known as artificial media containing some natural components.

Natural media are made up entirely of biological fluids that occur spontaneously. Natural media are ideal for a wide range of animal cell culture applications. Natural media has a significant disadvantage in terms of repeatability, which is owing to a lack of knowledge about the specific composition of these natural media. Artificial media consists of nutrients (both organic and inorganic), vitamins, salts, O_2 and CO_2 gas phases, serum proteins, carbs and cofactors. Various artificial media have been created to cater to one or more of the requirements of industry such as instant survival which offers specified pH and osmotic pressure in an optimised solution. Need for prolonged survival provides an optimised salt solution formulated with various organic components (Arora, 2013).

8.6   Types of Cell Detachment Techniques

When a monolayer is given treated with trypsin or any other enzyme, certain cells are removed quickly in comparison to others. Trypsin was used to eliminate fibroblasts from cultures of foetal human intestines and skin regularly. Lasfargues et al. (1973) discovered that exposing breast tissue cultures to collagenase enzyme for several days eradicated fibroblasts while leaving epithelial cells (Lasfargues et al., 1973). EDTA may be more likely to release epithelial cells than fibroblasts. Dispase II specifically disperses epithelial sheets from human cultures that have been maintained on monolayers of 3 T3 cells while leaving the 3 T3 cells intact (Hynds et al., 2018). This method could be useful for subculturing epithelial cells via different sources while excluding stromal fibroblasts. The initial stage in subculturing of adherent culture is to physically or enzymatically separate them from the culture vessel's surface. The processes for dissociating cells are listed in the table below.

Examine the cells for detachment using a microscope. If the cells are not fully detached after a few minutes, extend the incubation period by a few minutes, and check for detachment every 30 s. You can also use your finger to touch the vessel to hasten cell dissociation. Rotate the vessel for a short time after 90% of the cells have detached to enable the cells to flow. Add 2 volumes of pre-warmed complete media. Pipette the medium on the cell layer surface many times to disperse it. Several culture factors have a significant impact on cell behaviour and cell adherence to the culture substrate. There are certain things to take into account a variety of parameters and needs that determine the selection of the most appropriate cell detachment procedure. Below mentioned parameters must be taken care of.

8.6.1   Degree of Cell Adhesion

Strong cell adhesion is required for cell culture methods, such as microcarriers, which must be balanced by the severity of the detaching step. Cell culture in t-flasks, on the other hand, is better for cells that have poor connections (Merten, 2015).

8.6.2   Use of Detached Cells

Cell viability and membrane constituent stability are more significant than cell harvesting with proceeded production of extracts of cells when future reattachment is predicted from detached cells. Similarly, while obtaining a cell sheet, cell-to-cell interactions need to be preserved, but not while taking a suspension culture. A cell's ideal functionality/metabolic state is also desirable, in addition to its physical integrity.

8.6.3   Process Compatibility

There are chemicals used in specific cell detachment methods that cause hindrance with downstream processing that is not overcome and treated by other approaches. As a result, physical detachment is desirable to chemical treatments, and it can help alleviate problems that may influence the additives to the final product (Aijaz et al., 2018).

8.6.4   Culture Support

Detachment of cell processes such as cell scraping, which need direct access to the support, is incompatible with cell culture substrates including hollow fibres or microcarriers.

8.6.5   Process Scale

Manual detachment of cell approaches includes cell scraping that is sufficient at the laboratory level but is impractical at the scale-up level due to the laboriousness of the method.

8.6.6   Regulatory Constraints

Some cell detachment techniques may be excluded from evaluation in specific applications due to regulatory constraints, or rigorous evaluation may be required before use approval is granted.

8.6.7   Temporal Resolution

Some of the stimuli used to encourage cell separation are harmful to cells. Controlling the transient resolution of the cell detachment approach is critical in these situations.

8.6.8   Compatibility with Sterilisation Methods

If the cell culture equipment must be sterilised before use, care must be taken to ensure that the sterilisation approach does not alter the cell detachment technique's features.

8.6.9   Production Costs

High-cost cell detachment methods have a detrimental impact on the entire expense of the cell culture procedure and, in addition, limit the technique's applicability (Merten, 2015).

8.7   Different Cell Detachment Methods

A rubber or plastic spatula is being used to separate the cells from the culture substrate physically in cell scraping. When used for a few cells, the approach is rapid and easy but may disrupt the cells, further resulting in considerable cell death.

A combination of vigorous shaking and a modest trypsinisation treatment serves as an alternative for microcarrier-based cell culture. The mechanical and chemical stimuli are combined in this cell detachment process. The method has scalability possibilities as well. This method is simple to apply in humongous microcarrier cell culture, notwithstanding the issues posed by the employment of proteolytic enzymes. A variant of this technology takes advantage of the high shear forces found in microfluidic systems. Cells are cultivated on the inside microfluidic chip channel walls. When a strong liquid flow passes via these channels, the cells are exposed to high shear strain, which causes the cells to detach efficiently. This approach is regarded as a rigorous detachment procedure, however, as it frequently causes significant cell damage and even death.

Alternative proteolytic enzymes to trypsinisation include collagenase and pronase, a commercially available combination of multiple endo- and exo-proteases generated by *Streptomyces griseus*. These enzymes degrade proteins on the cell surface, causing the cells' interactions with the culture substrate to be distorted. Enzymatic activity is harmful to cells, and other proteases may cause irreparable damage, i.e. apoptotic impact caused by trypsin treatment. Because of the method's inadequate temporal resolution, the enzyme's activity on the cells is prolonged, resulting in cell damage.

Chemical treatment with a non-enzymatic approach can also be used. Enzymatic therapy is frequently paired with EDTA that aids trypsin's function while inhibiting the binding of several proteins. When cells are weakly adherent to the culture support, EDTA can be used alone. Cell detachment has also been observed to respond well to the citric saline approach. A gentle cell detachment approach is provided by non-enzymatic chemical treatment.

Under a transitory electrical potential, advanced gold-coated electro-responsive materials were used to hydrolyse the ether bond that held the cell-adhesive compounds. Furthermore, because cell-adhesive compounds stay linked to the cells after the ether bond is dissolved, the reaction is irreversible. Furthermore, parallel electrochemical processes may be induced by the applied electric potential. Despite the disadvantages listed above, the system's potential automation may lead to future efforts to construct electro-responsive systems (Forum, 2016).

8.8 Role of Cellular Density in the Subculturing Procedures

When the cells cover the plate or the medium's capacity is exceeded, the cells must be passaged or subcultured. This will maintain cells at a healthy density, allowing them to continue to develop and proliferate. The volume of healthy cells in your experiment will be maximised if you maintain log phase growth. Following a fixed timetable for cell passaging ensures repeatable behaviour and tracks their health state. Change the seeding density of cell cultures until a consistent growth rate and yield from a certain seeding density for your cell type. Deviations from the established development patterns usually signal that the culture is unhealthy or that a component of your culture system is broken (Segeritz & Vallier, 2017).

8.9 Determining the Time Interval and Schedule in Subculturing Process

Passaging your cells at a strict time assures repeatable behaviour and helps you to keep track of their health. There is a modification of the seeding density of cultures until it reaches a steady growth rate and yields from a certain seeding density for your cell type. Deviations from the established development patterns usually show that the culture is unhealthy or contaminated. Keep a complete cell culture scheduling check, including subculturing and adding media timings, types of medium used, dissociation processes, split proportions, morphological characteristics, levels of seeding and use of antibiotics.

It depends on the stage in which are you working, the species, the system that are you using as well as the health of plant materials, among the most important. The length of subculture is normally higher during in vitro establishment than in proliferation. In semisolid media it is also usually higher than in liquid media. In temporary immersion, you can enlarge more the subculture than with static systems. While, when decay symptoms begin to appear, it normally means that subculture is necessary.

8.10 Cellular Synchronisation

Cell synchronisation is the method of bringing cells in a culture that are at different phases of the cell cycle into their respective phases. Cell synchronisation is an important mechanism in the study of cells passing through the cell cycle because it permits data to be acquired from a population rather than just single cells. Physical fractionation and chemical blockage are the two methods of synchronisation that are commonly used. Physical fractionation is a technique for separating continuously dividing cells into phase-enriched pools based on specific characteristics like these:

- Density of cells.
- Dimensions of the cell.
- The presence of antibodies that recognise cell surface epitopes.
- Dispersion of light.
- Labelled cells emit fluorescent light.

8.10.1 Centrifugal Elutriation

Centrifugal elutriation is a technique for the separation of cells in different stages of the cell cycle's characteristics, i.e. size and sedimentation rate. It can separate cell culture into different phases of the cell cycle, i.e. G1, S, G2 and M phases with the help of increasing size and the sedimentation velocities (Keng et al., 1980).

8.10.2   Chemical Blockade

The exogenous outer substrates can be utilised to stop cells at specific stages of the cell cycle, and they are widely employed to target cell cycle checkpoints. The chemical blockage is usually better, effective and accurate in comparison to physical separation; some approaches can be flawed for a variety of problems, including the following:

- Chemical manipulations may disturb cell processes.
- Treatment may be hazardous.
- Treatment is not suitable in vivo.
- Insufficient proportions of synchronised cells (Banfalvi, 2011).

8.10.3   Arrest in the City of M

The mitotic arrest is accomplished in M phase in a variety of ways and at various points, including the G2/M transition, meta–/anaphase and end of mitotic division.

8.10.4   Nocodazole

Nocodazole is a microtubule polymerisation inhibitor that can be used to stop cells from entering anaphase in the metaphase/anaphase transition (Zieve et al., 1980).

8.10.5   Inhibition of CDK1

CDK1 inhibition is required for the shift from the G2/M phase. RO-3306 can cause cells to reversibly arrest during the G2/M transition, being an effective CDK1 inhibitor. RO-3306 synchronised >95% of cycling cells, allowing them to enter mitosis quickly (Vassilev, 2006).

8.10.6   Colchicine

Colchicine, like nocodazole, freezes arrested cells in metaphase and acts as a microtubule poison, blocking the development of mitotic spindles. It depolymerises tubulin into the microtubules, preventing anaphase progression by causing a sustained stop at the checkpoint of spindle assembly.

8.10.7   Arrest in S (Arrest in G1/S)

During the replication of the genome, arrest in the S phase usually entails the suppression of DNA synthesis. The majority of procedures are washable and reversible.

8.10.8   Double Thymidine Block

Through competitive inhibition, high amounts of thymidine disrupt the deoxynucleotide metabolism pathway, preventing DNA replication. Because a single thymidine treatment arrests cells through the S phase, a double treatment induces a much more consistent blockage in the early S phase. The procedure starts with a thymidine therapy, followed by washing the culture and another thymidine treatment (Kalyan Dulla, 2011).

8.10.9   Other Methods of Cell Synchronisation

8.10.9.1   Mitotic Selection

Mitotic selection is a drug-free method of selecting mitotic cells from an exponentially growing monolayer. Mitotic selection makes the use of morphological changes that occur during mitosis in adherent cells cultured on a monolayer. Mitotic cells can thus be entirely separated and retrieved from the supernatant by thoroughly mixing (Banfalvi, 2011).

8.10.9.2   Nutrient/Serum Deprivation

When serum is removed from the culture media for around 24 h, cells in the junction between G0 and early G1 aggregate. This stoppage is easily reversible with the reintroduction of serum or the vitamin that has been depleted. Following the release, cell cycle progression is varied, with some cells remaining dormant while others progressing at different rates (Davis et al., 2001).

8.10.9.3   Contact Inhibition

When cells are left to expand to high confluence, maximising cell-to-cell contact, contact inhibition develops. Normal cells are arrested in early G1 as a result of this. Replacing cells at a lower density reverses the arrest (Zeng & Hong, 2008).

8.11 Cellular Interaction and Signalling Mechanism

The majority of solid-tissue cells grow in monolayer cultures unless the cell lines are externally influenced to grow as anchorage-independent. Cell-cell interaction along with cell-matrix interaction greatly influences the in vivo cell growth. Further, nutritional- and hormonal-induced signals are considered important parameters in cell growth. Some signalling is conducted by cell adhesion molecules, although soluble, diffusible substances can also cause signalling. Cell signalling plays a crucial role in cell growth and is of multiple types.

Endocrine signals refer to signals reaching the cell from another tissue via the systemic vascular system. Meanwhile, the paracrine signals are signals which are generally generated from nearby cells as a direct or indirect interaction without entering the bloodstream. The production of double signals is referred to as homotypic paracrine or holocrine signalling. Further, autocrine signalling occurs when a cell interacts with its own produced signals.

Although discussed types of signalling occur in vivo except for autocrine and holocrine signalling, which occur in vitro conditions with a growth medium. Many cell cultures fail to grow effectively due to low seeding density which may occur due to dilution to reduce essential signalling factors. The use of cell line-specific media or feeder layers is to increase plating efficiency, as paracrine and endocrine factors may be involved in the growth of cultured cell lines, as well as differentiation stimulation.

However, the mechanism of action of these conditions is complex as it is essential for the synergistic effect of multiple signalling factors. The by-product factors produced over a specific time period have a significant impact on the development of the cell culture. This is another important factor in cellular phenotype interaction that aids in facilitating cell-cell communication through the administration of exogenous paracrine substances (Bhatia et al., 2019).

8.12 Attaining Standard Growth Cycles and Split Ratios

Just after the initial subculture, or passage, the main culture transforms into a cell line that may exhibit multiple subculture growth. With each successive subculture, slow or non-proliferating cells will be diluted and removed, and the population with the highest proliferative potential will progressively predominate and flourish. Disparities in proliferative capabilities after the first subculture are exacerbated by varying capacities to resist the shock of trypsinisation and transfer. Despite some phenotypic changes based on selection, the culture becomes more stable by the third passage and is distinguished by a durable, rapidly reproducing cell.

An overview of tissue culture since its start frequently aids in the prevention of expansion of more delicate or slow-growing specific cells, such as liver parenchyma and epidermal keratinocytes. Poor culture conditions are a key contributor to this issue, and substantial progress has been achieved in the use of selective media and substrates for the preservation of many specialised cell lines (Geraghty et al., 2014).

The Number of Passengers and the Rate of Population Doubling
The subculturing should be maintained at a 1:2 ratio; generally, this ratio is utilised in the majority of continuous cells as they multiply at a faster rate. While having a substantially greater split ratio when subcultured, the numerical frequency of sub-culturing is known as the passage number. Furthermore, the number of populations doubling since the culture began, also known as the population doubling level (PDL), is approximately equal to the passage number for diploid cultures. Whereas, continuous cell lines are passaged at greater split ratios. As a result, the PDL for continuous cell lines has not been determined. In most cases, the PDL is an estimate since it does not account for cells that perished from necrosis or apoptosis or cells that are reaching senescence and no longer divide.

8.13   Using Antibiotics for Contamination Control

Antibiotics cannot be used regularly in cell culture even though they foster the growth of antibiotic-resistant strains and allow less contamination to endure, which can progress to comprehensive contamination once the antibiotic is eliminated from the media, and may conceal mycoplasma infections (Nikfarjam & Farzaneh, 2012). Furthermore, some antibiotics may have a cross-reaction with cells, interfering with the biological processes being studied. Antibiotics must only be administered as a very last alternative and for a short period, and they will be eliminated as quickly as possible from the culture. Antibiotic-free cultures should be kept in succession as a control for contamination and infections if they are used for a long time (Fair & Tor, 2014).

Antibiotics are a useful strategy in cell culture when used correctly, but antibiotics can be quite harmful when overused or carelessly handled. Antibiotics should never be used frequently in culture media, according to experienced cell culture users. In a large study, Barile discovered that mycoplasma-contaminated 72% of cultures were grown continuously in antibiotics, but only 7% of cultures produced without antibiotics, a tenfold difference. Workers who use antibiotics in their media regularly and continually have more contamination problems, including myco-plasma. Antibiotic overuse leads to sloppy aseptic techniques.

Antibiotics should not be used in place of a good aseptic technique, although they can be strategically utilised to prevent the loss of crucial experiments and cultures. The trick is to only employ them for limited periods: in the first week, during the early stages of hybridoma creation, and research in general when the cultures will be discontinued at some point. Antibiotics should be demonstrated to be effica-cious, noncytotoxic and stable in whatever application they are used for (Suvikas-Peltonen et al., 2017).

8.14 Conclusion

The chapters discuss all the aspects of inoculating the adherent and suspension cell cultures covering the subculturing of monolayer cultures, media composition, cell detachment techniques, cellular synchronisation and the usage of antibiotics for controlling contamination in the culture. The proper maintenance of cell cultures must be followed in a time management process and must be performed in sterile conditions.

References

Aijaz, A., Li, M., Smith, D., Khong, D., LeBlon, C., Fenton, O. S., & Parekkadan, B. (2018). Biomanufacturing for clinically advanced cell therapies. *Nature Biomedical Engineering, 2*(6), 362–376.

Arora, M. (2013). *Cell culture media: A review*. Labome.

ATCC. (2022a). *ATCC animal cell culture guide: Tips and techniques for continuous cell lines*. ATCC.

ATCC. (2022b). *Maintaining cells: Protocols to ensure healthy cell growth*. ATCC.

Banfalvi, G. (2011). Overview of cell synchronization. *Methods in Molecular Biology, 761*, 1–23.

Bhatia, S., Naved, T., & Sardana, S. (2019). Introduction to animal tissue culture science. In *Introduction to pharmaceutical biotechnology* (Vol. 3, p. 1-1-1-30). IOP Publishing.

Davis, P. K., Ho, A., & Dowdy, S. F. (2001). Biological methods for cell-cycle synchronization of mammalian cells. *BioTechniques, 30*(6), 1322–1331.

Dorfman, N. A., Civin, C. I., & Wunderlich, J. R. (1980). Susceptibility of adherent versus suspension target cells derived from adherent tissue culture lines to cell-mediated cytotoxicity in rapid 51Cr-release assays. *Journal of Immunological Methods, 32*(2), 127–139.

Fair, R. J., & Tor, Y. (2014). Antibiotics and bacterial resistance in the 21st century. *Perspectives in Medicinal Chemistry, 6*, 25–64.

Forum, B. D. (Producer). (2016). *Adherent cell culture in biopharmaceutical applications: The cell-detachment challenge*. Biopharm International

Freshney, I. (1983). Subculture and cell lines. In: *Culture of animal cells*. A.R Liss.

Geraghty, R. J., Capes-Davis, A., Davis, J. M., Downward, J., Freshney, R. I., Knezevic, I., et al. (2014). Guidelines for the use of cell lines in biomedical research. *British Journal of Cancer, 111*(6), 1021–1046.

Hynds, R. E., Bonfanti, P., & Janes, S. M. (2018). Regenerating human epithelia with cultured stem cells: feeder cells, organoids and beyond. *EMBO Molecular Medicine, 10*(2), 139–150.

Kalyan Dulla, A. S. (2011). Large-scale mitotic cell synchronization. In: *Cell cycle synchronization*. Humana Press, Hungary.

Kaur, G., & Dufour, J. M. (2012). Cell lines: Valuable tools or useless artifacts. *Spermatogenesis, 2*(1), 1–5.

Keng, P. C., Li, C. K., & Wheeler, K. T. (1980). Synchronization of 9L rat brain tumor cells by centrifugal elutriation. *Cell Biophysics, 2*(3), 191–206.

Lasfargues, E. Y., Coutinho, W. G., Lasfargues, J. C., & Moore, D. H. (1973). A serum substitute that can support the continuous growth of mammary tumor cells. *In Vitro, 8*(6), 494–500.

Merten, O.-W. (2015). Advances in cell culture: Anchorage dependence. *Philosophical Transactions of the Royal Society of London. Series B, Biological Sciences, 370*(1661), 20140040–20140040.

Nikfarjam, L., & Farzaneh, P. (2012). Prevention and detection of Mycoplasma contamination in cell culture. *Cell Journal, 13*(4), 203–212.

Segeritz, C. P., & Vallier, L. (2017). *Cell culture: Growing cells as model systems in vitro.* Academic Press.

Suvikas-Peltonen, E., Hakoinen, S., Celikkayalar, E., Laaksonen, R., & Airaksinen, M. (2017). Incorrect aseptic techniques in medicine preparation and recommendations for safer practices: A systematic review. *European Journal of Hospital Pharmacy: Science and Practice, 24*(3), 175–181.

Vassilev, L. T. (2006). Cell cycle synchronization at the G2/M phase border by reversible inhibition of CDK1. *Cell Cycle, 5*(22), 2555–2556.

Zeng, Q., & Hong, W. (2008). The emerging role of the hippo pathway in cell contact inhibition, organ size control, and cancer development in mammals. *Cancer Cell, 13*(3), 188–192.

Zieve, G. W., Turnbull, D., Mullins, J. M., & McIntosh, J. R. (1980). Production of large numbers of mitotic mammalian cells by use of the reversible microtubule inhibitor Nocodazole: Nocodazole accumulated mitotic cells. *Experimental Cell Research, 126*(2), 397–405.

Chapter 9
Counting of Cells

Divya Jindal and Manisha Singh

9.1 Introduction

Cell counting is the process of calculation of viable cells in a given solution, a major application in research, diagnosis and treatment of a disease. Animal cell growth is proportional to the mass of the cells, which is directly associated with production and increase in the volume of various cell organelles. The concentration of cells and their components account for numbers. Cell counting is essential to observe and keep track of cell health and their differentiation rate, evaluating their perpetuation and transformation for seeding in subsequent experiments. The consistency in cell concentration ensures reproducibility and accuracy (Ongena & Das, 2010). Successful cell cultures grown in vitro always rely on cell concentration; a greater number of viable cells, the higher the cell density. A correct estimation of cell plating concentration is an essential method for any type of cell culture, say, somatic, plant (Kobayashi et al., 1999), protoplast (Schween & Hohe, 2003) or microspore culture. Every organism has its optimal culture density which may vary from $4 * 10^4$ microspores per ml for *Brassica naups* (Huang & Bird, 1990), $8 * 10^4 – 10 * 10^4$ for pepper (Kim & Jang, 2008), $8 * 10^4$ for rice (Raina & Irfan, 1998) and $140 * 10^4$ for apple (Hofer, 2004). Consistency in cell concentration marks with efficient reproducibility and accuracy. The account of cell concentration is important for cell health and proliferation rate monitoring, which is essential for immortalisation and transformation for further studies and experiments. With the advancement in technology, there exist many methods depending upon direct and indirect cell counting. The conventional method of calculating cell numbers is using a "hemocytometer". Initially, a hemocytometer was devised by Louis-Charles and Malassez, and later it

D. Jindal · M. Singh (✉)
Department of Biotechnology, Jaypee Institute of Information Technology (JIIT), Noida, Uttar Pradesh, India

© The Author(s), under exclusive license to Springer Nature Switzerland AG 2023
S. Mani et al., *Animal Cell Culture: Principles and Practice*, Techniques in Life Science and Biomedicine for the Non-Expert,
https://doi.org/10.1007/978-3-031-19485-6_9

was modified by Neubauer. Another method is automated cell counting using micro counter 3100, which works on the system of an inverted microscope, digital camera and an image analyser that automatically counts the number of cells in a picture frame (Camacho-Fernández & Hervas, 2018). Nowadays, flow cytometry is used for the fast detection and calculation of fluorescent cells when tagged with fluorescent light sources (McKinnon, 2018).

9.2   Cell Growth, Cell Propagation, and Cell Viability

Within the intact animal cell tissues, the majority of cells are tightly attached to cells via different types of cellular junctions. The cells connect extracellular metric (ECM), which is a network of different proteins and other biomolecules with cells of varied origins. Most animal cells grow only on special solid surfaces, and they require characterised components to facilitate their propagation and viability. Cell growth is a process of production of cells and increases in physical size, in between 5–20 µm. Cell growth follows a series of events for the overall synthesis of cellular components initiating with the log phase, lag phase, stationary phase and finally death phase, which is regulated by many growth factors and nutrients (Guertin et al., n.d., p. 2015).

For propagation of cells, it requires optimum media, proper environment and regular passaging, which is essential for cellular function and biological component production. Media components are important for the proper functioning and growth of cells, containing essential and non-essential amino acids, vitamins, salts, glucose as an energy source, growth factors, antibiotics and organic supplements. An optimum environmental condition of 37 °C temperature, 4–10% CO_2, 7.4 pH and relative humidity of 95% are required (Butler, 2005; Spier, 1997). In addition, passaging in regular intervals depending upon the cell line is essential as it enables the culture enough for cells according to the requirement of an experiment and cell line maintenance (Lin & Wwigel, 1998).

In order to offer information about the integrity of the cell membrane, the viability of cells is calculated in a solution. This is the most common method for estimating cell viability in which an aqueous solution of dye is mixed with the cells, followed by viewing them in a light microscope. Here, the dead cells become permeable to dye and hence get coloured, whereas the live cells are not coloured as they can't take up the dye. Eventually, by counting the number of dead and live cells, the percentage of viable cells can be easily estimated. The most common dyes used are istryphan blue and erythrosine B, which are preserved with methyl p-hydroxybenzoate (Kim & Kim, 2016). The number of cells taking up the dye depends upon the pH, maximum at 7.5 for tryphan blue and 7.0 for erythrosine B. Tryphan blue has more affinity towards proteins in the cell suspension than protein in non-viable cells, while erythrosine B has more affinity towards non-viable cell's protein. Erythrosin B creates noise-free backgrounds by forming outcomes that are significantly more accurate and have less false-negative and false-positive results (Franke & Braveman, 2020). Cell count is directly related to cell viability. Cell viability is the number of active cells in the population, which can be assessed by several methods. Cell proliferation plays an important role as an indicator in assessing growth or cell survival. Proliferation assay is an analysis of DNA biosynthesis during the S phase of the cell cycle. Another criterion is the ATP measurement assay to quantify the cell's ATP content, which is proportional to the number of viable and metabolically active cells (Kumar & Nagarajan, 2018).

9.3   Earlier Methods of Cell Counting

The conventional method of cell counting that was initially employed for analysis of RBCs, WBCs and platelets used hemocytometers in pathology and bacterial cells. Out of many researchers working in animal cell cultures (Ongena & Das, 2010), 71% of them are using counting chambers to accomplish cell counting calculations. Mostly, microspore cultures are calculated using a hemocytometer, out of which Neubauer Chamber (Castillo, 2000; Zhang & Gu, 2020; Corral-Martínez et al., 2013; Nelson & Marson, 2009; Kim & Park, 2013) Fuch-Rosenthal (Simmonds & Long, 1991), Burker Chamber (Lantos & Juhasz, 2012) are used. Cell counting can be automated. Manual cell counting was primitive and required no special equipment, easily assessed in any biology laboratory. In the counting chamber method, a microscope slide is used as a counting chamber to enable cell density calculation. Hemocytometer comprises two gridded chambers in the centre and is covered with a glass slide when assessing the cell counting with calibrated area and depth. Cell culture is dropped on the space in the chamber below the coverslip, filling it by capillary action, as shown in Fig. 9.1. The hemocytometer slide surface must be cleaned with 70% ethanol and needs to be pressed firmly in the centre above the cell culture sample. The cell number can be analysed under the microscope by forming grids to manually count in the areas of known size. The space's distance between the chamber and the upper layer of the cover is predefined; hence, the volume of the cell culture is calculated, also determining the viability of cells. A transverse groove is dividing the cell plate platform into two chambers, consisting of a 3 * 3 mm grid, and is subdivided into equal 1 * 1 mm squares. This is then divided into 16 squares and later into 2 tertiary squares to facilitate counting (Phelan & Lawler, 2001). The sample is prepared with the treatment of enzymes on cell monolayers, followed by resuspension in a culture medium. The cell density has to be monitored, if concentrated, then there is a need to dilute the suspension culture before loading and the sample to be mixed properly for proper cell dispersion. Cells are stained for proper visualisation and counted with tryphan blue in an equal volume of cells. Tryphan blue differentiates between dead and live cells in the suspension. Generally, the Pasteur pipette is used for loading the sample into the hemocytometer Neubauer chamber. The cells are left undisturbed for a few minutes. If the aim is to determine cell viability, statins can be incorporated with cell

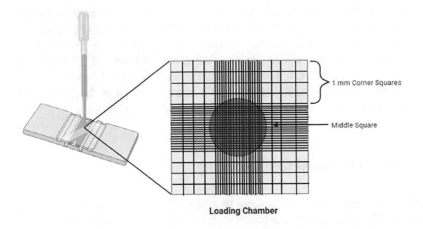

Fig. 9.1 Hemocytometer for quantification of cells in a medium

suspension, which unveils the percentage the viable cells (Absher, 1973). This method is generally favoured in many biological research when it is necessary to assess whether the cell culture is growing well and as planned. It is also relatively quick and inexpensive. Hemocytometers usually need samples to be diluted, or else highly densified cell counting could happen. While diluting, it may hamper the count and confer inaccurate cell count.

Another method of manual cell counting is plating and CFU counting (Sanders, 2012). The cells are plated on Petri dishes along with their respective growth mediums, as shown in Fig. 9.2. It is assumed that if cells are distributed evenly on a Petri dish, they form a single colony or Colony-forming unit (CFU). These colonies can be counted based on the volume of culture spread on the plate (which is already known), assessing the calculation of cells. In comparison to counting chambers, colony-forming unit cell counting requires time to grow a colony, which might take a few hours to days along with their overlapping nature of colonies, which forms a "lawn" of cells. This method can estimate the number of viable cells even though they are time-consuming and hence can be used for research aiming for the quantification of cells that can resist drugs or any extrinsic factor.

Coulter counter is another method for electronic particle estimation which comes as an alternative to a hematocytometer for the evaluation of cell concentration (Rodríguez-Duran et al., 2017), as shown in Fig. 9.3. In this, the suspended cells

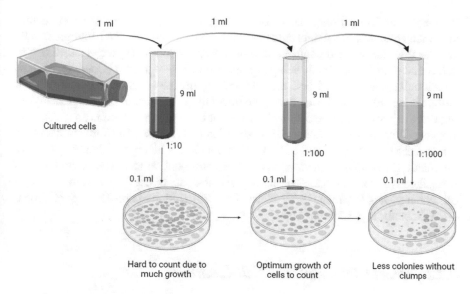

Fig. 9.2 Manual cell counting is Plating and CFU counting

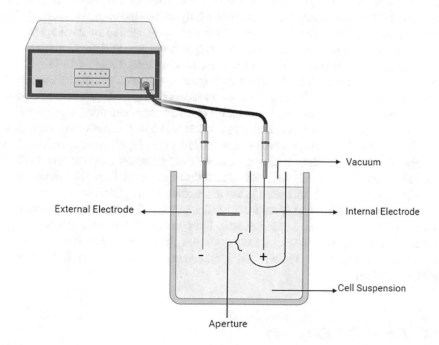

Fig. 9.3 Simplified diagram of cell Coulter counter

surpass through an aperture, upon which electric current flows. The electrical resistance of the medium is amended with their presence and causes a transformation in current and voltage flow. The cell concentration is directly related to the magnitude of this transformation and then electronically converted to particle counts. Cells can be counted as 500 particles per second individually, despite their shape, structure and orientation. Mainly, a threshold control is used to eradicate the counting of cells smaller than the selected control cell size, thus reducing inaccuracy caused due to debris and other components, being non-essential. Zap-O-globin (Coulter) is added to cell culture, which will lyse the residual erythrocytes. The vial containing the culture is placed below the aperture tube under the vacuum control instrument passing electric current, followed by calculation of cell number. There is an automated sceptre cell counter which provides automation and accuracy to Coulter cell counting at a cheap cost. After the cell count completion, a histogram detailing the cell distribution is visualised on the screen of the instrument, followed by applying gates using the toggle button, which eventually sets the gate or uses them from the previous count (Davis & Green, 1967; Mei & Cho, 2012; Richardson-Jones & Hellman, 1985; Schonbrun & Caprio, 2015).

9.3.1 Recent Techniques in Cell Counting

Cell enumeration techniques have been developed from the usage of manual calculation from the use of a hematocytometer to a fully automated counting with reliable counts, precision and high throughput. This approach has led to development in areas like micro- and nanotechnology to analyse multivariate data. Commercially available high throughput cell counters are classified into direct and indirect cell counting methods; these are flow cytometry, spectrophotometry or electrical impedance. Precision in the determination of cell numbers is essential for a broad range of applications, as used in tissue culture studies by researchers and disease propagation in diagnostic laboratories (Oyama & Eagle, 1956). The initial automated technique was done using electrical properties as a feature to distinguish cells, followed by impedance properties. Recent and advanced colter counters can measure cells of less than 1 μm size (Verso, 1964). An optical flow cytometer is established on optical cells detection, commonly used in the diagnosis and prognosis of highly chronic diseases like leukaemia and lymphoma (Jelinek & Bezdekova, 2017). In the flow cytometer, there is the illumination of laser light on the cell's stream, which will get scattered according to their fluorescent properties, followed by their detection using optical detectors (Cho & Godin, 2010; Sandhaus & Ciarlini, 2015; Krediet & DeNofrio, 2015).

9.4 Direct Cell Counting

(a) Electrical Resistance Cell Counting
     A Coulter counter is used which counts and differentiates cells based on the size which is suspended in electrolytes (Cho & Godin, 2010). This counter works on the technique of resistive pulse or electrical zone sensing, containing one or more

microchannels (separate chambers of electrolyte solutions). This works on the principle that the cell is drawn from the orifice, with a simultaneous electric current, which emits a charge in impedance and is proportional to the volume of cells. Platinum electrodes are submerged in the electrolyte solution and a glass tube to which there is an orifice to pass cells. The pulse changes in impedance are derived from electrolyte displacement triggered by cell particles. The Coulter principle is coined after its inventor Dr. Wallace H. Coulter and is nowadays widely used in the medical industry for cell counting. This system can acclimate cell counting rates to 4000 per second. Each cell is passed through an aperture that modifies the electric current by briefly dispersing a solution of conductive liquid equal to their volume. Recent studies have indicated that conductive metals act like non-conductive particles when in an orifice. The rest of the Coulter system comprises pulse amplification and distinction of cell size by the operator (Kinsman, 1969). Further, these Coulter is tagged with ImageJ Plugins for high-quality micrographs with minimal background noise (O'Brien & Hayder, 2016). Along with the cell counting, electrical Coulters also facilitate insights to sustain the interpretation of complex physiological conditions of the cell, after being infected with bacteria that have altered activities of ion channels (Kunzelmann, 2005), conductivity, resistance in the cell cytoplasm (Huang et al., 2014; Jang & Wang, 2007) and deformability (Abdolahad & Zeinab, 2012).

Before performing the counting, the Coulter has to be warmed up, which is followed by trypsinisation of cells. The cells are then resuspended in PBS solution to wash and transfer to a cuvette. The cuvette is placed on the aperture platform, an isotonic solution is poured, and the aperture tube is slowly lowered into the cuvette with the other two electrodes already submerged. A blank reading is taken to calibrate the instrument, and then the reading of the sample is taken. Now, the cell number is calculated by multiplying it with the average of two readings by 20 (if the volume taken is 10 ml), proving the total volume of the cuvette. If a 200 µl sample is taken, then it is multiplied to obtain the cell number per ml.

(b) Flow Cytometry

This technology provides instant analysis with a parametric approach of a cell in a suspended buffered salt solution. They employ photodiodes, fluorescent light, laser light sources for scattering, and signals for detection. The visible scattered light is calculated in two different directions, forward (Forward scatter, FSC) which shows a relative estimation of cell size, and another at 90°, which is a side scattered, SSC and calculates the complexity of cell internally. The light scattered from it is independent of fluorescent light. Samples are prepared via transfection and their expression of respective fluorescent proteins (GFP), fluorescent dyes (Propidium Iodide) and specific fluorescent conjugated antibodies. The signals formed are further transformed into electric signals that can be easily analysed on a computer and saved as a .fcs standardised data file. The cell quantification is purified based on their pattern of scattering and characteristics of fluorescent light, as shown in Fig. 9.4.

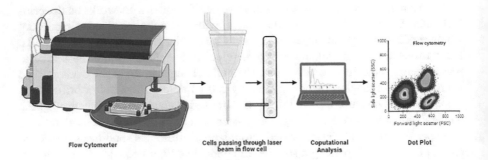

Flow Cytomerter   Cells passing through laser   Coputational   Dot Plot
                              beam in flow cell   Analysis

Fig. 9.4 Brief overview of a flow cytometry experiment and its computational analysis

With the advancement in technology, this can be used in multiple areas of biomedical sciences like immunology, molecular biology, monitoring of infectious diseases and cancer biology as it allows simultaneous description of numerous populations of cell types. This is used in cell sorting for further analysis of experiments, with evolution in instrumentation. Different laser techniques have been proposed due to their specificity in solving various areas of biomedical sciences. For example, for bead analysis, 96-well loaders are deliberated or can be a combination of two systems, like mass spectrometry and FACS. Flow cytometry not only sorts cells but also quantifies cell attributes, DNA/RNA concentration, intracellular receptors and expression of genes (Macey, 2007; Mach & Thimmesch, 2010; McCoy, 2002). The whole flow cytometer comprises four systems. The first system in use is fluidics, in which fluorochrome monoclonal Ab is used to label cellular suspensions before being put into the equipment for cell characterization. The antibodies are specific to the antigen against which the testing is being done, present on the cell surface. Fluorochromes are compounds that can emit light on excitation, bound to antibodies, and the cells get labelled with fluorochrome and can pass individually through a sensing area of the laser beam (Ibrahim, 2007). The second system is optics, which consists of a laser, a light amplification source, emitting light of a particular wavelength. Various laser types are used with their ability to excite specific fluorochromes and trigger them to emit light (Jaroszeski & Radcliff, 1999). The suspension solution then passes the laser, directed in two detections, that is forward scatters (responsible for cellular morphology and cell size) and side scatters, which are scattered at 90° from the laser direction (responsible for cell granularity) (Picot & Guerin, 2012; Herzenberg & Tung, 2006). The third system is electronics, which is involved in converting photos into understandable data. There are photomultiplier tubes (PMTs), composed of semiconductors that can produce current on the intensity of light detection. The last step is data analysis, in which the computer is regulating the functioning of the flow cytometer and representing a graphical analysis of the sample in 1-D, 2-D and 3-D images. Flow cytometry demands designing their instrumentation based on the sample being used (Betters, 2015).

There can be two forms of cell sorters, one is quartz cuvette, which has fixed alignment of laser localisation, and another is jet-in-air, which needs to be aligned on daily basis for small cell detection. They both differ in the interrogation of laser point localisation (McKinnon, 2018). The combination of methods can also be used for rapid and specific analysis, TOF mass spectrometry and flow cytometry. In this technique, the cells are not labelled with fluorescent Abs but tagged with heavy metal ion-tagged Abs, followed by detection via the TOF MS method. This technique doesn't work on the principle of FSC and SSC light detection and hence does not allow the detection of cell aggregates and is overcome by using cell barcoding (Leipold & Newell, 2015).

(c) Electric Cell Sizing

This is used to assist cell counting based on size distribution by achieving the frequency of pulses obtained on their passage via the aperture. The Coulter produces a "channeliser", which connects counter with pulse frequency. This is then coupled with a system that evaluates and converts them into a readable format. Since channelisers are expensive, the counter can be directly connected to a computer with optimised software on which peak frequency can be measured and the respective cell size evaluated.

9.5 Indirect Cell Counting

(a) Spectrophotometer

Cultures cells become turbid and able to absorb light of specific intensity. Cell concentration is directly proportional to the turbidity formed. The approach of the spectrophotometer is being used to estimate the culture turbidity accurately, being simple and insensitive. The culture is filled in a translucent cuvette, which is placed in a spectrophotometer, and finally, the turbidity is measured by using mathematical calculation, as shown in Fig. 9.5. The wavelengths of absorbance are taken at 550–610 nm. The readings implicate the relative estimation of cells in the solution. In general, cell concentration and absorbance have a linear relationship. However,

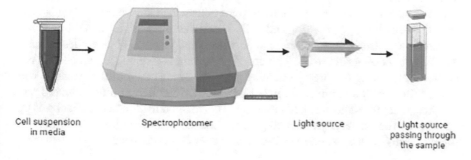

Cell suspension in media   Spectrophotomer   Light source   Light source passing through the sample

Fig. 9.5 Spectrophotometer for quantification of cells in a media by passing a light source

spectrophotometers don't estimate the cell number but rather calculate the absorbance. This technique does not need that the cells be diluted with water, even if the cell density is beyond a particular threshold. Serial dilution of suspended cells containing a quantified number of cells is absorbed and read to generate a standard curve. This method is easy to use as several samples can be loaded at a time, which reduces the workload and time. In addition to this, they require a very little volume of sample, which can be as low as 1 μl (Mohler & Charlton, 2020; Aijaz & Trawinski, 2020; Gimpel & Katsikis, 2021).

(b) Protein Content

The protein content can also determine the number of cells, by using serial dilution method. Within the limitation of an assay that follows a linear relationship of total protein content cell number being evaluated ensures a cell suspension that has been assayed in the linear range. There are many methods assessed for the quantification of protein content, but the most efficient method is by estimating the absorbance at 280 nm of the lysed solution of suspended culture. Lysis can be done via sonication, chemical or enzymatic, out of which the chemical method by using a detergent is most common as it can easily get absorbed at 280 nm and generate a high and good background (Lodish et al., 2000). Almost every detergent gets absorbed, except Empigen BB from Albright and Wilson Ltd (Høj & Moller, 1986). Special assistance is required while performing a protein assay on cell suspension being separated as it may interfere in the quantification process. These contaminants can be removed by extensive washing with PBS solution.

(c) Assays Involving Radioactive Labels:

Labeling cells with radioactive materials before separation is one of the sensitive techniques for determining cell counts. The application of radioactive labels gives very specific activities, allowing for the accurate counting of cells with counts as low as 100 (Wilson & Lukowicz, 2017). The selection of the radio-labelled compound is essential for ensuring cell incorporation in the same amount.

Generally, 14C or 3H thymidine is labelled with DNA but is not recommendable to ensure the cells are equally labelled and incubated for an exact time corresponding to a single growth cycle (Elkind et al., 1976), or else the cell proportion will be processed to labelled DNA synthesis. Protein can also be labelled with radioactive amino acids if cells are projected to be synthesised at the same rate. The has been used to count a small number of human bone cells which have been fractionated. In this, the cell culture of fractionated cells is differentiated from labelled cells and then removed. This is further precipitated with 10% TCA (w/v) by centrifugation of the cells at 1000× g for around 15 min. 1% Albumin is also added, which acts as a carrier for ensuring the precipitation of proteins. This is then further washed with TCA, followed by solubilisation in 0.1 M NaOH. Scintillation fluid is mixed with the precipitate for radio-labelled counting. This method has some drawbacks as the cells have to be actively synthesised proteins and may cause hindrance while assessing radioactivity in basic extracts because of chemiluminescence.

The most efficient technique of radio-labelled counting is the use of chromate 51Cr, described by Walter and Krob in 1983 to enable an analysis of separated cells population. Sodium Chromate 51Ce is diluted with a buffer so that 10micro Ci is assembled in 100 μl. 1.5 ml of cells are incubated in plasma with 15 micro Ci at room temperature for half an hour. Then, an ascorbate solution of 500 mg/ml is added to the buffer solution and separated. These cells are now lysed, and the cells are counted by a liquid scintillation counter. Every cell in the suspension culture need not be labelled; only a small amount is separated and labelled and then remixed with the stock solution. These can be now used as markers for unlabelled cells. The reaction parameters for labelled cells vary depending on the cell type being used (Burdon et al., 1988).

9.6 Errors in Cell Counting

1. **Errors in Earlier Methods of Cell Counting**

    (a) Time-consuming: The Animal cultures require time to grow and multiply to visualise colonies in the earlier method of cell counting.
    (b) Limited accuracy: Accuracy is a key regulator in most highly regulated pharmaceutical industries, and the assessment of manual cell counting can concede a great extent of human error.
    (c) Transfer errors: Some experiments require the transfer of cultures from different temperatures at a particular time. The manual cell counts demand additional human involvement, every time there is a need to change the environmental conditions. This will eventually cause an out-of-specification event. Majorly, the errors can occur during sample preparation and pipetting of the cell on the loading chamber due to fibres or bubbles in them.
    (d) Human handled: Since the hemocytometer is a manual method of cell counting, the protocol depends upon individualised approach, which leads to variation and is prone to becoming a non-standardised protocol. This may lead to bias and misuse of cell counting and eventually result in inaccurate results. There can be user-to-user variation with variation in the hematocytometer filling rate.
    (e) Limited surface areas: The cell culture is unevenly and limit distributed in the loading chamber, with too many or too less cells in the culture sample, which leads to subjective counting (Ongena & Das, 2010).
    (f) Cell visualisation: Due to the lack of visualisation technology in a hemocytometer, inadequate counting of the cells may hamper the cell density calculation. This improper visualisation can occur due to cell aggregation, debris formation and even due to eyesight problems, leading to significant fluctuation in cell number.

(g) Cell density: Suspension with too high or too low cell density is more susceptible to inaccurate results. Too low density indicates that the cells in the chamber may not be an exact source from the stock solution, and too high density indicates aggregation of cells (Chan & Laverty, 2013).

2. **Errors in Recent Methods of Cell Counting**

(a) Coincidence error in electrical resistance Coulter method: Coincidence errors depend upon the length of the sensing zone and the cell concentration with their flowing rate and particle size. If there is some change in the parameters, it may lead to improper cell counting. The coincidence error can also occur due to the use of an anticoagulant-induced sample (Lombarts & Koevoet, 1986).

(b) Flow cytometry: Use of expired or degraded antibodies is more susceptible to causing cell counting error and may happen due to improper handling with varied temperatures. Another reason could be their low concentration for detection. With time, the fluorescence of fluorochrome may get faded due to storing issues or light exposure. When the specific antibody interacts, expression of the target antigen could be low due to their sub-optimal binding, old isolated cells of non-optimised cell culture or maybe due to less knowledge on the antigen expression. The fluorescence on the cells stained may have beached in the absence of fixatives like PFA in samples.

(c) Spectrophotometer: One of the relative errors in spectrophotometer depends on the absolute value of the variables measured, which accounts for the reproducibility of measurements in the equipment for all the operations. Common errors overall can be due to sample handling or poorly manufactured cells from the operator. Any problem associated with instruments, such as instability, aging, incorrect signal amplification, electrical circuitry fluctuations, incorrect calibration or improper slit width, may result in fluctuation in cell numbers. Another common issue is concerned with errors arising from chemical reactions which are associated with improper thermodynamic and optical properties of absorbing components or varied scattering of radiation (Lantos, 1989).

9.7   Conclusion

The chapter discussed various methods of cell counting, from hemocytometer to flow cytometer, which have been used to date. Any method for quantitative estimation of the cell requires certain procedural calibration and noise removal for correct counting. The choice of counting of cells depends upon various factors, like cell viability, instrumentation, operator handling, time consumption, sample preparation, etc.

References

Abdolahad, M., & Zeinab, S. (2012). Vertically aligned multiwall-carbon nanotubes to preferentially entrap highly metastatic cancerous cells. *Carbon*.

Absher, M. (1973). *Hemocytometer counting. Tissue culture*. Academic.

Aijaz, A., & Trawinski, D. (2020). Non-invasive cell counting of adherent suspended and encapsulated mammalian cells using an optical density. *BioTechniques, 68*, 35.

Betters, D. (2015). *Use of flow cytometry in clinical practice*. Journal of the Advanced Practitioner in Oncology (Vol. 6, pp. 435–440).

Burdon, R. H., et al. (1988). Chapter 2: Methods of cell counting and assaying cell viability. In *Laboratory techniques in biochemistry and molecular biology* (p. 18). Elsevier.

Butler, M. (2005). Animal cell cultures: Recent achievements and perspectives in the production of biopharmaceuticals. *Applied Microbiology and Biotechnology, 68*, 283.

Camacho-Fernández, C., & Hervas, D. (2018). Comparison of six different methods to calculate cell densities. *Plant Methods, 14*–30.

Castillo, A. M. (2000). Comparison of anther and isolated microspore cultures in barley. Effects of culture density and regeneration medium. *Euphytica, 113*, 1–8.

Chan, L. L., & Laverty, D. (2013). Accurate measurement of peripheral blood mononuclear cell concentration using image cytometry to eliminate RBC-induced counting error. *Journal of Immunological Methods, 388*, 25.

Cho, S. H., & Godin, J. (2010). Recent advancements in optofluidic flow cytometer. *Biomicrofluidics, 4*, 43001.

Corral-Martínez, P., Parra-Vega, V., & Segui-Simarro, J. (2013). Novel features of Brassica napus embryogenic microspores revealed by high pressure freezing and freeze substitution: evidence for massive autophagy and excretion-based cytoplasmic cleaning. *Journal of Experimental Botany*, 3061–3075.

Davis, R. E., & Green, R. (1967). Automatic platelet counting with the Couler particle counter. *Journal of Clinical Pathology*, 777–779.

Elkind, M. M., et al. (1976). Spurious photolability of DNA labeled with [14C]-thymidine. *Biochemical and Biophysical Research Communications, 68*, 691–698.

Franke, J. D., & Braveman, A. (2020). Erythrosin B: A versatile colorimetric and fluorescent vital dye for bacteria. *BioTechniques, 68*, 7.

Gimpel, A. L., & Katsikis, G. (2021). Analytical methods for process and product characterization of recombinant adeno-associated virus-based gene therapies. *Molecular Therapy – Methods and Clinical Development, 20*, 740–754.

Guertin, D. A., et al. (n.d.). Chapter 12 – Cell growth. In Mendelsohn (Ed.), *The molecular basis of cancer* (4th ed.). W.B. Saunders.

Herzenberg, L. A., & Tung, J. (2006). Interpreting flow cytometry data: A guide for the perplexed. *Nature Immunology, 7*, 681–685.

Hofer, M. (2004). In vitro androgenesis in apple–improvement of the induction phase. *Plant Cell Reports, 22*, 365–370.

Høj, P. B., & Moller, B. (1986). The 110-kDa reaction center protein of photosystem I, P700-chlorophyll a-protein 1, is an iron-sulfur protein. *Journal of Biological Chemistry, 261*, 14292.

Huang, B., & Bird, S. (1990). Effects of culture density, conditioned medium and feeder cultures on microspore embryogenesis in Brassica napus L. cv. Topas. *Plant Cell Reports, 8*, 594–597.

Huang, S.-B., et al. (2014). A clogging-free microfluidic platform with an incorporated pneumatically driven membrane-based active valve enabling specific membrane capacitance and cytoplasm conductivity characterization of single cells. *Sensors and Actuators B: Chemical, 190*, 928.

Ibrahim, S. F. (2007). Flow cytometry and cell sorting. In *Advances in biochemical engineering/biotechnology*. Springer.

Jang, L. S., & Wang, M. (2007). Microfluidic device for cell capture and impedance measurement. *Biomedical Microdevices, 9*, 737–743.

Jaroszeski, M. J., & Radcliff, G. (1999). Fundamentals of flow cytometry. *Molecular Biotechnology, 11*, 37–53.

Jelinek, T., & Bezdekova, R. (2017). Current applications of multiparameter flow cytometry in plasma cell disorders. *Blood Cancer Journal, 7*, e617.

Kim, M., & Jang, C. (2008). Embryogenesis and plant regeneration of hot pepper (Capsicum annum L.) through isolated microspore culture. *Plant Cell Reports*, 425034.

Kim, S. I., & K. Kim H-J. (2016). Application of a non-hazardous vital dye for cell counting with automated cell counters. *Analytical Biochemistry, 492*, 8.

Kim, M., Park, E.-J. (2013). High-quality embryo production and plant regeneration using a two-step culture system in isolated microspore cultures of hot pepper (Capsicum annuum L.). *Plant Cell, Tissue and Organ Culture (PCTOC)*.

Kinsman, S. (1969). *Electrical resistance method for automated counting of particles* (Vol. 158, p. 703). Annals of the New York Academy of Sciences.

Kobayashi, T., et al. (1999). Physiological properties of inhibitory conditioning factor(s), inhibitory to somatic embryogenesis, in high-density cell cultures of carrot. *Plant Science, 144*, 69–75.

Krediet, C. J., & DeNofrio, J. (2015). Rapid, precise, and accurate counts of Symbiodinium cells using the Guava flow cytometer, and a comparison to other methods. *PLoS One, 10*, e0135725.

Kumar, P., & Nagarajan, A. (2018). *Analysis of cell viability by the MTT assay*. Cold Spring Harbor protocols.

Kunzelmann, K. (2005). Ion channels and cancer. *The Journal of Membrane Biology*.

Lantos, C.(1989). Chapter 3 – Errors in spectrophotometry. In *Sommer, studies in analytical chemistry* (pp. 78–94). Elsevier.

Lantos, C., & Juhasz, G. (2012). Androgenesis induction in microspore culture of sweet pepper (Capsicum annuum L.). *Plant Biotechnology Reports, 6*, 123.

Leipold, M. D., & Newell, E. (2015). Multiparameter phenotyping of human PBMCs using mass cytometry. In *Methods in molecular biology*.

Lin, J. H., & Wwigel, H. (1998). Gap-junction-mediated propagation and amplification of cell injury. *Nature Neuroscience*, 1, 494.

Lodish, H., Berk, A., Zipursky, L., Matsudaira, P., Baltimore, D., & Darnell, J. (2000). *Purifying, detecting, and characterizing proteins*. W.H. Freeman.

Lombarts, A. J., & Koevoet, A. L. (1986). Basic principles and problems of haemocytometry. *Annals of Clinical Biochemistry, 23*, 198607.

Macey, M. G. (2007). Principle of flow cytometry. In *Flow cytometry: Principles and applications*. Springer.

Mach, W. J., & Thimmesch, A. (2010). Flow cytometry and laser scanning cytometry, a comparison of techniques. *Journal of Clinical Monitoring and Computing, 24*, 251–259.

McCoy, J. P. (2002). Basic principles of flow cytometry. *Hematology/Oncology Clinics of North America, 16*, 229–243.

McKinnon, M. (2018). Flow cytometry: An overview. *Current Protocols in Immunology, 20*, 5.1.-5.1.11.

Mei, Z., & Cho, S. H. (2012). Counting leukocytes from whole blood using a lab-on-a-chip Coulter counter. In *Annual International conference of the ieee engineering in medicine and biology society IEEE engineering in medicine and biology society annual international conference* (pp. 6277–6280).

Mohler, W. A., & Charlton, C. A. (2020). Spectrophotometric quantitation of tissue culture cell number in any medium. *BioTechniques, 21*, 260–262.

Nelson, M. N., & Marson, A. S. (2009). Microspore culture preferentially selects unreduced (2n) gametes from an interspecific hybrid of Brassica napus L. × Brassica carinata Braun. *TAG Theoretical and Applied Genetics Theoretische und angewandte Genetik, 119*, 497–505.

O'Brien, J., & Hayder, H. (2016). Automated quantification and analysis of cell counting procedures using ImageJ Plugins. *Journal of Visualized Experiments: JoVE, 17*, 54719.

Ongena, K., & Das, C. (2010). Determining cell number during cell culture using the Scepter cell counter. *Journal of Visualized Experiments: JoVE, 52*, 45.

Oyama, V. I., & Eagle, H. (1956). Measurement of cell growth in tissue culture with a phenol reagent (folin-ciocalteau). In *Proceedings of the society for Experimental Biology and Medicine Society*. Blackwell Science

Phelan, M. C., & Lawler, G. (2001). Cell counting. *Current Protocols in Cytometry, 00*, A.3A.1–A.3A.4.

Picot, J., & Guerin, C. L. (2012). Flow cytometry: retrospective, fundamentals and recent instrumentation. *Cytotechnology, 64*, 109–130.

Raina, S. K., & Irfan, S. (1998). High-frequency embryogenesis and plantlet regeneration from isolated microspores of indica rice. *Plant Cell Reports, 17*, 957–962.

Richardson-Jones, A., & Hellman, R. (1985). The coulter counter leukocyte differential. *Blood Cells, 11*, 203–240.

Rodríguez-Duran, L. V., Torres-Mancera, M. T., & Trujillo-Roldan, M. A. (2017). Standard instruments for bioprocess analysis and control. In *Current developments in biotechnology and bioengineering*. Elsevier.

Sanders, E. R. (2012). Aseptic laboratory techniques: Plating methods. *Journal of Visualized Experiments: JoVE*.

Sandhaus, L. M., & Ciarlini, P. (2015). Automated cerebrospinal fluid cell counts using the Sysmex XE-5000: Is it time for new reference ranges? *American Journal of Clinical Pathology, 134*, 34–38.

Schonbrun, E., & Caprio, G. D. (2015). Differentiating neutrophils using the optical coulter counter. *Journal of Biomedical Optics, 20*, 111205.

Schween, G., & Hohe, A. (2003). Effects of nutrients, cell density and culture techniques on protoplast regeneration and early protonema development in a moss, Physcomitrella patens. *Journal of Plant Physiology, 160*, 209–212.

Simmonds, D. H., & Long, N. E. (1991). High plating efficiency and plant regeneration frequency in low density protoplast cultures derived from an embryogenic Brassica napus cell suspension. *Plant Cell, Tissue and Organ Culture, 27*, 231.

Spier, R. E. (1997). Factors limiting the commercial application of animal cells in culture. *Cytotechnology, 23*, 113–117.

Verso, M. L. (1964). The evolution of blood-counting techniques. *Medical History, 8*, 149–158.

Wilson, C., & Lukowicz, R. (2017). Quantitative and qualitative assessment methods for biofilm growth: A mini-review. *Research & Reviews Journal of Engineering and Technology, 6*.

Zhang, M., & Gu, L. (2020). Improvement of cell counting method for Neubauer counting chamber. *Journal of Clinical Laboratory Analysis, 34*, e23024.

Chapter 10
Cryopreservation of Cell Lines

Vinayak Agarwal and Manisha Singh

10.1 Introduction

Metabolic activities have been observed to drastically reduce or cease to exist at extremely low temperatures in all living systems including humans; however such phenomenon if applied judiciously and effectively can result in ensuring long-term preservation of viable cells and tissues. Nevertheless, one of key and scientifically proven facts is that freezing alters the chemical profile inside the cells which summons distinct cellular mechanical constraints and damages which are fatal to almost all living organisms. One of the primary and the most crucial hurdles for animal cells at low temperatures is to overcome the phasic transition of water-to-ice which paves the way for several cellular injuries. Moreover there are several attributes which contribute to these injuries such as fast cooling rates leading to intracellular ice crystal formation, whereas slow cooling rates inflict osmotic damage due to concentration alteration in intracellular and extracellular solutions or hampered mechanical interaction between ice crystals and cells. Cryopreservation aids in maintaining biological samples in an animated suspended state at cryogenic temperatures for extended spells of time and simultaneously preserving cellular components. Several research studies have investigated and observed that cryoprotective agents (CPA) have distinct attribute which aids in altering cellular behaviour in cryogenic temperatures by influencing water transport, growth of the ice crystal and nucleation as cell survival in these freezing states depends upon several biophysical properties during the cooling and warming cycles of the cryopreservation. Apart from copious of such attributes, CPA also caters to several vital requirements necessary for ensuring optimum cell vitality under cryogenic environment, and some of the most common as

V. Agarwal · M. Singh (✉)
Department of Biotechnology, Jaypee Institute of Information Technology (JIIT), Noida, Uttar Pradesh, India

well as successful CPAs employed in research facilities are dimethyl sulphoxide (DMSO), glycerol, protein and specific polymers. Modus of operandi for conventional cryopreservation has been found to be relatively simple which can be directly translated into several lab–/industry-level applications where large cell volumes can be subjected to cryopreservation in one vial however exhibit substandard survival rates. Therefore, despite having the potential of being an exceptional preservation method, it's slightly more complex than what it seems. Thus necessity to employ programmed freezing modules operating at specific and sustained cooling rates becomes imperative in ensuring cell vitality. Moreover, exploring novel cryopreservation techniques such as vitrification has also been developing in order to reduce cost and optimise several process parameters along the way. Though various evidence suggests that cryopreserved cells/tissues reap more benefits in research and future application as sustained availability of these cells ensures extensive quality control testing and reproducible results in cases of transplantation without the requirement to obtain fresh organ/tissue samples (Jang et al., 2017).

10.2   Fundamentals of Cryopreservation

The basic principle of cryopreservation deals with the utilisation of extremely low temperatures for the preservation and maintenance of cell's mechanical integrity for long storage applications. There are several cellular moieties which are classed depending upon their respective species, thereby imparting their influences on a distinct biological and survival response of the cells with respect to the cryogenic conditions and its subsequent thawing cycle. Moreover, for quality and reproducible results, it's necessary to identify some the characteristic attributes associated with the process of cryopreservation which are stated as follows:

- Prior to cooling, ensure proper mixing of CPAs with cells or tissues.
- Freezing and storage of cells or tissues at cryogenic temperatures.
- Thawing or warming cycle of cells or tissues.
- Efficient separation of CPAs from cell or tissues subsequently after thawing.

Amidst various extremities playing their part, the cooling rate of the process dominates all the biological effects as 80% of the cell's mass is occupied by water which gets converted to ice crystals. Such crystal formation generally employs either of the two processes heterogeneous or homogeneous ice nucleation as freezing has been found to be a nucleation induced event. Heterogeneous ice nucleation (generally at $-35\ ^\circ C$) has been reported to be more prevalent in the process as compared to its counterpart, due to the higher thermodynamic stability offered by the random stacking of water molecule which further reinforces the stability than to cause thermodynamic decay. Homogeneous ice nucleation requires extremely low temperature $(> -35\ ^\circ C)$ to induce ice crystallisation catalysed by appropriate liquid or solid interface adjacent to the sample solution. Subsequent ice crystal formation has been observed to exhibit a pure crystalline water profile with no trace of dissolved solute which encompasses the solute particles along with cells in a liquid phase at cryogenic temperatures. To ensure the viability of the cell samples under such temperatures reminds us the importance of CPAs in cryopreservation even more (Gupta et al., 2017).

10.3 Cryoprotective Agents

Copious research studies have been conducted to explore as well to elucidate the chemical and physicochemical profile of various cryoprotective agents. Detailed assessment of these parameters deduced a high biological acceptance rate towards CPAs having liquids or aqueous profile as such attributes impart least strain, high penetration along with stunted toxicity under cryogenic environment. There have been several neoteric research approaches selected to devise an optimised form CPAs which enables to reduce the ice crystal formation at lower temperatures depending upon various factors such as warming rate, cooling rate and cell types. To accomplish the optimum survival rate of cells, cooling and warming rates, volume and CPA concentrations ought to be enhanced relying upon the distinctive cell types and setting of tissues (Jesus et al., 2021). It ought to be referenced that the macroscopic components of the tissue is a significant highlight be elucidated in cryopreservation convention due to warmth and mass exchange impediments in these mass frameworks. CPAs can be isolated and classed into two primary classes: first and foremost, cell layer saturating cryoprotectants (cell membrane permeating), for instance, dimethyl sulphoxide (DMSO), glycerol and propylene glycol (1, 2-propanediol). There are certain non-permeating cryoprotectants as well such as

Table 10.1 Various attributes of commonly employed cryoprotective agents

Cryoprotective agents	Toxicity profile	Potential applied cryoprotective application
Dimethyl sulphoxide (DMSO)	Cell membrane toxicity	Adipose tissue, bone marrow, amniotic fluid, umbilical cord, hepatocytes, platelets
Glycerol	Renal toxicity/failure	Amniotic fluid, red blood cell, spermatozoa
Ethylene glycol	Pulmonary oedema and gastrointestinal irritation	Amniotic fluid, dental pulp
Cell banker series	Low cell membrane toxicity	Adipose tissue-derived stem cells, bone marrow, synovium, amniotic fluid
Propylene glycol	Impairment in developmental potential of germ cells	Embryo, hepatocytes

2-methyl-2,4-pentanediol and polymers which have been intricately classed in accordance with their applied cryopreservation application (Table 10.1). In contrast to engineered synthetics and biomaterials, this can be further utilised to obstruct ice crystal development, alongside conventional little particles. The immediate restraint of ice gem arrangement and utilisation of cancer prevention agents and different mixes have been utilised to endeavour to diminish cell passing from cycles leading to apoptosis during the freezing and defrosting cycle (Taylor et al., 2019).

10.4   Polymers

Amidst several modus of operandi, resuspending encapsulated CPA within the capsule in the wake of cryopreservation tends to serve as a complex technique in manipulating the cell's location in the solution. Recent developments in modern science has offered several exemplifying materials which could be potentially be engineered to synthesis non-infiltrating polymers which tends to provide protection to cells moieties within the scaffold at cryogenic temperatures and thereby eluding several constraints associated with diffusion at higher-dimensional cryopreservation applications. However, there are also various conventional polymer class vinyl-determined   polymers   which   further   encompasses   polyethylene   glycol ($C_2nH_4n+2On+1$), polyvinyl liquor [(C_2H_4O)n] and hydroxyethyl starch with subatomic load of 200–9500 Da, 30–70 kDA and 130–200 kDa, respectively. These listed polymers have reported to exhibit extensive capabilities in reducing the crystal size, thereby offering better alternatives than the existing polymers employed as CPAs (Fuller & Paynter, 2007).

10.5   Glycerol

Observations of glycerol having cryoprotectant credits were first expressed by Polge et al. in 1949. Even after decades of research and development, there have been quite few compounds which can serve as a potential alternative of glycerol. The significance as well as usefulness of glycerol is second to none and has remained as highly effective CPA until DMSO was assessed and exhibited in 1959 (Lovelock & Bishop, 1959). Several studies have revealed that glycerol comes across as a non-electrolyte, and thereby its expression is associated with a decreased electrolyte concentration of the remaining aqueous regions of the freezing solution. Such phenomenon is more frequently and consistently observed near as well as around all the sample cells at all temperature levels. Its primary and widely accepted application has been identified in chronic storage of animals as well as bacterial cell samples in medical facilities. The defensive attributes of glycerol have been observed to be restricted to substances that themselves don't exhibit toxic profile and have a low atomic weight, a high solvency in fluid electrolyte arrangements and a capacity to penetrate living cells. The quantity of solutes equipped for satisfying these conditions is little; thus far glycerol most intently moves towards the ideal cryoprotectant agent (Hubálek, 2003).

10.6   Dimethyl Sulphoxide (DMSO)

Several studies have reported and evaluated the application of DMSO in cryopreservation of animal cells as a result of its minimal effort and relatively low cytotoxic profile which was originally reported by a Russian scientist Alexander Zaytsev in 1866. DMSO functions have been closely observed to be analogues to that of glycerol as both of the CPA facilitates their expression by lowering the electrolyte concentration in the remaining thawed solution arrangement in and around a cell at any random temperature. Moreover, a reduced viability rate along with acceptance of cell separation brought about by DNA methylation and histone change has been accounted for. Adipose-derived mesenchymal stem cells (ASCs) in recent research have exhibited attributes which contributes in the preservation of the properties of subcutaneous fat tissue especially effectively; thus cryopreservation is right now preceded as a standard strategy for safeguarding ASCs to securely gain enormous quantities of cells. Notwithstanding, numerous investigations have revealed that cell movement subsequent to freezing and defrosting might be influenced by the arrangements utilised for cryopreservation. Besides copious complex attributes offered by DMSO, the primary characteristic of DMSO-mediated cryopreservation is that it diffuses across the plasma membrane into the cell and shields the cells from any kind of impairments/damages caused in a cryogenic environment (Ishizuka & Bramham, 2020).

10.7   Proteins

Among others one the most promising recently discovered proteins which exhibited attributes similar to that of CPA is a sticky hydrophilic protein sericin having a molecular weight of 30 kDa. Sericin was derived from the cocoon of silkworm. Wide range of extensive applications has seen to be associated with this distinct protein; however, its potential as a CPA specifically for human adipose tissue-derived stem and hepatocytes is quite valued. Apart from sericin, a distinct protein harnessing attributes similar to antifreeze compound has been extracted from salt water teleosts/fishes which have been garnering a lot of attention due to its potential of being quality CPAs; however, there hasn't been enough detailed exploratory studies to state anything conclusive (Naing & Kim, 2019).

10.8 Primary Mechanism of Action of Cryoprotectants

Several research studies have been conducted aimed to explore as well as elucidate the basic functioning of CPAs. Glycerol and DMSO are two of the most popular and widely used CPAs categorised as penetrating cryoprotectants among the many others. They have been found to function by and large with small non-ionic particles with elevated levels of water solubility which facilitate easy diffusion across the plasma membrane and position themselves into the intracellular compartment. The preferred mode of their action was explored when the total concentration of the solute particles was observed to have remained unfrozen and possessed an independent profile of a partially frozen solution. Nevertheless various deleterious impacts of salts are straightforwardly identified with concentration, thereby diluting the concentration and subsequently supplanting a segment of the harming solutes in the partially frozen solution with a CPA, which will bring about less cryoinjuries at cryogenic temperatures. Reduction in the extent of ice crystal formation in the solution due to the presence of CPA would further enable the cell structure to be equipped when exposed to lower temperature let's say $-20\ °C$ than solution without CPAs. The cryoinjuries caused by the salts have been reported to be connected not exclusively to the concentrations yet additionally to the varying cryogenic temperatures. Accordingly, within the sight of a CPA, the cell will be presented to a similar toxic salt concentration, yet exposure will happen at a lower freezing temperature where its harming impact is diminished (Mandawala et al., 2016). Nonetheless, during the freezing cycle, salts are not by any means the only solutes to be concentrated – the CPA goes through a comparable cryogenic concentration and has appeared to add to the haemolysis of frozen defrosted red cells. Hence, insurance includes some significant downfalls in those convergences of CPA that in general have an advantageous defensive impact can likewise contribute to the "solution effect" injury at moderate cooling rates. Moreover, the proof for such effect has been documented in several prior studies, and different systems of activity have been proposed including adjustment of the plasma film by the CPA. Non-penetrating CPAs are for the most part can be referred to as long chained polymers along with heavy molecular weight that don't cross the plasma layer and stay in the extracellular compartment during freezing. The preferred mechanism of action for operation is most probably up for discussion; however, it has been credited to both a colligative impact and the pre-freeze drying out of the cell; diminishing intracellular water molecules apart from improving the probability of the sustaining cell's equilibrium by accepting loss of water molecules rather than having ice crystal formation is even worse with respect to cell survival. Elevated levels of compound viscosity due to addition of CPAs further helps to inhibit the ice nucleation during the cooling cycle and recrystallisation during the warming cycle (Li et al., 2010) (Fig. 10.1).

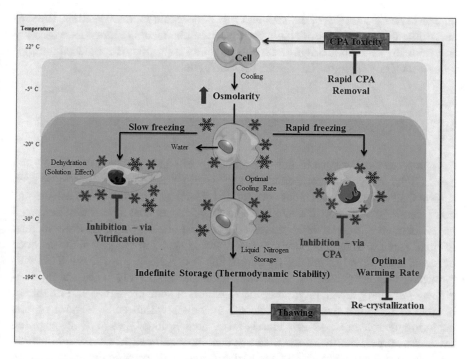

Fig. 10.1 Schematic representation of several facets employed in cryopreservation

10.9   The Cell Banker Series

One the recently devolved technology which is generally referred to as cell banker series (CBS) (established by Nippon Zenyaku Kogyo Co., Ltd., Fukushima, Japan) which was aimed to expedite the process of cryopreservation of cell samples at −80 °C and has been reported to accomplish better endurance and viability rates after subsequent freezing and defrosting cycles. This specific cell banker series of cryopreservation media holds distinct composition of glucose, an endorsed high-grade polymer, pH modulator and 10% DMSO. Moreover, their certain variations in distinct cell banker series on the basis of their constituents, for instance, CBS 1 and 1+ have been reported as serum-containing variants which can be utilised for the cryopreservation of a broad range of animal cells. To be sure, regular media employed for cryopreservation include the employment of foetal bovine serum which serves as a pool of copious cytokines, growth factors and vague substances (encompassing bovine exosomes) which made it illegal and unsafe in the foundation of a normalised cryopreservation convention for any kind of clinical application/experimentation associated with human subjects. While CBS2 serves as a potential alternative to CBS1 as it employs non-serum-type media for cryopreservation in a culture condition specifically engineered to compliment CBS2 which is a serum free environment. CBS3 which is more commonly also known as stem cell banker is made out of 10% DMSO and other inorganic mixes (US20130198876) and fulfils the model of artificially characterised chemical components which are free from any xenobiotic origin and is subsequently appropriate for induced pluripotent stem cells and protection of somatic stem cells.

10.10   Cryoinjury

Cryopreservation indulges in the phase change of water within intracellular, and extracellular at cryogenic temperature leads to serious cell damage. There have been several damaging factors such as cooling and thawing velocities which are the primary cause in inflicting biophysical and physicochemical interaction-mediated impairment and further deteriorate the cell survival rates. Apart from these factors, osmotic rupture in the cell tends to allow ice crystal formation which has been observed to be directly regulated by the cooling rate. Additionally retention of semipermeable attributes exhibited by plasma membrane tends to aid in illustrating comprehensively the cell viability; however, retention of such properties still might not be able to influence the survival rate of organelles within cells. Several reports exploring the role of CPAs have clearly articulated various types of damages posed by the cryoprotectant themselves. The cyclic entry and exit of CPAs significantly cause the cell structure to expand and shrink sequentially causing the damage to the cells (McGann et al., 1988).

Clinically, DMSO has appeared to trigger several extreme allergic responses in patients which were reported to be treated with CPA-infused haematopoietic stem cell samples and also has additionally appeared to incite apoptosis machinery in specific locations of the central nervous system in mouse models. Numerous facilities have preferred to employ a relatively diluted form of DMSO before reinfusion of stem cells into beneficiaries, especially youngsters. Cells also can be influenced by varying molar concentrations of CPA which in general serves as a protecting shield for cells from freezing and defrosting cycle associated with cryopreservation. Differentiation process is strongly induced by DMSO on haematopoietic stem cells along with several other cells in vitro; furthermore, this process is associated with triggering apoptosis. The primary cause for initiation of both these processes is employment of DMSO at low concentrations for long spells of time frames at raised temperatures (+37 °C). Henceforth variations were reported with respect to the epigenetic profile of stem cells, and moreover the concentration ranging from 0.5 to 2 M has been seen to function at cryogenic temperatures during cryopreservation which offers apt exposure time and optimum survival chances along with temperatures which are limited to those adequate to permit equilibration of the cell. An elevated level of concentration utilised to accomplish vitrification tends to exhibit a serious toxic and osmotic profile which further requires various ways to deal with their presentation and expulsion (Yavin & Arav, 2007).

10.11 Varied Application of Cryopreservation Technique on Special Cell Cultures/Cells

Contemporary past has witnessed the evolution of copious neoteric applications of cryopreservation with respect to varying research field areas encompassing biochemical molecular biology, food technology, cryosurgery and ecology and plant anatomy along with several biomedical purposes (blood transfusion, artificial insemination, in vitro fertilisation (IVF) and bone marrow transplantation) with the primary application being cryopreservation of cells. Scientific communities in coherence have comprehensively observed the potential of cryopreservation in developing elaborate cell bank facilities for better and rapid analysis for human leukocyte antigen (HLA) typing during the organ transplantation process (Keck et al., 2011). This further enhances the distribution process efficiency of viable cell samples amidst various medical facilities. One of chief growing areas of application is the long-term storage of stem cells which further caters to several ancillary applications such as tissue engineering and cellular therapies. Besides that, there are various animal cell types associated with their specific application in relation to cryopreservation in modern medicine and are mentioned as follows.

10.12 Female Germ Cells (Oocytes and Embryos)

The principal instance associated with human embryo cryopreservation aimed to safeguard the fertility was reported in early 1996, which employed in vitro fertilisation on a female subject which was diagnosed with breast cancer, thus operating before undergoing chemotherapy. In the wake of 1996 success, cryopreservation of mature female germ cells has been viewed as a way of safeguarding the reproductive capacity of female subjects. Moreover, several retrospective exploratory studies have been conducted during the time frame of 1986 to 2007, investigating the efficacy and viability of the technique (Mandawala et al., 2016). The researchers didn't observe any conclusive evidence to support that the process of cryopreservation has affected or has any significant influence over pregnancy, childbirth, labour, birth rate or even miscarriages in comparison to that of natural or IVF. However, the results are highly promising; still scope of improvement and modification remains (Chang et al., 2017).

10.13 Male Germ Cells (Sperm, Testicular Tissue and Semen)

Last decade witnessed and faced several issues regarding the germ cell applications, not due to the lack of scientific resources but always depleting sample germ cells due deficient techniques which were unable to guard the sample cells from toxicity (chemical and physical) and infection/contamination during the course of experimentation, thus, the suboptimal success rate. However, nowadays cryopreservation has resolved the related concerns and ensured fertility as well quality of patients undergoing cancer treatment which involves radio and chemotherapies. Research has shown that proper cryopreservation of sperm and semen samples can preserve the vitality along with viability for unlimited periods of time. Moreover, novel trials encompassing testicular tissue-centric cryopreservation have been reported under early clinical phases. The trial focuses on achieving cell suspension of testicular tissues which further been subjected to cryopreservation, although in early stages yet trials hold huge potential for offering as well as safeguarding fertility of copious men facing therapeutic altercations which elevates the probability to compromise their fertility (Alotaibi et al., 2016).

10.14 Hepatocytes

Fundamentally there have been several significant discoveries in the field of research and medicine focused on hepatocytes centred application which have been documented in the last four decades, amidst various few of the primary applications encompassing varying physiological inspection of liver-related attributes such as metabolism, drug toxicology profile and organ/cell preservation along with other clinical approaches. Moreover, recent times have garnered a coherent support for developing cryopreservation facilities for hepatocytes in wake of soaring interest in liver-specific research along with dedicated employment of biotechnological aid to further expedite the related clinical applications (Ibars et al., 2016).

10.15 Stem Cells

Copious amounts of evidence support the potential application of cryopreservation specifically which has been reported to be associated with adult stem cells. Extraction of these cells has been primarily subjected to locations such as the umbilical cord, adipose tissues, bone marrow and amniotic fluids apart from several other sources of extraction. The primary functional profile of most stem cell types (such as stromal, embryonic, mesenchymal, etc.) makes them ideal candidates which cater to the requirements necessary to be employed as therapeutic altercations in regenerative therapy. Although it's not just regenerative medicine; besides this there have been several others highly expensive as well as exclusive therapeutic medical applications which have been found to be closely associated with stem cells such as tissue grafting and engineering, gene therapy and organ transplantation. Thereby making a sustained and quality supply of stem cells even more important and therefore cryopreservation of such cells is one of the most valued applications in recent times (Li et al., 2010).

10.16   Other Frequently Cryopreserved Cells

Amidst several animal cells, there are some distinct classes of cells which extensively witness cryopreservation due to their lack of availability, and thus different neural cells as well as cell lines along with cardiomyocytes are being routinely utilised at medical and research facilities across the globe. Yet a best-quality level methodology for safeguarding their vitality has not been elucidated. Neoteric studies dealing with CPAs have revealed certain alternative therapeutic regimes avoiding any involvement of glucocorticoid immunosuppressive agents; however, a potential transplantation of pancreatic islet may serve as an ideal treatment of type I diabetes. Therefore, the improvement as well development in islet cryopreservation strategies has been progressing; however, the results reported weren't that promising, and one of the factors contributing to that was the suboptimal survival rate of the treated cells (Taylor et al., 2019).

10.17   Optimum Cooling Rate

The expressions moderate as well rapid are generally considered as relative and closely identified with the permeability of water along with the surface/volume proportion of the varied class of cell which is subjected to the process. Cooling operated at a moderate rate has been observed to be characterised as cooling at suboptimum pace beneath which cell could sustain a constant responsiveness with respect to steady change in temperature together with the resulting expansion in extracellular ice crystal formation, exclusively caused by efflux of water which further contributed to dwindling of cell. Operating under the influence of such conditions harm is inferable from arrangement impacts caused straightforwardly or in a roundabout way by the raised solute concentration. The impacts of moderate cooling injury aggregate with expanding openness time to harming solute focuses. Accordingly, expanding the cooling rate will bring about more limited spells of exposure that further elevated the endurance rates. Fast cooling might be characterised as any pace of cooling above which the cells neglect to keep up synthetic harmony by diffused water flow along with the balance re-establishment via nucleation further leading to ice crystal formation inside the cells. These newly formed ice crystals would rely directly upon the subjected warming rate yet tend to prompt deadly occasions and diminished endurance. Endurance will keep on being diminished as cooling rates keep on expanding. The ideal rate of cooling is subsequently a trade-off amidst the two contending factors, primarily when the cooling is sufficiently fast to decrease harm caused by the solute impacts however tapered down to lessen rate in the cell populace.

10.18 Influence of Warming Rate

Warming rate can significantly influence the survival rate of the cells which were subjected to cooling at elevated rates post-defrost. Here, fast rewarming considerably elevated endurance contrasted with rewarming gradually. Several studies explain distinct facets of a relatively rapid warming rate which limits ice recrystallisation, consequently ensuring at the same time negating the conversion of tiny intracellular ice crystals to develop into sizable crystals which could result in cellular damage; sluggish rate of warming generally offers additional time at raised cryogenic temperatures which aids in the process of recrystallisation. Rates of warming have been observed to impact the cooled cells which is much more complex as a quick rewarming tends to exhibit analogous expression to that of cells which were cooled quickly, i.e. expanded endurance when operated at moderate rewarming rates. Studies have revealed the application of specific models where the influence imparted by the warming rate on the cryopreserved cells tends to exhibit more or less null effect which eventually proves to be extremely helpful in preserving cell's attributes (Jang et al., 2017). Cells which have been subjected to a steady rate of cooling have been observed to experience antagonistic effects of warming rates on the cell's survival rate as the damages or injuries incurred by the cell during its time at cryogenic temperatures get re-exposed during thawing. Thus, the process becomes extremely time-sensitive, and any kind of delay due to the sluggish rate of warming could further deteriorate and damage the cellular conformation of the cryopreserved cells (Hunt, 2017).

10.19 Limitation of Cryopreservation

Albeit various uses of the cryopreservation methods exist, both in fundamental and clinical examination, a few limits actually exist. At cryogenic temperatures, the cell tends to cease all metabolic activities which has inescapable results which eventually pave the way for several associated complexities encompassing specific genetic alterations correlating to impairment to the cell's structural as well as functional conformation (Li et al., 2010). Generally, it has been witnessed that CPAs themselves can be harmful to cells, particularly when utilised in soaring levels of concentrations. Many exploratory studies have firmly suggested the likelihood of DMSO expression may change chromosome steadiness, which can prompt a danger of tumor pro-genesis. Besides copious unwanted endogenous alterations in cell, the conceivable disease or contamination at the cellular level leads to several discrepancies such as tumours which ones ought to be forestalled (Luhur et al., 2019).

10.20   Vitrification

Although cryopreservation is one of the most preferred and favoured means of cell preservation, however, there are certain limitations such as ice crystal formation which significantly damage cell moiety and reduce cell viability; therefore an effective alternative to this method is vitrification. This is a process where the sample solution gets solidified without ice crystal formation. Broadly vitrification requires two primary criteria for successful attempt towards cryopreservation which includes highly concentrated solute systems and elevated cooling rates which evade ice crystal formation via ice nucleation inhibition due to raised viscosity levels. Sustained cooling rate leads to a steady increase in the solution's viscosity until the intermolecular motion is halted and solution turns into glass which displays all the prominent attributes of solids, yet retention of all molecular structures of liquid has been observed to be ensured (Reubinoff et al., 2001). Vitrification caters to several needs of the researchers by ensuring limited or no intracellular damage of ice crystals, henceforth subsequently evading varying types of solution effect injuries. Process is relatively cost-effective as no expensive and intricate equipment is needed if compared against traditional cryopreservation processes. Numerous mathematical models were utilised, and numerous rich sluggish cooling conventions were intended to make these stepwise systems successful for the cryopreservation of various kinds of cells and tissues. Be that as it may, ice ought to never be permitted to show up and develop inside the cells or tissue as this prompts harm and demise of the living framework. Realise that the last objective of both lethargic cooling and vitrification is the equivalent: to actuate a glasslike cementing inside cells to shield them from harm by ice gems at all phases of cryopreservation, which further guides in investigating and broadening the extent of its application at both lab and industrial scale. Nevertheless, there are copious serious complications associated with the process too such as its high toxic profile along with critical osmotic damage to the cellular structures which reinforces the apprehensions regarding vitrification and limits its application (Hunt, 2017).

10.21   Conclusion

Cryopreservation has been viewed as a relatively simple yet significant process which primarily aids to sketch a comprehensive exploratory analysis with respect to extraction and efficient employment of cryopreserved cells along with several ancillary applications such as optimising the delivery of mechanism of cryopreserved cells and establishing facilities like cell banks to ensure quality and sustainability of clinical research approaches. Contemporary times have necessitated as well as validate the employment of optimised cryopreservation strategies along with the optimised logistics facilities. Vitrification has been viewed as a chief candidate which in the future can be the potential alternative for cryopreservation. The process has reported to negate all the major limitations associated with cryopreservation and also exhibit a phenomenal recuperative rate, yet there are still no concrete/standard protocols to optimise vitrification. Another major concern associated with vitrification is its inability to withstand scale processes which drastically limit its application to laboratory scale. Importance of optimising vital factors influencing the process of cryopreservation is necessary to attain higher levels of cell survival and cell vitality. Factors such as CPAs, toxicology and osmotic profile along with cooling and warming rates are the chief modulators operating the proper functioning required to achieve quality clinical applications. This chapter discusses several facets of cryopreservation encompassing copious strategies employed and role CPAs. Sustained efforts in exploring novel CPA were made which significantly contribute in determining numerous inalienable toxic profiles of many known specialists. Superior comprehension of the chemical as well as biochemical profile dictating the cryopreservation and warming rates would be essential for any kind of modern future advancements. Effectiveness of the process protocols highly influenced by the employed animal cell sample which is an essential part in examination associated with clinical utility for a wide range of human preliminaries. All in all, unmistakably all the future endeavours aimed with respect to cryopreservation must be zero in on the improvement in facilities along with application which would significantly influence the reproducibility, efficacy and popularity of cryopreservation not only in scientific but also amidst leman population.

References

Alotaibi, N. A. S., Slater, N. K. H., & Rahmoune, H. (2016). Salidroside as a novel protective agent to improve red blood cell cryopreservation. *PLoS One, 11*, e0162748. https://doi.org/10.1371/journal.pone.0162748

Chang, A., Kim, Y., Hoehn, R., Jernigan, P., & Pritts, T. (2017). Cryopreserved packed red blood cells in surgical patients: Past, present, and future. *Blood Transfusion, 15*(4), 341–347. https://doi.org/10.2450/2016.0083-16

Fuller, B. J., & Paynter, S. J. (2007). Cryopreservation of mammalian embryos. *Methods in Molecular Biology (Clifton, N.J.), 368*, 325–339. https://doi.org/10.1007/978-1-59745-362-2_23

Gupta, V., Sengupta, M., Prakash, J., & Tripathy, B. C. (2017). Animal cell culture and cryopreservation. *Basic and Applied Aspects of Biotechnology*, 59–75. https://doi.org/10.1007/978-981-10-0875-7_3

Hubálek, Z. (2003). Protectants used in the cryopreservation of microorganisms. *Cryobiology, 46*(3), 205–229. https://doi.org/10.1016/S0011-2240(03)00046-4

Hunt, C. J. (2017). Cryopreservation: Vitrification and controlled rate cooling. *Methods in Molecular Biology, 1590*, 41–77. https://doi.org/10.1007/978-1-4939-6921-0_5

Ibars, E. P., Cortes, M., Tolosa, L., Gómez-Lechón, M. J., López, S., Castell, J. V., & Mir, J. (2016). Hepatocyte transplantation program: Lessons learned and future strategies. *World Journal of Gastroenterology, 22*(2), 874–886. https://doi.org/10.3748/wjg.v22.i2.874

Ishizuka, Y., & Bramham, C. R. (2020). A simple DMSO-based method for cryopreservation of primary hippocampal and cortical neurons. *Journal of Neuroscience Methods, 333*, 108578. https://doi.org/10.1016/J.JNEUMETH.2019.108578

Jang, T. H., Park, S. C., Yang, J. H., Kim, J. Y., Seok, J. H., Park, U. S., Choi, C. W., Lee, S. R., & Han, J. (2017). Cryopreservation and its clinical applications. *Integrative Medicine Research, 6*(1), 12–18. https://doi.org/10.1016/J.IMR.2016.12.001

Jesus, A. R., Meneses, L., Duarte, A. R. C., & Paiva, A. (2021). Natural deep eutectic systems, an emerging class of cryoprotectant agents. *Cryobiology, 101*, 95–104. https://doi.org/10.1016/J.CRYOBIOL.2021.05.002

Keck, M., Haluza, D., Selig, H. F., Jahl, M., Lumenta, D. B., Kamolz, L. P., & Frey, M. (2011). Adipose tissue engineering: Three different approaches to seed preadipocytes on a collagen-elastin matrix. *Annals of Plastic Surgery, 67*(5), 484–488. https://doi.org/10.1097/SAP.0B013E31822F9946

Li, Y., Tan, J. C., & Li, L. S. (2010). Comparison of three methods for cryopreservation of human embryonic stem cells. *Fertility and Sterility, 93*(3), 999–1005. https://doi.org/10.1016/J.FERTNSTERT.2008.10.052

Lovelock, J. E., & Bishop, M. W. H. (1959). Prevention of freezing damage to living cells by dimethyl Sulphoxide. *Nature, 183*(4672), 1394–1395. https://doi.org/10.1038/1831394a0

Luhur, A., Klueg, K. M., Roberts, J., & Zelhof, A. C. (2019). Thawing, culturing, and cryopreserving drosophila cell lines. *Journal of Visualized Experiments, 146*, 59459. https://doi.org/10.3791/59459

Mandawala, A. A., Harvey, S. C., Roy, T. K., & Fowler, K. E. (2016). Cryopreservation of animal oocytes and embryos: Current progress and future prospects. *Theriogenology, 86*(7), 1637–1644. https://doi.org/10.1016/J.THERIOGENOLOGY.2016.07.018

McGann, L. E., Yang, H., & Walterson, M. (1988). Manifestations of cell damage after freezing and thawing. *Cryobiology, 25*(3), 178–185. https://doi.org/10.1016/0011-2240(88)90024-7

Naing, A. H., & Kim, C. K. (2019). A brief review of applications of antifreeze proteins in cryopreservation and metabolic genetic engineering. *3 Biotech, 9*(9), 1–9. https://doi.org/10.1007/S13205-019-1861-Y

Reubinoff, B. E., Pera, M. F., Vajta, G., & Trounson, A. O. (2001). Effective cryopreservation of human embryonic stem cells by the open pulled straw vitrification method. *Human Reproduction, 16*(10), 2187.

Taylor, M. J., Weegman, B. P., Baicu, S. C., & Giwa, S. E. (2019). New approaches to cryopreservation of cells, tissues, and organs. *Transfusion Medicine and Hemotherapy, 46*(3), 197–215. https://doi.org/10.1159/000499453

Yavin, S., & Arav, A. (2007). Measurement of essential physical properties of vitrification solutions. *Theriogenology, 67*(1), 81–89. https://doi.org/10.1016/J.THERIOGENOLOGY.2006.09.029

Chapter 11
Resuscitation of Frozen Cell Lines

Vinayak Agarwal and Manisha Singh

11.1 Introduction

Most academic, biotechnical and clinical facilities routinely propagate cultivated tissue cells. Most labs keep their own cryopreserved cell banks even though it is impractical to maintain all cells in culture at the same time; to safeguard cells from contamination, genetic drift and loss of function owing to over-passaging; and to conserve cells with poor propagation potential. Frozen aliquots of cells are very useful in the field of cell therapy whenever there is a paucity of freshly donated cells or when cells derived from a specific blood sample must be kept for several years to treat a patient. To sustain cell viability while keeping all genetic markers and functional features of the original tissue, sufficient attention must be taken. To reduce the possibility of cell injury during the cryopreservation process, adequate freezing conditions, proper storage and proper thawing techniques are required. These parameters are especially important for delicate cells like stem cells, where inappropriate freezing and thawing conditions would almost certainly have a negative influence on cell viability and activity. Currently, the most popular recommendation for freezing cells is slow cooling at a controlled pace (usually $-1\ °C\ min^{-1}$) to $-80\ °C$ using either a passive freezing container or a programmable controlled-rate freezer, followed by storage in liquid nitrogen at $-160\ °C$ or lower. Recent research indicates that even short-term temperature variations (i.e. slow thawing) during transport or storage, as well as repeated freeze–thaw cycles of cryopreserved cells, are detrimental to cell viability. As a result, it is suggested that cells be thawed, or raised to slightly over $0\ °C$ from cryogenic conditions, as soon as feasible. To mitigate any potential deleterious effects of dimethyl sulfoxide (DMSO) or other cryoprotectants,

V. Agarwal · M. Singh (✉)
Jaypee Institute of Information Technology (JIIT), Noida, Uttar Pradesh, India

© The Author(s), under exclusive license to Springer Nature Switzerland AG 2023
S. Mani et al., *Animal Cell Culture: Principles and Practice*, Techniques in Life
Science and Biomedicine for the Non-Expert,
https://doi.org/10.1007/978-3-031-19485-6_11

prompt washing out of the freezing medium and transfer of thawed cells into optimal medium for recovery and growth is generally recommended (Thompson et al., 2014).

11.2 Freezing Cell Lines (Cryopreservation)

A frozen bank of cells for any cell type should be established as soon as there are enough cells to store; the fewer times the cells have been passaged (grown to optimal confluency and split to a larger number of culture plates or flasks), the more likely they will retain the characteristics and function of the tissue from which they were obtained. To achieve the greatest outcomes during freezing and post-thaw recovery, cells should be actively developing at a high density (preferably just sub-confluent). Cell cultures should be checked for obvious and detectable contamination (e.g. bacteria, mycoplasma, fungus and viruses) before freezing. Furthermore, cells should be screened to ensure that no different cell types are growing in the culture (e.g. fibroblasts in epithelial cultures). The cells to be frozen are split from the cell culture substrate (for adherent cells) physically, enzymatically (usually with trypsin) or chemically, depending on the cell type. Cells are pooled before the freezing step begins, and an aliquot is taken to calculate cell count and viability (e.g. trypan blue staining). Damaged cells are less likely to withstand the mechanical stress of the freezing and thawing process; hence it is vital that cells are handled delicately throughout these preparation processes https://currentprotocols.onlinelibrary.wiley.com/doi/abs/10.1002/0471142735.ima03gs99.

11.3 Thawing Cell Lines

Despite the fact that it has been clearly demonstrated that the rate at which cells are thawed is just as important to cell survival as the rate at which they are frozen, no clear guidelines on the optimal rate and temperature for cell thawing have been developed (Yokoyama et al., 2012). There are few, if any, automated temperature control methods for the thawing phase of any cryopreservation technique. Furthermore, current accepted thawing cell methodology is not standardised and generally relies on manually holding frozen cell vials in a 378 °C or hotter water bath (or inserting the vials into a floating rack in a water bath) and then periodically inspecting the vial until the last ice crystal has dissolved before diluting the thawed cells into tissue culture medium for centrifugation to remove the cryoprotective agents. This procedure not only introduces the possibility of contamination, but it is also subjective and can result in significant diversity in the results produced. DMSO, in particular, is a very effective cryoprotectant and, therefore, commonly used in freezing media; however, it is known to be toxic to mammalian cells at higher concentrations and is linked to some serious side effects when used to cryopreserve cells intended for use in cell therapy (Chua & Chou, 2009). Some protocols seek to lower the concentration, and thus the toxicity, of DMSO in the freezing medium, but this is not always possible, and other common cryoprotectants like glycerol can also have toxic effects on cells. DMSO, in example, is a very effective cryoprotectant and is thus widely employed in freezing media; however, it is poisonous to mammalian cells at higher doses and has been related to some major adverse effects when used to cryopreserve cells for use in cell therapy. Some techniques attempt to reduce the quantity of DMSO in the freezing solution, and therefore its toxicity, although this is not always practicable, and other frequent cryoprotectants, such as glycerol, can also be harmful to cells (Armitage & Mazur, 1984).

Cryoprotectant toxicity isn't the only risk that cells confront as they shift from a cryogenic to an active metabolic state. The number of freeze–thaw cycles, cooling and thawing speeds and absolute temperatures to which a cell is subjected all have an impact on whether the cell survives the cryogenic storage procedure. Cells that have been thawed and then refrozen are more vulnerable to damage than cells that have just gone through one freeze–thaw cycle. Cells are most vulnerable when stored in frequently accessed freezers, during transport from one facility to another, or during cryogenic storage when frozen cells held in the vapour phase of liquid nitrogen tanks may experience temperature changes depending on their storage position in relation to the liquid nitrogen level, or during prolonged exposure to room temperature air, such as when storage box racks are removed from the storage tank to access other samples. Thus, evidence indicates that even brief exposure to cryopreserved cells to extreme temperatures can drastically reduce their post-thaw viability (Norkus et al., 2013). As a result, thawing cryopreserved cells is done as rapidly as possible to minimise exposure of the resuscitated cells to the environment.

11.4   Cell Thawing Techniques

11.4.1   Hand-Warming

Warming cryovials by hand can sometimes be appealing as a cell thawing approach since it does not necessitate any equipment, body temperature is comparable to a water bath and the user can continuously check and agitate the vial (see Fig. 11.1). It is important to keep in mind, too, that heat exchange when warming a hand-filling a vial is likely to be slower and less consistent because it is substantially less efficient and much more unpredictable than a vial immersed in water. Hand-warming is also inconvenient when thawing more than one item, two vials at once, that can result in even slower cooling rates. Additionally, even though nitrile, exposing flesh to temperatures around 196 °C while wearing gloves might cause serious harm. Thus, it's not recommended in common lab practice (Strober, 2015).

Fig. 11.1 Graphical representation of hand thawing

11.4.2   Water Bath

The communal water bath is probably the most often used approach in the lab for thawing frozen cell vials. Water's high conductivity ensures rapid heating even while preventing localised overheating within the vial. A water bath is also part of the regular equipment used in the cell culture lab on a daily basis, so no additional preparation or investment is required. It is critical not to expose cells to temperatures higher than 37 °C when reheating them in a water bath or any other form of thawing. Even if the total temperature rise during warming can be even more than 200 °C, reaching 37 °C localised inside the vial might soon result in undesirable effects, including cell death (see Fig. 11.2). However, one significant disadvantage of water baths is the possibility of contamination due to the contact between both the water and the specimen. Due to the water level and the requirement to examine the vial, maintaining the tip of the vial dry throughout thawing can be problematic, constantly, as well as the requirement to stir the sample to prevent gradients in temperature (Hunt, 2019).

11.4.3   Specialised Devices

Numerous manufacturers have developed specialised equipment during the last decade that can provide the requisite rapid warming rate without the use of a water bath. Most are built for cryovials only, but bigger volume systems are also available. Dedicated devices perform the thawing phase in a reproducible manner, which has significant advantages in cases where standardisation and validation of the thawing process are required (see Fig. 11.3). They are also useful in situations when water baths are not feasible. The most significant disadvantage of specialised thawing equipment is the initial cost required. Whereas water bath thawing normally does not necessitate any additional expenditure, specialist thawing equipment adds additional costs to the cell culture process. Furthermore, some devices can only thaw one vial at a moment, making them potentially incompatible with certain cell culture techniques (Arai et al., 2020).

Fig. 11.2 Cellular thawing
by water bath

Fig. 11.3 Specialised
equipment to facilitate cell
thawing

Fig. 11.4 Cellular thawing
employing bead thawing
method

11.4.4 Bead Bath

Bead baths can be used instead of water baths to warm up cultured cells media or
maintain cultures warm outside of the incubator. However, beads, like hand-
warming or air warming (e.g. in an incubator), do not provide the efficient heat
transmission of a water bath since there is less surface area in touch with the vessel,
which makes it highly unlikely for cells and can indeed be thawed reliably or rapidly enough. Bead baths are frequently advocated since they do not require refilling;
nonetheless, dust and spillages may require cleaning and refilling over time to prevent contamination (see Fig. 11.4) (Arai et al., 2020).

11.5 Material and Reagents Employed for Resuscitation

It is critical to properly defrost frozen cells in order for the culture to recover fast
and produce the best viability and functionality feasible. Because some cryoprotectants, like as DMSO, are poisonous, it is critical that cells be thawed immediately
and not left in the freezing medium for any longer than necessary.

11.5.1   Materials

Frozen cryogenic vial containing cell line, 70% ethanol, medium (used for growth of cell line), dry ice, CoolRack CFT30 thermo-conductive cryogenic tube module (or equivalent), 37 °C water bath, ThermalTray HP platform (optional), sterilised tissue culture hood, 1 ml pipet, sterile, 12 ml tube with cap, sterile, Beckman TH-4 rotor (or equivalent), tissue culture flasks, inverted microscope.

11.5.2   Reagents and Solvents

Freezing medium (specific for cell type) and trypsin EDTA solution.

11.6   Methodology for Cell Revival

11.6.1   Equipment Setup

1. As needed, label new cell culture vessels with cell type information.
2. Fill a 15 mL conical tube with 9 mL of acceptable recovery media, and set it in a CoolRack1 15 mL module that has been preheated to room temperature. The DMSO and other cryoprotective chemicals will be washed out of this tube.

11.6.2   Thawing Protocol

1. Remove the cryogenic vial containing the frozen cells from the storage compartment, and immediately place it in a ThawSTAR1 Equilibration Module that has been preconditioned. Before proceeding to the following stage, let at least 10 min for equilibration to 278 °C.
2. Place the frozen cryogenic vial in the Thaw STARTM thawing device. To start the thawing cycle, press down on the vial. Retrieve the thawed vial as soon as the auditory and visual alert signals indicate completion.
3. Wipe the outside of the vial with 70% ethanol to prevent contamination before placing it within a sterile tissue culture hood in a prechilled 54 °C CoolRack1 CFT30, using an ice-free cooling base if necessary to maintain 54 °C. Because some cells and freezing medium are temperature-sensitive, this phase is very dependent on cell type.
4. Carefully open the vial and pipette out the cells with a transfer pipette with a big opening to avoid harming the cells.

5. Fill a sterile 15 mL conical tube with 9 mL of room temperature thawing medium (original culture medium containing at least 10% foetal bovine serum or equivalent).
6. Close the 15 mL tube and centrifuge at 300 g for 5 min.
7. To avoid cell injury, the cell pellet should be loose and easy to resuspend with extremely low pipetting velocity.
8. Remove supernatant and resuspend cells in culture media to the desired concentration before transferring to a tissue culture flask or other preferred culture vessel.
9. A tiny sample of this cell solution can be obtained for rapid cell counting, viability measurement or phenotyping.

For adherent cell lines, adjust the medium volume and, if necessary, flask size to meet the cell seeding density specified on the cell line spreadsheet for adherent cell lines. A pre-centrifugation procedure to removing cryoprotectant is usually unnecessary because the initial medium change removes any remaining cryoprotectant. This will then be specified on the data sheet. If the cells are to be employed immediately rather than subculture, a pre-centrifugation process to remove cryoprotectant may be necessary.

For suspension cell lines, a pre-centrifugation step is recommended to remove cryoprotectant, i.e. pellet the cells by centrifuged at 150× g for 5 min, and resuspend the cell pellet in new media using the required volume to obtain the correct seeding density (Thompson et al., 2014).

11.7 Critical Parameters for Resuscitation

The most important criteria to consider for optimal cryopreservation are starting with cells in the logarithmic development phase nearing confluency and freezing at a high cell density ($10–30 \times 10^6$ mL^{-1}). Selecting an optimum freezing medium and optimal cryo-preservative and adhering to a controlled pace of freezing and quick thawing are also crucial aspects. After resuspension in their freezing medium, cells should be frozen as quickly as feasible. Frozen cells are stable at −80 °C for a few weeks (recovered viability gradually diminishes over months); however, it is not recommended to store cells at −80 °C for long periods of time (i.e. more than 6 months). Cryopreserved cells, on the other hand, are practically endlessly stable in a liquid nitrogen storage system. As a result, it is advisable to move the cryogenic vials from the 280 °C freezer to the liquid nitrogen storage tank within 24 h for maximum stability and recovery after long-term storage.

During the transfer, care should also be taken to minimise sample thermal cycling (transient warming episodes), which might have a negative impact on stability during frozen storage (Cosentino et al., 2007). Thermo-conductive cryogenic tube module has been prechilled on dry ice when transferring cells from 280 °C to liquid nitrogen. When exposed to room temperature air, cryogenic vial contents can rise from 280 °C to above 250 °C in less than 1 min. To avoid cell damage from delayed thawing after removal from cryogenic storage, the same conductive cryogenic tube module can be utilised to keep the sample from warming over 275 °C until the required rapid thawing process is commenced.

The rate of thawing is critical for maintaining good cell viability. To avoid ice recrystallisation and osmotic stressors, cells should be thawed as soon as possible. Passive thawing results in lower cell viability and recovery than active thawing at 37 °C (Klossner et al., 2007). To achieve the proper thawing rate, a common practice is to manually submerge the vial in a water bath only to the point where the bath level is above the level of the vial contents, being very careful not to submerge or splash bath liquid onto the vial–vial cap interface, especially for external thread vials, as capillary action will retain the liquid in the thread region and greatly increase the likelihood of contamination of the vial contents upon opening. Periodic withdrawal of the vial from the bath to visually examine the thawing status allows the user to cease the thaw when the last vestige of the solid is visible, but it also slows down the thawing process, which may damage cells.

After thawing and washing, cells should be seeded into a tissue culture vessel large enough for the number of cells. Prior to freezing, cells should be cultured in the same medium in which they were developing (Akhtar et al., 1979).

Finally, it is critical to remember that timing is critical during the cryopreservation operation. The most time-consuming aspect of this technique should be the preparation of materials and reagents for both freezing and thawing. One flask or multi-well plate of adherent cells takes around 30 min to process; if more than one flask or plate needs to be processed, they should be processed individually to avoid stressing cells unduly. When thawing cells, do not melt more than two vials at once. Thawing time can be reduced to 15 min or less using this guidance. Staggering the technique for numerous vials also helps the researcher to strictly stick to overall timing for each cell vial (Mazur et al., 1974).

11.8 Conclusion

It is critical to thaw cells carefully in order to sustain culture viability and allow the cell to recover quickly. Several cryoprotectants, including DMSO, are hazardous at temperatures over 4 °C. As a result, cultures must be frozen promptly and dilute in growth media to reduce harmful effects. Furthermore, it's vital to highlight that proper thaw and post-thaw handling of a cryopreserved clinical specimen is just as critical as the initial cold process for good results. Immersion of cryo-containers in a warmer water bath is the tried-and-true method of thawing cryopreserved biological samples. Dry thawing methods have grown more widespread in the last decade due to decreased contamination hazards and more consistency. To maintain cell viability while conserving all specific genes and functional properties of the original cell, careful consideration must be given. Adequate freezing settings, correct storage and appropriate thawing techniques are necessary to decrease the chance of cell harm during the cryopreservation process.

References

Akhtar, T., Pegg, D. E., & Foreman, J. (1979). The effect of cooling and warming rates on the survival of cryopreserved L-cells. *Cryobiology, 16*(5), 424–429.

Arai, K., Murata, D., Takao, S., Verissiomo, A. R., & Nakayama, K. (2020). Cryopreservation method for spheroids and fabrication of scaffold-free tubular constructs. *PLoS One, 15*(11), e0230428.

Armitage, W. J., & Mazur, P. (1984). Toxic and osmotic effects of glycerol on human granulocytes. *American Journal of Physics, 247*(5 Pt 1), C382–C389. https://doi.org/10.1152/ajpcell.1984.247.5.C382

Chua, K. J., & Chou, S. K. (2009). On the study of the freeze–thaw thermal process of a biological system. *Applied Thermal Engineering, 29*(17), 3696–3709.

Hunt, C. J. (2019). Technical considerations in the freezing, low-temperature storage and thawing of stem cells for cellular therapies. *Transfusion Medicine and Hemotherapy, 46*(3), 134–150.

Klossner, D. P., Robilotto, A. T., Clarke, D. M., VanBuskirk, R. G., Baust, J. M., Gage, A. A., & Baust, J. G. (2007). Cryosurgical technique: Assessment of the fundamental variables using human prostate cancer model systems. *Cryobiology, 55*(3), 189–199.

Mazur, P., Miller, R. H., & Leibo, S. P. (1974). Survival of frozen-thawed bovine red cells as a function of the permeation of glycerol and sucrose. *The Journal of Membrane Biology, 15*(1), 137–158.

Norkus, M., Kilmartin, L., Fay, D., Murphy, M. J., Olaighin, G., & Barry, F. (2013). The effect of temperature elevation on cryopreserved mesenchymal stem cells. *Cryo Letters, 34*(4), 349–359.

Cosentino, L. M., Corwin, W., Baust, J. M., Diaz-Mayoral, N., Cooley, H., Shao, W., et al. (2007). Preliminary report: evaluation of storage conditions and Cryococktails during peripheral blood mononuclear cell cryopreservation. *Cell Preservation and Technology, 5*(4), 189–204.

Strober, W. (2015). Trypan blue exclusion test of cell viability. *Current Protocols in Immunology, 111*, A3.B.1–A3.B.3. https://doi.org/10.1002/0471142735.ima03bs111.

Thompson, M. L., Kunkel, E. J., & Ehrhardt, R. O. (2014). Cryopreservation and thawing of mammalian cells. *eLS, 9*, 3417.

Yokoyama, W. M., Thompson, M. L., & Ehrhardt, R. O. (2012). Cryopreservation and thawing of cells. *Current Protocols in Immunology, 99*(1), A.3G.1–A.3G.5.

Chapter 12
Isolation and Culturing of Cells from Different Tissues

Sakshi Tyagi and Shalini Mani

12.1 Introduction

Tissues are made up of different types of cells. To understand the function/defect of any cell, we first need to isolate the cell from the tissue sample provided. Cells isolated from tissues are not immortal so we have to design specific methods to isolate, culture them and perform assays related to such cell types. Also, they are susceptible to different contamination hence need to be careful while working with them (Rodríguez-Hernández et al., 2014).

Primary culture is that phase of the culture after cell isolation from the tissue, which is proliferated under the suitable conditions till they engross all of the accessible substrate (i.e., reach confluence). At this phase, the cells have to be passaged (i.e. subcultured) by moving them to a new flask/vessel with fresh growth media to offer more space for continuous growth (Freshney, 2005). Subculturing, also known as passaging, is the exclusion of the media and relocation of cells from a former culture into fresh growth media, a process that allows the further proliferation of a cell line or strain.

Primary culture is the foremost culture or a culture that is directly obtained from animal or human tissue by enzymatic or mechanical methods. These cells are generally grown at a slower rate, heterogeneous in nature and convey all the characteristics of the tissue of their origin. The main aim of this culture is to sustain the growth of cells on a suitable substrate, existing in the form of plastic or glass vessels, under regulated environmental conditions. As they are directly accessible from original tissue, they have a similar karyotype (number and appearance of chromosomes in the nucleus of a eukaryotic cell) as the original tissue. Once passaged, primary cell

S. Tyagi · S. Mani (✉)
Centre for Emerging Diseases, Department of Biotechnology, Jaypee Institute of Information Technology, Noida, India

© The Author(s), under exclusive license to Springer Nature Switzerland AG 2023
S. Mani et al., *Animal Cell Culture: Principles and Practice*, Techniques in Life Science and Biomedicine for the Non-Expert,
https://doi.org/10.1007/978-3-031-19485-6_12

cultures can generate cell lines, which may either decease after numerous passages (often called as finite cell lines) or may endure to grow indeterminately (such cell lines are known as continuous cell lines). Typically, normal tissues produce finite cell lines, while cancerous cells/tissues (characteristically aneuploid) generate continuous cell lines (Rodríguez-Hernández et al., 2014).

For many reasons cells attained from primary cultures have a restricted life span, i.e., the cells cannot be sustained indeterminately. An upregulated cell number in a primary culture leads to consumption of the substrate and nutrients, which can impact cellular activity and thus causing the build-up of high levels of lethal metabolites in the culture. This may eventually lead to inhibited cell growth. This stage is known as the confluence stage (contact inhibition), when a secondary culture or a subculture requires to be established to assure constant cell growth (Verma et al., 2020).

12.2   General Methods to Isolate Cells from Tissues

Numerous techniques have been developed for the disaggregation of tissue isolated for primary culture. Dissociation/disaggregation is an essential technique that permits researchers to mark and isolate the types of cells they are trying to study. There are certain approaches for cell dissociation such as mechanical, enzymatic, primary explant and chemical (Fig. 12.1) as mentioned below:

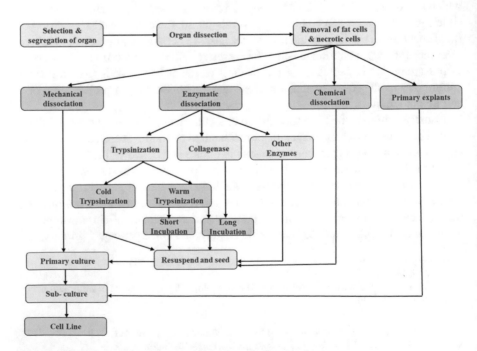

Fig. 12.1 Different methods for the preparation of primary culture

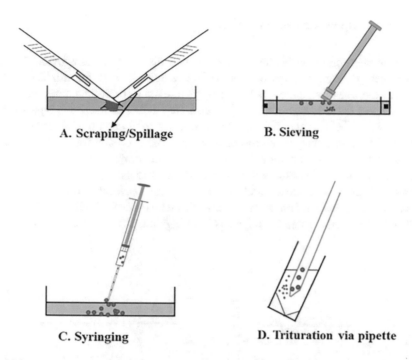

A. Scraping/Spillage   B. Sieving

C. Syringing   D. Trituration via pipette

Fig. 12.2 Mechanical disaggregation. (**a**) Scraping or "spillage": chopping action, or scraping of cut surface, freed cells. (**b**) Sieving: compelling tissue via sieve with syringe piston. (**c**) Syringing: collecting tissue into a syringe with the help of a wide bore needle or canula and expressing. (**d**) Trituration by pipette: pipetting tissue remains upward and downward via a wide bore pipette

12.2.1   Mechanical Dissociation

It is one of the simplest techniques amongst the above-mentioned methods for tissue dissociation. It includes breaking down tissue by the means of physical force, like crushing, scraping or cutting (Fig. 12.2). This method is generally applied for the dissociation of soft tissues (such as brain, spleen, soft tumors, embryonic liver). It involves the usage of a tool like mortar and pestle to disrupt and demolish the extracted sample, leaving behind leftover pieces. Among these remains, the antici-pated cells should be lightly floating for more segregation. This method is fast and thus easier in the case of loosely linked samples. This is a preferable process in the case of tissues such as lymph nodes, bone marrow, spleens and the tissues that are loosely bound (Freshney et al., 1987).

12.2.2 Enzymatic Dissociation

This is another common process to achieve cell dissociation. Enzymatic dissociation utilises definite proteins to disassemble samples of cell culture. The method involves the use of enzymes including trypsin, dispase or collagenase, which dissolve parts of tissue to release the target cells. Depending on the tissue type, the type of enzyme is decided, and discovering the correct combination leads to better outcomes. Choosing the right cell dissociation enzyme is crucial for the efficiency of the process. However, it is more time-consuming in comparison to mechanical dissociation. It is more beneficial for more compact tissues, which may have a large amount of debris. It is most widely used when high retrieval of cells is obligatory from a tissue. Dissociation of embryonic tissues is more effective with a higher yield of cells via the usage of enzymes (Cunningham, 2010).

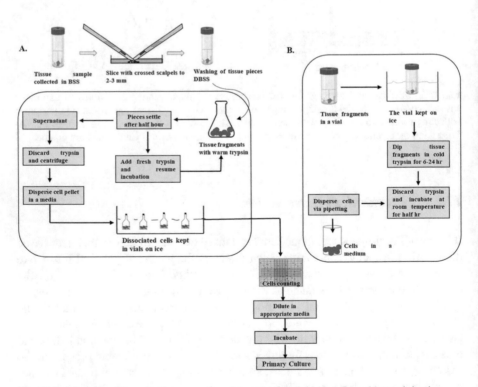

Fig. 12.3 Methods of trypsin disaggregation: (**a**) warm trypsinization (**b**) cold trypsinization

12.2.2.1 Disaggregation by Trypsin

The disaggregation of tissues with the help of trypsin (generally crude trypsin) is commonly termed trypsinization. Crude trypsin is preferred usually as this trypsin comprises other proteases. Moreover, cells can accept crude trypsin and the eventual effect of crude trypsin can easily be counteracted by trypsin inhibitor (in case of serum-free media) or serum. However, pure trypsin can also be used for cells dissociation, on the condition that it is less toxic and action-specific (Gori, 1964). There are two techniques for trypsinization: (1) warm trypsinization and (2) cold trypsinization, which have been discussed below and also summarised in Fig. 12.3.

Warm Trypsinization

1. This technique is helpful for the dissociation of large amounts of tissue in a comparatively short time, especially for minced whole mouse embryos or embryos of a chick.
2. There should be minimum exposure of cells to activate trypsin to maintain maximum viability. Therefore, when whole tissue trypsinization is being done at 37 °C, disaggregated cells should be collected every half hour, and the trypsin should be eliminated via centrifugation and nullified with the media's serum.

Cold Trypsinization

1. Immersing the tissue in trypsin for 6 to 18 hours at 4 °C will allow the enzyme to perforate the tissue and cause minimal tryptic activity, which is a straightforward method of mitigating cell damage after exposure.
2. The cold trypsin process normally gives a higher recovery of viable cells, with amended survival after 24-h culture, and maintains more diverse cell types than the warm method.

Disadvantages of Trypsin Disaggregation

Trypsinization of cells may impair some cells, like epithelial cells, and occasionally it is not operative for certain tissues, which include fibrous connective tissue, thereby other enzymes like collagenase and pronase are also suggested for cell dissociation.

12.2.2.2 Disaggregation by Collagenase

Collagenase is an enzyme that is involved in peptide bonds cleavage in collagen. Collagen is a structural protein, which is profusely found in higher animals, chiefly in the extracellular matrix of connective tissue and muscle. Collagenase, predominantly crude collagenase, can be effectively used for the dissociation of many tissues (adult, normal, malignant, embryonic) that may or may not be trypsin sensitive. During trials also, crude collagenase has shown better outcomes in comparison to purified collagenase. Collagenase disaggregation has been experimented on

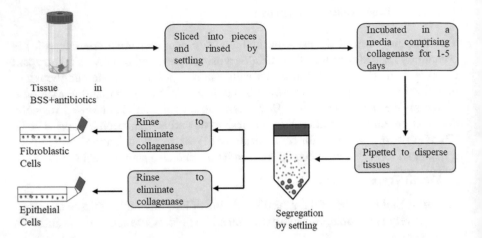

Fig. 12.4 Tissue dissociation with the help of collagenase

numerous human tumors, the brain, lungs, epithelial tissues and other mammalian tissues (Cunningham, 1994).

This technique includes a primary transfer of the chosen tissue into a basal salt solution (BSS) that includes antibiotics. This is accompanied by washing with subsiding and then allocation into a medium comprising collagenase. The solution is incubated for 1–5 days and trailed by constant pipetting for even dispersal of cells as depicted in Fig. 12.4.

12.2.2.3 Other Enzymes

Additionally, to the above-mentioned enzymes, there are certain other enzymes like bacterial proteases (e.g., pronase, dispase) that have been tried, but regrettably have not shown noteworthy results. Immersing the tissue in trypsin for 6 to 18 hours at 4°C will allow the enzyme to perforate the tissue and cause minimal tryptic activity, which is a straightforward method for minimising cell damage after exposure.

12.2.3 Primary Explant Technique

The first illustration of the primary explant method was provided by Harrison in 1907; however, it underwent many changes afterwards. In this technique, tissue is firstly suspended in BSS and then sliced properly and washed by subsiding. Tissue fragments are evenly spread over the growth surface followed by the incorporation of an appropriate medium and then incubated for a time period of 3–5 days (as depicted in Fig. 12.5). Old media is substituted by fresh media as long as anticipated

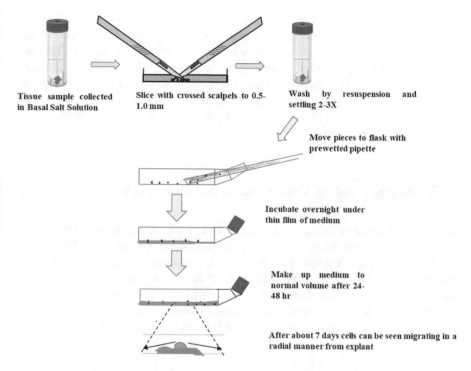

Tissue sample collected in Basal Salt Solution

Slice with crossed scalpels to 0.5–1.0 mm

Wash by resuspension and settling 2-3X

Move pieces to flask with prewetted pipette

Incubate overnight under thin film of medium

Make up medium to normal volume after 24–48 hr

After about 7 days cells can be seen migrating in a radial manner from explant

Fig. 12.5 Schematic representation of primary explant culture

growth or significant growth of the cells is not attained. Once optimal growth is attained, the explants are segregated and moved to new culture vessels comprising fresh media as shown in Fig. 12.5 (Masters, 2000).

This process is mostly used for dissociation of small amount of tissue. Mechanical and enzymatic disaggregation are not advisable for small quantities of tissues because of the risk of cell damage, which can eventually affect the viability of the cell. A major limitation of this technique is the poor gumminess of some tissues on the growth surface, which can hinder the choice of cells for required outgrowth. Despite that, this process has been used typically for culturing embryonic cells, in specific glial cells, myoblasts, fibroblasts and epithelial cells (Masters, 2000).

12.2.4   Chemical Dissociation

The third technique that has been utilised to disaggregate cells is termed chemical dissociation. This method takes benefit of the cations that grip together intracellular bonds. Cations are the atoms bearing the positive charge, and adjoining a chemical compound with a likeness to them leading to the dissolution of bonds within the tissue. Egtazic acid (EGTA) is an instance of a compound bearing this kind of property (Cunningham, 2010).

This process does not change the cell's surface as enzymatic dissociation does, thus making it a very safe and gentle alternative. It also produces high viability and cell number post-expansion since the cultured cells are usually healthy. This is a time-taking process, and similar to enzymatic disaggregation produces uncertain results. The overall functional capability of this technique can be affected by the environment, the kind of chemicals used, the amount and the concentration. This process is generally used for more tender forms of dissociation, such as embryonic cells (Cunningham, 2010).

12.3   Segregation of Viable and Non-viable Cells

After the expansion of primary cell culture, it is crucial to eliminate the non-viable cells from the dissociated cells that can be accomplished by constantly changing the media. With primary cultures sustained in suspension, non-viable cells are slowly diluted out when the viable cells initiate proliferating. Another method of centrifugation, mingling cells with ficoll and sodium metrizoate, can also be used to eliminate dead cells from the primary cell culture. Non-viable cells form a pellet at the bottom, which can simply be eliminated from the solution (Freshney, 2005).

12.4   Summary

In the cell culture, tissue disaggregation and grounding of the primary culture with definite functions make up the primary, and conceivably the most essential, stage. If the obligatory cells are lost at this period, then the loss is irreversible. By using the right techniques, it is possible to cultivate a wide variety of cell types. Trypsin generally causes more damage than collagenase, but it can occasionally produce a single-cell suspension more successfully. Collagenase does not detach epithelial cells readily, but this characteristic can be beneficial for segregating the epithelial cells from stromal cells. Mechanical disaggregation is much faster than the technique using collagenase, but compensates more cells. If none of those approaches are effective, try using supplementary enzymes, like DNase, Pronase, Accutase and Dispase.

References

Cunningham, R. E. (1994). Tissue disaggregation. In L. C. Javois (Ed.), *Immunocytochemical methods and protocols* (Methods in molecular biology, 34). Humana Press.
Cunningham, R. E. (2010). Tissue disaggregation. *Methods in Molecular Biology, 588*, 327–330.
Freshney, R. I. (1987). *Animal cell culture: A practical approach*. IRL Press.

Freshney, R. I. (2005). *Culture of animal cells: A manual of basic technique* (5th ed.). Wiley.
Gori, G. B. (1964). Trypsinization of animal tissues for cell culture: Theoretical considerations and automatic apparatus. *Applied Microbiology, 12*(2), 115–121.
Masters, J. R. W. (2000). *Animal cell culture: Practical approach* (3rd ed.). Oxford University Press.
Rodríguez-Hernández, C. O., Torres-Garcia, S. E., Olvera-Sandoval, C., Ramirez-Castillo, F. Y., Muro, A. L., & Avelar-Gonzalez, F. J. (2014). Cell culture: History, development and prospects. *International Journal of Current Research and Academic Review, 2*, 188–200.
Verma, A., Verma, M., & Singh, A. (2020). Animal tissue culture principles and applications. In *Animal biotechnology* (pp. 269–293). Academic.

Chapter 13
Co-culture Techniques

Vinayak Agarwal, Manisha Singh, and Vidushi Srivastava

13.1 Introduction

Over the last few decades there has been a lot of development in cell culture techniques, co-culturing technique happens to be one of them. Co-culturing specifically provides a detailed insight regarding cell-cell interaction (Tan et al., 2019). Fundamentally co-culturing involves a specific culturing set-up encompassing two or more cells which are permitted to be cultivated with a degree of physical contact amidst them (Kirkpatrick et al., 2011). Such techniques amassed plenty of interest from the field of synthetic biology for exploring novel complex multicellular synthetic systems. Co-culturing enables us to explore cellular interactions, assessing natural interaction between different cellular populations along enhancing successful culturing with respect to specific cell populations (Bogdanowicz & Lu, 2013). However, there are several external as well as internal variables influencing the success of the culture, thus making it a complex process. Neoteric drug research and testing is one of the most promising avenues for scouting applications for co-culturing (Miki et al., 2012). It offers a comprehensive in vivo tissue model system which permits high-throughput testing along with a more in-depth analysis of the drug's expression concerning cell to cell interaction, which were limited in case of animal models (Hendriks et al., 2007). Study conducted by Holmes et al., 2009 suggested that essential aspects of human pathophysiology and physiology were frequently lacking in animal models which eventually lead to false positive or false negative human responses during clinical trials (Holmes et al., 2009). Therefore, culturing human cells could significantly elevate our potential insight into diseased and healthy human systems. Besides aiding in drug testing, co-culture techniques have been observed to be highly effective in enhancing the success of cell cultivation (Tan et al., 2019). Prior scientific literature enlists some cell types as hard to

V. Agarwal · M. Singh (✉) · V. Srivastava
Jaypee Institute of Information Technology (JIIT), Noida, Uttar Pradesh, India

© The Author(s), under exclusive license to Springer Nature Switzerland AG 2023
S. Mani et al., *Animal Cell Culture: Principles and Practice*, Techniques in Life
Science and Biomedicine for the Non-Expert,
https://doi.org/10.1007/978-3-031-19485-6_13

cultivate in in vitro models or exhibit negligible desired physiological expression in in vivo models; however, this can be changed by cultivating different populations of cells with them. Presence of another cell in close proximity tends to impart positive expression on the former cell's growth behaviour, leading to improved overall cell cultivation (Zhang et al., 2018). Moreover, to complement such culturing, three dimensional (3D) models are employed for fabricating specific cellular microenvironments closer to its actual natural counterpart. This drastically improves the significance of several physiological responses as well as scales the level of complexity (Mountcastle et al., 2020). For instance, Zhang et al., 2019 studied osteocytes with the help of different 3D in vitro models. Their research indicated that mimicking the native cellular microenvironment enables us to attain superior morphology along with better expression in contrast to the traditional monolayer cultures (Zhang et al., 1914). Co-culture techniques present a broad scope of application which stretches from synthetic ecology to industrial, medical as well as environmental biotechnology; simply due to its superior productivity and beneficial attributes over monoculture (Jessup et al., 2004). This chapter focuses on copious facets influencing the co-culture system, probable drawbacks along with scope for future application and potential.

13.2   Co-cultures

Cell culture techniques have been in practice for well past a century; however, the frequent development and advancement has only aided modern science. Amongst the recent developments in this methodology is the co-culture techniques, which involves the culturing of different cell types in a single culture well or petri plate (Shafiee et al., 2021). Several research articles have suggested that this technique of culturing can be directly applied to most of the cell types such as stromal–mesenchymal interaction, analysing the interaction between different immune cells as well as culturing embryonic stem cells on feeder cells. One of the easiest ways to design a co-culture system that enables physical contact is by fabricating a mixed monolayer of desired cell types. To create such a monolayer, simply combine both cell suspensions according to the necessary seeding ratio for the co-culture (IntanHasbullah et al., 2021). The mixed monolayer models would now offer the highest local heterotypic interaction, and furthermore the relative levels of hetero- and homotypic interactions can be modulated by altering the seeding density ratio for the co-culture model (Park et al., 2022).

Co-culturing techniques have already shown potential to have a plethora of applications in biotechnology which are aimed to explore the interaction between divergent cell types. Synthetic biology and synthetic ecology also find their fair share of serviceability of such a technique for creating a multispecies cell consortium along with foundational research concerning novel environmental, medical and industrial application with several proof-of-principle studies (Soriano et al., 2021).

13.3   Extracellular Microenvironment

There have been several reports suggesting that different cell populations employed for co-cultures are quite vulnerable to extracellular environments. It is important to understand that cellular phenotypes are expressed because of an intricate interplay between the environment and genotype pertaining to the cells employed. Consequently in regards to co-culture studies, it simply means that interaction of microenvironment with genotype is not only responsible for altered expression but the population structure of the cells can also fundamentally tweak the ways in which the genotype would interact with the environment ideally (Mcglothlin et al., 1982). For instance, certain cellular populations tend to form cooperative relationships under specific environmental conditions, which ultimately alter the desired and predicted expression of the cell population. Environmental influence is closely associated with evolutionary trajectory is a well-known fact and with respect to the co-culture population it requires a watchful consideration, at the initial stages of gene circuit designing (Lee et al., 2010). Cellular population offers a higher probability of witnessing a mutation due to larger cell aggregation or elevated levels of stress. Therefore, controlling extracellular microenvironment is highly essential for a successful co-culturing; however, this condition also presents us with an opportunity for genetic modification or control over the expression of cell population (Williams & Wick, 2005).

13.4   Variables in Co-cultures

Apart from the extracellular microenvironment, co-cultures are subjected to several variables that influence cellular growth. Generally, these variables have separate identities, yet they tend to exhibit consolidated interlinked effects causing irregularities in the behaviour of the cell population. Some of the usually observed variables in a co-culture system are cellular communication, culture medium, seeding cell population of the co-culture as well as varying volumes of co-cultures (Rollié et al., 2012; Basu et al., 2005).

13.4.1   Cell Communication During Co-cultures

Human and animal systems require a smooth function of the signalling cascades, which are spread throughout the body. These signalling cascades are further fractionated based on the signalling pathways and the type of signalling involved, such as synaptic signalling (facilitated by neurons), endocrine signalling (facilitated through bloodstream), juxtacrine signalling (facilitated by cell - cell interaction), paracrine and autocrine signalling (via hormones) or communication via gap junctions (facilitated by intracellular spaces between cells) (Bogdanowicz & Lu, 2013). All things considered, all types of signalling can be broadly classified under heterotypic and homotypic interactions. Co-culture technologies serve as an apt tool for investigating these distinct cellular interactions along with analysing geometry and cellular responses by utilising the in vitro tissue model systems (Fey et al., 2012).

Co-culture presents the opportunity for different types of cells to communicate and interact, employing various pathways and signalling types based on their affinity and cellular response towards one another. One of the common interaction modes observed by Stagg et al., 2006 in co-cultures is via gap junctions which not only facilitates heterotypic interaction between different types of cells but also allows homotypic interaction from cells of non-human origin. Nonetheless, besides the gap, endocrine signalling also plays a vital role in cell signalling in co-cultures as it lacks the necessity of physical contact (Stagg et al., 2006). Alberts et al., 2003 in their research study observed that paracrine, autocrine or endocrine signalling employ extracellular secretions which regulate the interaction between the cells (Hartung, 2014).

In co-cultures copious scientific literature observed that gap junction or endocrine signalling supports communication and interaction in cells from intracellular origin. Furthermore, no conclusive evidence was reported indicating correlation between the cell's development stage and type of communication. Although divergent cells are cultured together in co-culture, there is not any significant influence over each other's differentiation profile or cellular morphology. However cellular interaction in co-cultures have been reported to impart their expression leading to tissue formation in vitro models. Thereby it offers an opportunity to identify and study specific regulators and stimulators involved in the co-culture for this phenomenon (Eagle, 1959).

13.4.2   Cell Culture Medium

Universally culture media is the most important component for cell culture technology irrespective of the specific culture technologies employed. Primary function of any cell culture medium is to provide nourishment for the cells, usually medium are added in liquid state consisting of base medium, various modulatory factors along with serum. Going back in time the first culture medium was developed in 1959 by Eagle, 1959 where they designed a minimal essential medium (MEM) having vitamins, amino acids and ionic species in the ratio of (8:13:6) respectively along with glucose. Apart from base medium, serum (predominantly fatal bovine serum (FBS)) and modulating factors (such as growth factors and hormones) were also added due to their key role in guiding the cellular expression towards the desired cell behaviour. Establishing an optimum cell culture medium is also important for creating a good in vitro model (Polizzi et al., 2014).

First and foremost, it is essential to engineer culture medium specific for certain cell types due to their different functionality and varying nutrition requirement during co-cultures. However, this is not an easy task as finding the correct balance in the medium is difficult because of the involvement of more than one cell type as reported by Polizzi et al., 2014. Based on such criteria there are several types of media encompassing partitioned culture environments, supplemented media and mixed media (Langenbach & Handschel, 2013).

In one of the research studies conducted by Langenbach and Handschel, 2013 they designed a co-culture incorporating osteoclasts and osteoblasts. The culture medium used had osteogenic supplements like β-glycerophosphate and dexamethasone which are essential for maturation and differentiation of osteoblast cells. Although these supplements were added to support co-culture, in spite of imparting any positive value, they tend to exhibit inhibitory expression towards the differentiation of monocytes into osteoclasts (Zhu et al., 2018). Later another team took the mantle of designing an alternate approach without imparting any negative effect. Whereas, Zhu et al., 2018 used a simple base medium and added the liquid supplement to it in a sluggish manner to avoid any negative effects (Vis et al., 2020).

This challenge tends to multiply when multi-organ systems are involved, as searching for optimum balance for cell culture media is even harder in such cases. For instance, a medical device that practically integrates adipose tissue, liver, lung and kidney exhibited significant inhibition in liver cells due to transforming growth factor β1 (TGF-β1), which is essential for lung cells (Zhang et al., 2009). Zhang et al. (2009) used gelatine microspheres that tend to release TGF-1 locally, supporting lung cell proliferation. As a result, TGF-1 levels were compartmentalized, preventing its inhibitory expression over the liver cells (Kaji et al., 2011).

13.4.3 Cell Population in Co-cultures

Cell number is exceedingly important for any type of culturing technique; while talking specifically about co-cultures the relative seeding cell count not only decides the probability of its survival but also indicates whether the co-culture would reach its physiological equilibrium or not. Although both cell types are seeded with the intention of being grown together, the importance of the seeding order cannot be overstated (Jensen & Therkelsen, 1981). Seeding order is decided based on the required microenvironment for the cell along with its associated phenotypic behaviour. Besides seeding order, several researchers have found simultaneous planting to be an effective alternative. Albeit there are not any specific protocols for calculating optimum seeding ratios, cells having a more proliferative expression are taken in lower proportion as compared to their less fastidious counterparts. Usually employed seeding ratios for co-culturing based on the above stated concept are 1:2 and 1:3 (Alberts et al., 2003).

13.4.4 Volume of Co-cultures

Several studies in the past have been carried out investigating the varying volume ranges employed in the co-culture system, which suggested that the choice of volume solely depends upon the purpose for which the co-culture set-up has been designed. Potentially co-culture techniques can be operated at any volume range logistically possible. Although volume does not influence much cellular interaction directly, biometrics of the in vitro have been observed to be highly sensitive towards culture density and volume. Typically, the volume ranges on the equipment are in millilitres (mL) in animal tissue culture laboratories. Carrying out co-culturing at a small scale (lab scale) requires either 96-well plate, microfluidic devices or micro-droplets on petri plates, which all come in with this volume range only (Park et al., 2011). At industrial scale, bioreactors are employed for carrying out co-culturing and the volume ranges are in litres (L) (Banic et al., 2012).

13.5 Model Systems in Co-cultures

Establishing a quality model system is imminent for evaluating cellular interaction in the culture population. Generally, 2D and 3D models are designed by rendering tissue engineering technology. These model systems exhibit a wide variety of applications besides co-culture technology, such as computational biology (Polizzi et al., 2014), understanding microphysiology of human organ systems (Wang et al., 2017) as well in ecological studies (Jessup et al., 2004; Polizzi et al., 2014; Wang et al., 2017). In vitro 3D models are designed with respect to specific tissues which are associated to specific organ systems like, ovary, lung, bone, kidney and liver. Apart from normal physiological tissues these models in co-culture can also be used to investigate certain pathological tissues in diseased state. Traditionally, co-culture encompasses different types of cell populations requiring specific and controlled microenvironments mimicking human systems, attaining which is a complex process; thus, designing computational as well as in vitro models becomes necessary (Ahluwalia et al., 2017).

13.5.1 Computational Models

Synthetic or computational biology is based on the basic principle of analysing and evaluating engineered systems for predicting behaviour in a controlled environment. In terms of computational biology, co-culture has several facets such as population dynamics and cellular interaction among others which are crucial for its success. Accounting for all these variables in a single model is a tough task to accomplish and translating such data for a co-culture model system is even more arduous. Contemporary times have witnessed studies focused on exploring the theoretical along with experimental outlooks for carrying out co-cultures (Jones et al., 2016).

Due to its unique parameter space, computational modelling is most often used to simulate model systems that are challenging to replicate in lab settings, such as microenvironments, long-term cellular behaviour, and population ratios (Bersini et al., 2016). A research study conducted by Kerr et al., 2002 reported that diversity in the ecosystem elevates in a co-culture system as compared to a static environment (Kerr et al., 2002). Another study by Byun et al., 2013 explained that co-culture systems usually cause oxygen depletion, which can be controlled by splitting or colony delocalisation (Byun et al., 2013).

These model systems highly depend upon a variety of factors such as rate constant, which is difficult to determine as parameters obtained are from monocultures which cannot be translated for co-culture techniques. In simple terms, system biology helps to integrate computational modelling with experimental systems. Such modelling aids in identifying several facets which tend to influence co-cultures during experimental set-up, thereby providing valuable insight regarding the designed system's response in the environment (Scott-Drechsel et al., 2012).

13.5.2   In Vitro Tissue Models

One of the chief applications associated with co-culturing techniques involves in vitro tissue modelling to explore the interaction profile of the cells. These in vitro models are synthetic replicas of their natural counterparts which enables us to investigate several human processes including homeostasis, development, disease and regeneration. They are hugely valuable in experimental research as they enable us to understand various regulatory and biochemical mechanisms of the human system; moreover, they provide us with in-depth crosstalk analysis at cellular level of the interactions. Here are some of the widely employed in vitro tissue models in co-culture along with their specific organ (Unger et al., 2002).

13.5.2.1   Skin

Studies investigating the tissue damage and wound healing process of human skin was conducted by Bishop et al. 1999; Donovan et al., 2001. Their study primarily focused on the identification of various cell types involved in the process of wound healing at different time points which cause a change in signal transduction subsequently after the injury. Initial observation suggested a close association between endothelial and fibroblast cells. Based on these findings they employed umbilical endothelial cells and dermal fibroblasts cells for developing in vitro tissue models for co-culturing (Bishop et al., 1999; Donovan et al., 2001). The crosstalk between the cells induced endothelial cell migration which further led to the synthesis of microcapillaries or tubules. Furthermore, the importance of growth factors, particularly vascular endothelial growth factor, in microcapillaries creation, differentiation, and stability studies was shown using 3D scaffolds of endothelial and fibroblast co-culture (Tremblay et al., 2005).

13.5.2.2 Liver

Liver cells such as hepatocytes and sinusoidal epithelial cells (EC) are some of the most important cells in human physiology due to their role in hepatic regeneration along with devising extracorporeal liver systems. A few of the initial studies conducted by Goulet et al. 1988 incorporated adult hepatocytes cells along with epithelial rat liver cells for exploring and understanding hepatic systems. The combinatorial expression of these cells leads to the formation of extracellular matrices, besides prominent expression in gap junctions over 28 days of the co-culture (Goulet et al., 1988). Furthermore, Bader et al. 1996 established a novel 3D co-culture model derived by sandwiching monolayers of primary hepatocytes discovered in collagen sponge. This newly synthesised in vitro model was covered by non-parenchymal liver cells reported to induce certain liver cells (epithelial cells, Ito cells and Kupffer cells) (Canová et al., 2004).

13.5.2.3 Blood Vessels

Blood capillaries, more commonly known as blood vessels, remain in constant contact with blood flowing throughout the body. Amidst several, ECs have been observed to constitute the endothelial layer which is in direct contact with the blood. However, blood vessels based upon their functionality show divergence, for instance, multi-layered smooth muscle cells are highly populated in the blood vessels catering to arteries specifically in tunica media (Hellström et al., 2001). Several research groups have explored the combination of epithelial and mural cell co-cultures rendered from differing sources like human mesenchymal stem cells. This specific co-culture aids to get a better understanding of the notch signalling, improve shear stress response, elucidate heterotrophic occurrences along with the expression of angiopoietin/Tie2. Furthermore, apart from natural sources, there also has been a synthetic alternative involving (poly (ethylene glycol)) (PEG)-based biomimetic hydrogels to formulate an EC-MSC co-culture for emulating the formation of vascular networks (Moon et al., 2010).

13.5.2.4 Blood Brain Barrier

Blood brain barrier (BBB) is the most intricate vasculature present in the human system whose primary function is to ensure brain protection; moreover, due to its specific functionality all the microcirculation of the central nervous system (CNS) had to pass through it. Any attempt to design a BBB co-culture must incorporate three vital cell types, which are primarily astrocyte, brain pericyte and brain capillary endothelial cell (BCEC). There has been a lot of literature pertaining to in vitro BBB-based tissue models suggesting that the employment of BCEC cells for monoculturing leads to inferior results and is not advisable (Fischer et al., 2005). Thereby the application of co-culture finds its place regarding BBB. Dehouck et al. 1990 designed a co-culture where they rendered the BCEC out of bovine and astrocytes from new born rat's cerebral cortex. The cellular population of the latter co-culture exhibited tight gap junctions, electrical resistance and gamma-glutamyl transpeptidase expression which indicated proper cerebral functioning (Dehouck et al., 1990).

13.5.2.5 Bone

Copious research studies in the past have explored the bone's healing potential in the absence of any kind of damage, yet the cellular and molecular aspects of this phenomenon is relatively unexplored. Co-culture technology offers the means to understand bone cell behaviour and expression at cellular level. Moreover, specific in vitro models enable us to offer comprehensive insights regarding the associated drug delivery therapies (Gupta et al., 2006). Research led by Sebrell et al., 2019 designed a spheroid co-culture model with the help of human umbilical vein endothelial cells (HUVEC) along with osteoblast cells. The study was able to conclude the presence of bi-directional influence of cells on the gene expression corresponding to angiogenesis and osteogenesis. Another combination of HUVEC with MSC cell-based co-culture was also observed in several prior research literature. Co-culture of these cells exhibited complete formation of an in vitro network, emulating vascular cells, mostly observed after 10 days of seeding. Subsequently, after 14 days, anastomosis of the HUVEC was observed in the subcutaneous region of the nude mice employed for the study (Stahl et al., 2004).

13.6 Potential Challenges with Co-cultures

Although co-culture offers numerous beneficial applications, there are certain drawbacks associated with this specific technique. The most prominent challenges observed with co-cultures is that they are quite laborious and time-consuming, cost inefficient and there is a high risk of failure during the culturing due to lack of knowledge and modulation of microenvironment. Certain research articles have also reported higher likelihood of bacterial and viral infection especially in Vero cells. Co-culturing necessitates the maintenance of ideal culture conditions, which are very variable (El-Ali et al., 2006).

13.7 Future Perspective

Recent decades have witnessed the development of various culture techniques that offer numerous benefits and latest advancements, yet co-culture technology holds most potential. This technique specifically serves as the link between complex prospective systems and genetic engineering. Applications or avenues that co-culture caters exist multifariously in the biological hierarchy of the human system, which comprises of tissue organelles as well as microenvironment. Several industrial concerns such as resistance to toxic waste, metabolic stability, tolerance to stress or additives and composition in feedstock can be dealt with ease with the help of co-cultures. It permits the inclusion of most of the desired characteristics in the cellular population, as there are two types of cells present which make genetic engineering relatively straightforward. In the near future co-culture holds the key to answer copious necessary biological queries (Tan et al., 2019).

13.8 Conclusion

All in all, co-cultures exhibit significant adventitious facets which make them superior to the existing monoculture technology because of several chief aspects like stability, reciprocity, modularity along with industrial scalability. Besides drug testing, it also aids in inspecting intercellular interaction, which necessitates specific tissue culture in vitro model systems along with the participation of different cell types in the culture. Co-cultures are operated under extremely specific and controlled microenvironments, which need plenty of optimisation for attaining desired goals. Its active role in modernising in vitro tissue models not only assists physiological research studies but also helps in drug development based on the cellular interaction in a pathological state.

References

Ahluwalia, A., Mattei, G., Sartori, S., Caddeo, S., & Boffito, M. (2017). Tissue engineering approaches in the design of healthy and pathological in vitro tissue models. *Frontiers in Bioengineering and Biotechnology, 5*, 40. https://doi.org/10.3389/fbioe.2017.00040

Alberts, B., Johnson, A., Lewis, J., Raff, M., & Roberts, K. (2003). Molecular biology of the cell. *Annals of Botany, 91*(3), 401. https://doi.org/10.1093/aob/mcg023

Bader, A., Knop, E., Böker, K. H. W., Crome, O., Frühauf, N., Gonshior, A. K., Christians, U., Esselmann, H., Pichlmayr, R., & Sewing, K-F. (1996). Tacrolimus (FK 506) biotransformation in primary rat hepatocytes depends on extracellular matrix geometry. *Naunyn-Schmiedeberg's Archives of Pharmacology, 353*, 461–473. https://doi.org/10.1006/excr.1996.0222

Banic, M., Banic, B., Franceschi, F., Babic, Z., Babic, B., & Gasbarrini, A. (2012). Extragastric manifestations of helicobacter pylori infection. *Helicobacter, 17*, 49. https://doi.org/10.1111/j.1523-5378.2012.00983.x

Basu, S., Gerchman, Y., Collins, C. H., Arnold, F. H., & Weiss, R. (2005). A synthetic multicellular system for programmed pattern formation. *Nature, 434*(7037), 1130–1134. https://doi.org/10.1038/nature03461

Bersini, S., Gilardi, M., Arrigoni, C., Talò, G., Zamai, M., Zagra, L., Caiolfa, V., & Moretti, M. (2016). Human in vitro 3D co-culture model to engineer vascularized bone-mimicking tissues combining computational tools and statistical experimental approach. *Biomaterials, 76*, 157–172. https://doi.org/10.1016/J.BIOMATERIALS.2015.10.057

Bishop, E. T., Bell, G. T., Bloor, S., Broom, I. J., Hendry, N. F. K., & Wheatley, D. N. (1999). An in vitro model of angiogenesis: Basic features. *Angiogenesis, 3*, 335.

Bogdanowicz, D. R., & Lu, H. H. (2013). Studying cell-cell communication in co-culture. BTJ-COMMENTARY. *Biotechnology Journal, 8*, 395–396. https://doi.org/10.1002/biot.201300054

Byun, C. K., Hwang, H., Choi, W. S., Yaguchi, T., Park, J., Kim, D., Mitchell, R. J., Kim, T., Cho, Y. K., & Takayama, S. (2013). Productive chemical interaction between a bacterial microcolony couple is enhanced by periodic relocation. *Journal of the American Chemical Society, 135*(6), 2242–2247. https://doi.org/10.1021/JA3094923/SUPPL_FILE/JA3094923_SI_001.PDF

Canová, N., Kmoníčková, E., Lincová, D., Vítek, L., & Farghali, H. (2004). Evaluation of a flat membrane hepatocyte bioreactor for pharmacotoxicological applications: Evidence that inhibition of spontaneously produced nitric oxide improves cell functionality. *ATLA, Alternatives to Laboratory Animals, 32*(1), 25–35. https://doi.org/10.1177/026119290403200106

Dehouck, M.-P., Méresse, S., Delorme, P., Fruchart, J.-C., & Cecchelli, R. (1990). An easier, reproducible, and mass-production method to study the blood–brain barrier in vitro. *Journal of Neurochemistry, 54*(5), 1798–1801. https://doi.org/10.1111/J.1471-4159.1990.TB01236.X

Donovan, D., Brown, N. J., Bishop, E. T., & Lewis, C. E. (2001). Comparison of three in vitro human 'angiogenesis' assays with capillaries formed in vivo. *Angiogenesis, 4*(2), 113–121. https://doi.org/10.1023/A:1012218401036

Eagle, H. (1959). Amino acid metabolism in mammalian cell cultures. *Science, 130*(3373), 432–437. https://doi.org/10.1126/SCIENCE.130.3373.432/ASSET/3239A937-07CE-4D92-9 DBA-107399109FD5/ASSETS/SCIENCE.130.3373.432.FP.PNG

El-Ali, J., Sorger, P. K., & Jensen, K. F. (2006). Cells on chips. *Nature, 442*(7101), 403–411. https://doi.org/10.1038/nature05063

Fey, D., Croucher, D. R., Kolch, W., Kholodenko, B. N., & Barberis, M. (2012). Crosstalk and signaling switches in mitogen-activated protein kinase cascades. *Frontiers in Physiology*. https://doi.org/10.3389/fphys.2012.00355

Fischer, S., Wiesnet, M., Renz, D., & Schaper, W. (2005). H2O2 induces paracellular permeability of porcine brain-derived microvascular endothelial cells by activation of the p 44/42 MAP kinase pathway. *European Journal of Cell Biology, 84*(7), 687–697. https://doi.org/10.1016/J.EJCB.2005.03.002

Goulet, F., Normand, C. N., & Morin, O. (1988). Cellular interactions promote tissue-specific function, biomatrix deposition and junctional communication of primary cultured hepatocytes. *Hepatology, 8*(5), 1010.

Gupta, H. S., Seto, J., Wagermaier, W., Zaslansky, P., Boesecke, P., & Fratzl, P. (2006). *Cooperative deformation of mineral and collagen in bone at the nanoscale*. https://www.pnas.org

Hartung, T. (2014). 3D — A new dimension of in vitro research. *Advanced Drug Delivery Reviews, 69–70*, vi. https://doi.org/10.1016/J.ADDR.2014.04.003

Hellström, M., Gerhardt, H., Kalén, M., Li, X., Eriksson, U., Wolburg, H., & Betsholtz, C. (2001). Lack of pericytes leads to endothelial hyperplasia and abnormal vascular morphogenesis. *The Journal of Cell Biology, 153*(3), 543–554. https://doi.org/10.1083/JCB.153.3.543

Hendriks, J., Riesle, J., & van Blitterswijk, C. A. (2007). Co-culture in cartilage tissue engineering. *Journal of Tissue Engineering and Regenerative Medicine, 1*(3), 170–178. https://doi.org/10.1002/TERM.19

Holmes, A., Brown, R., & Shakesheff, K. (2009). Engineering tissue alternatives to animals: Applying tissue engineering to basic research and safety testing. *Regenerative Medicine, 4*(4), 579–592. https://doi.org/10.2217/RME.09.26

IntanHasbullah, N., Aminah Syed Mohamad, S., Iberahim, R., Hasan, A., Ahmad, N., Kheng Oon, L., Azfa Johari, N., Nuruddin Abd Manap, M., & Mohamed Boudiaf, O. (2021). A review on in vitro cell culture model for bacterial adhesion and invasion: From simple monoculture to co-culture human intestinal epithelium model. *Journal of Pharmaceutical Research International, 33*(43B), 97–106. https://doi.org/10.9734/JPRI/2021/V33I43B32530

Jensen, P. K. A., & Therkelsen, A. J. (1981). Cultivation at low temperature as a measure to prevent contamination with fibroblasts in epithelial cultures from human skin. *The Journal of Investigative Dermatology, 77*(2), 210–212. https://doi.org/10.1111/1523-1747.EP12479920

Jessup, C. M., Kassen, R., Forde, S. E., Kerr, B., Buckling, A., Rainey, P. B., & Bohannan, B. J. M. (2004). Big questions, small worlds: Microbial model systems in ecology. *Trends in Ecology & Evolution, 19*(4), 189–197. https://doi.org/10.1016/J.TREE.2004.01.008

Jones, J. A., Vernacchio, V. R., Sinkoe, A. L., Collins, S. M., Ibrahim, M. H. A., Lachance, D. M., Hahn, J., & Koffas, M. A. G. (2016). Experimental and computational optimization of an Escherichia coli co-culture for the efficient production of flavonoids. *Metabolic Engineering, 35*, 55–63. https://doi.org/10.1016/J.YMBEN.2016.01.006

Kaji, H., Camci-Unal, G., Langer, R., & Khademhosseini, A. (2011). Engineering systems for the generation of patterned co-cultures for controlling cell–cell interactions. *Biochimica et Biophysica Acta - General Subjects, 1810*(3), 239–250. https://doi.org/10.1016/J. BBAGEN.2010.07.002

Kerr, B., Riley, M. A., Feldman, M. W., & Bohannan, B. J. M. (2002). Local dispersal promotes biodiversity in a real-life game of rock–paper–scissors. *Nature, 418*(6894), 171–174. https:// doi.org/10.1038/nature00823

Kirkpatrick, C. J., Fuchs, S., & Unger, R. E. (2011). Co-culture systems for vascularization — Learning from nature. *Advanced Drug Delivery Reviews, 63*(4–5), 291–299. https://doi. org/10.1016/J.ADDR.2011.01.009

Langenbach, F., & Handschel, J. (2013). Effects of dexamethasone, ascorbic acid and β-glycerophosphate on the osteogenic differentiation of stem cells in vitro. *Stem Cell Research & Therapy, 4*(5), 1–7. https://doi.org/10.1186/SCRT328/FIGURES/1

Lee, J. H., Wang, H., Kaplan, J. B., & Lee, W. Y. (2010). Effects of Staphylococcus epidermidis on osteoblast cell adhesion and viability on a Ti alloy surface in a microfluidic co-culture environment. *Acta Biomaterialia, 6*(11), 4422–4429. https://doi.org/10.1016/J.ACTBIO.2010.05.021

Mcglothlin, J. W., Moore, A. J., Wolf, J. B., & Brodie, E. D. (1982). Interacting phenotypes and the evolutionary process. III. Social evolution. *Evolution, 64*, 2558. https://doi. org/10.1111/j.1558-5646.2010.01012.x

Miki, Y., Ono, K., Hata, S., Suzuki, T., Kumamoto, H., & Sasano, H. (2012). The advantages of co-culture over mono cell culture in simulating in vivo environment. *The Journal of Steroid Biochemistry and Molecular Biology, 131*(3–5), 68–75. https://doi.org/10.1016/J. JSBMB.2011.12.004

Moon, J. J., Saik, J. E., Poché, R. A., Leslie-Barbick, J. E., Lee, S. H., Smith, A. A., Dickinson, M. E., & West, J. L. (2010). Biomimetic hydrogels with pro-angiogenic properties. *Biomaterials, 31*(14), 3840. https://doi.org/10.1016/J.BIOMATERIALS.2010.01.104

Mountcastle, S. E., Cox, S. C., Sammons, R. L., Jabbari, S., Shelton, R. M., & Kuehne, S. A. (2020). *A review of co-culture models to study the oral microenvironment and disease*. https://doi.org/1 0.1080/20002297.2020.1773122

Park, J., Kerner, A., Burns, M. A., & Lin, X. N. (2011). Microdroplet-enabled highly parallel co-cultivation of microbial communities. *PLoS One, 6*(2), 17019. https://doi.org/10.1371/journal. pone.0017019

Park, H., Cooke, M. E., Lacombe, J.-G., Weber, M. H., Martineau, P. A., Nazhat, S. N., & Rosenzweig, D. H. (2022). Continuous two-phase in vitro co-culture model of the enthesis. https://doi.org/10.1101/2022.02.07.479445

Polizzi, K. M., Goers, L., & Freemont, P. (2014). Co-culture systems and technologies: Taking synthetic biology to the next level. *Journal of the Royal Society Interface, 11*, 20140065. https://doi.org/10.1098/rsif.2014.0065

Rollié, S., Mangold, M., & Sundmacher, K. (2012). Designing biological systems: Systems engineering meets synthetic biology. *Chemical Engineering Science, 69*(1), 1–29. https://doi. org/10.1016/J.CES.2011.10.068

Scott-Drechsel, D., Su, Z., Hunter, K., Li, M., Shandas, R., & Tan, W. (2012). *A new flow co-culture system for studying mechanobiology effects of pulse flow waves* (Vol. 64, p. 649). https://doi. org/10.1007/s10616-012-9445-2

Sebrell, T. A., Hashimi, M., Sidar, B., Wilkinson, R. A., Kirpotina, L., Quinn, M. T., Malkoç, Z., Taylor, P. J., Wilking, J. N., & Bimczok, D. (2019). A novel gastric spheroid co-culture model reveals chemokine-dependent recruitment of human dendritic cells to the gastric epithelium. *Cellular and molecular gastroenterology and hepatology, 8*(1), 157–171. https://doi.org/10.1016/j.jcmgh.2019.02.010

Shafiee, S., Shariatzadeh, S., Zafari, A., Majd, A., & Niknejad, H. (2021). Recent advances on cell-based co-culture strategies for Prevascularization in tissue engineering. *Frontiers in Bioengineering and Biotechnology, 9*, 1155. https://doi.org/10.3389/FBIOE.2021.745314/BIBTEX

Soriano, L., Khalid, T., O'brien, F. J., O'leary, C., & Cryan, S.-A. (2021). A tissue-engineered tracheobronchial in vitro co-culture model for determining epithelial toxicological and inflammatory responses. *Biomedicines, 9*, 631. https://doi.org/10.3390/biomedicines9060631

Stagg, S. M., Gürkan, C., Fowler, D. M., LaPointe, P., Foss, T. R., Potter, C. S., Carragher, B., & Balch, W. E. (2006). Structure of the Sec13/31 COPII coat cage. *Nature, 439*(7073), 234–238. https://doi.org/10.1038/nature04339

Stahl, A., Wenger, A., Weber, H., Stark, G. B., Augustin, H. G., & Finkenzeller, G. (2004). Bi-directional cell contact-dependent regulation of gene expression between endothelial cells and osteoblasts in a three-dimensional spheroidal coculture model. *Biochemical and Biophysical Research Communications, 322*(2), 684–692. https://doi.org/10.1016/J.BBRC.2004.07.175

Tan, Z. Q., Leow, H. Y., Lee, D. C. W., Karisnan, K., Song, A. A. L., Mai, C. W., Yap, W. S., Lim, S. H. E., & Lai, K. S. (2019). Co-culture Systems for the Production of secondary metabolites: Current and future prospects. *The Open Biotechnology Journal, 13*(1), 18–26. https://doi.org/10.2174/1874070701913010018

Tremblay, P.-L., Erie Hudon, V., & Fran, F., Berthod, F., Germain, L., & Auger, F. A. (2005). Inosculation of tissue-engineered capillaries with the Host's vasculature in a reconstructed skin transplanted on mice. *American Journal of Transplantation, 5*, 1002–1010. https://doi.org/10.1111/j.1600-6143.2005.00790.x

Unger, R. E., Krump-Konvalinkova, V., Peters, K., & James Kirkpatrick, C. (2002). In vitro expression of the endothelial phenotype: Comparative study of primary isolated cells and cell lines, including the novel cell line HPMEC-ST1.6R. *Microvascular Research, 64*(3), 384–397. https://doi.org/10.1006/MVRE.2002.2434

Vis, M. A. M., Ito, K., & Hofmann, S. (2020). Impact of culture medium on cellular interactions in in vitro co-culture systems. *Frontiers in Bioengineering and Biotechnology, 8*, 911. https://doi.org/10.3389/fbioe.2020.00911

Wang, Y. I., Oleaga, C., Long, C. J., Esch, M. B., McAleer, C. W., Miller, P. G., Hickman, J. J., & Shuler, M. L. (2017). Self-contained, low-cost body-on-a-chip systems for drug development. *Experimental Biology and Medicine, 242*(17), 1701–1713. https://doi.org/10.1177/1535370217694101

Williams, C., & Wick, T. M. (2005). Endothelial cell-smooth muscle cell co-culture in a perfusion bioreactor system. *Annals of Biomedical Engineering, 33*(7), 920–928. https://doi.org/10.1007/s10439-005-3238-0

Zhang, C., Bakker, A. D., Klein-Nulend, J., & Bravenboer, N. (1914). Studies on osteocytes in their 3D native matrix versus 2D in vitro models. *Current Osteoporosis Reports, 17*, 207–216. https://doi.org/10.1007/s11914-019-00521-1

Zhang, C., Zhao, Z., Abdul Rahim, N. A., van Noort, D., & Yu, H. (2009). Towards a human-on-chip: Culturing multiple cell types on a chip with compartmentalized microenvironments. *Lab on a Chip, 9*(22), 3185–3192. https://doi.org/10.1039/B915147H

Zhang, Y., Guo, W., Wang, M., Hao, C., Lu, L., Gao, S., Zhang, X., Li, X., Chen, M., Li, P., Jiang, P., Lu, S., Liu, S., & Guo, Q. (2018). Co-culture systems-based strategies for articular cartilage tissue engineering. *Journal of Cellular Physiology, 233*(3), 1940–1951. https://doi.org/10.1002/JCP.26020

Zhang, L., Yang, G., Johnson, B. N., & Jia, X. (2019). Three-dimensional (3D) printed scaffold and material selection for bone repair. *Acta Biomaterialia, 84*, 16–33. https://doi.org/10.1016/j.actbio.2018.11.039

Zhu, S., Ehnert, S., Rouß, M., Häussling, V., Aspera-Werz, R. H., Chen, T., & Nussler, A. K. (2018). From the clinical problem to the basic research-co-culture models of osteoblasts and osteoclasts. *International Journal of Molecular Sciences, 19*. https://doi.org/10.3390/ijms19080000

Chapter 14
3D Cell Culture Techniques

Madhu Rani, Annu Devi, Shashi Prakash Singh, Rashmi Kumari, and Anil Kumar

14.1 Introduction

The detailed studies that are conducted on the formation, function and pathogenesis of tissues and organs using cell culture systems and animal models are beneficial for managing any pathological condition like cancer and neurodegenerative diseases (Kapałczyńska et al., 2018; Koledova, 2017). Traditional two-dimensional (2D) cell culture models or animal model systems are widely used for better understanding the formation, function of tissue/organ under normal and diseased conditions. It has been inferred by various studies that the 2D cell culture models have helped in understanding the fundamental concepts of biological processes. The cells produced on a flat 2D surface (such as polystyrene substrates) are obserevd to be dramatically different from cells grown in 3D cell culture systems in terms of their shape, cell-cell interactions, cell-matrix interactions, and cellular differentiation (Freshney, 2015; Koledova, 2017). Experimental studies on animal models offer relevant information about specific molecules and biological processes. However, regular disparities have also been observed in results that were gathered from studies of desired gene and protein expression profiles. Besides this, animal model systems also could not reflect the appropriate features of human tumours, therapeutic drug responses, stem cell differentiation and autoimmune diseases (Yamada & Cukierman, 2007;

M. Rani · A. Devi
Genome Biology Laboratory, Department of Biosciences, Jamia Millia Islamia, New Delhi, India

S. P. Singh · R. Kumari
Special Centre of Molecular Medicine, Jawaharlal Nehru University, New Delhi, India

A. Kumar (✉)
Rani Lakshmi Bai Central Agricultural University, Jhansi, Uttar Pradesh, India

© The Author(s), under exclusive license to Springer Nature Switzerland AG 2023   197
S. Mani et al., *Animal Cell Culture: Principles and Practice*, Techniques in Life Science and Biomedicine for the Non-Expert,
https://doi.org/10.1007/978-3-031-19485-6_14

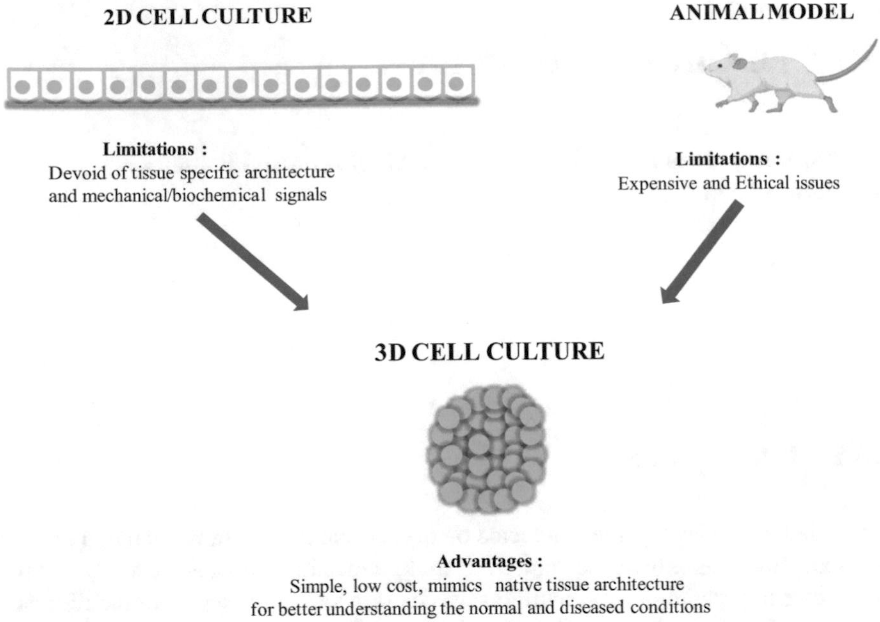

2D CELL CULTURE

Limitations :
Devoid of tissue specific architecture
and mechanical/biochemical signals

ANIMAL MODEL

Limitations :
Expensive and Ethical issues

3D CELL CULTURE

Advantages :
Simple, low cost, mimics native tissue architecture
for better understanding the normal and diseased conditions

Fig. 14.1 A comparison between two-dimensional cell culture, animal model system and 3D cell culture methods. On the one hand, the 2D cell culture model lacks tissue architecture as well as biochemical/mechanical signals, whereas in vivo models require high costs for maintenance and are also associated with ethical concerns. Therefore, 3D cell culture systems are recognised as a link between 2D cell culture and animal model systems. It also mimics native tissue architecture, which helps in better understanding the physiology of normal and diseased conditions

Justice et al., 2009). Henceforth, in vitro 3D cell culture models are recognised as a third approach that works efficiently over the pitfalls of traditional cell culture techniques and animal model systems (Fig. 14.1).

3D cell culture models have made a revolutionary path towards gathering information about biological processes in in vivo molecular mechanisms. Further, these models have also enabled advancement in cellular and molecular biological studies for the cellular growth, cell proliferation and profile expression studies of genes and proteins. The popularity of 3D models is due to their capability to mimic specific environments and tissues, which facilitated their use in applications like advances in tissue engineering, development and screening of new therapeutics against pathological conditions (Edmondson et al., 2014).

The rationale behind this book chapter is to discuss the general principles, ideas and cautions towards the use of in vitro 3D cell culture systems for enhancing knowledge in morphogenesis of tissues and pathological conditions, mainly carcinogenesis. In the present chapter, we deliberate the different types of 3D cell culture methods that help in gaining information about cell growth, cell proliferation, growth conditions and expression profiles of genes and proteins. We will also intend to compare the characteristics of the 3D cell culture and two-dimensional (2D) monolayer culture in order to perceive the advancement of 3D culture over the traditional cell culture. Finally, we will also connote the applications of 3D cell culture models and challenges faced by researchers and medical professionals for their use in personalised medicine.

14.2   3D Cell Culture Versus 2D Cell Culture Systems

Earlier experimental studies conducted on 2D cell cultures in vitro for disease model conditions, including cancer, inferred that the traditional cell cultures have many limitations like changes in morphology, polarity, a flaw in interactions between cellular and extracellular environments and a division method. These limitations make their limited use in regular experimental research in the area of drug screening, drug discovery etc. Therefore, 3D cell culture systems show up-gradation in the context of real reflection of cellular organ system and pathological conditions. These modern systems have also facilitated the study of biomarkers and targeted therapies against diseases significantly. In Table 14.1, we have compiled the comparison of properties of 3D versus 2D culture. The diagrammatical comparison of 3D versus 2D cell culture is also shown in Fig. 14.1.

Table 14.1 Shows the comparative analysis of 3D cell culture systems and the traditional 2D cell culture model

Type of culture	The 3D cell culture system	2D cell culture model
Time taken for culture formation	From a few hours to few days	From minutes to few hours
In vivo initiation	In vivo tissues and organs are in 3D form.	Not reflect the native structure of tissue or tumour mass.
Quality of culture	Lacks significant performance and reproducibility. So, it is difficult to interpret and also culturing is not easy.	Exhibits significant performance and reproducibility, long-term culture in maintenance. So, it is easy to interpret due to their simplicity.
Cell interactions	Exhibits significant interactions of cell-cell and cell-extracellular environment, environmental niches	Lack of inter-cellular and intra-cellular interactions and also in vivo-like microenvironment and niches
Cellular characteristics	Native morphology and way of division, diverse phenotype and polarity	Altered morphology and way of division; loss of diverse phenotype and polarity
Accessibility towards crucial compounds	Approachable contents such as oxygen, nutrients, metabolites and signalling molecules are inconsistent as in vivo.	Unlimited approachability towards the oxygen, nutrients, metabolites and signalling molecules (in contrast to in vivo)
Molecular mechanisms	Possesses expression of genes, mRNA splicing, topology and biochemistry of cells as in vivo	Exhibits alterations in gene expression, mRNA splicing, topology and biochemistry of cells (in contrast to in vivo)
Cost of Maintenance a culture	Expensive, time-consuming, Fewer commercially available tests	Cheap, commercially available tests and the media

Adapted and modified Kapałczyńska et al. (2018)

14.3   Overview of 3D Cell Culture Techniques

Various approaches were recognised like whole animals, organotypic explants culture, cell spheroids, tissue-engineered models and micro-carrier cultures for the use of 3D cell culture in many applications (Carletti et al., 2011). These models require both scaffolds for culturing and scaffold-free cultures. Organotypic explants are mainly used in those studies where complete information is required in whole animals such as *Drosophila melanogaster* and zebrafish. These models provide crucial data wherein the cells are physically located in their native niche. The growth conditions can be varied in cell culture studies on animal models like *Drosophila melanogaster* and zebrafish. However, those involving mouse embryos require highly controlled conditions such as the pH, temperature and oxygen levels should be highly precise (Corrò et al., 2020).

Organ explantation is mainly done for brain and neural tissues. The explanted tissue is grown on gels or semi-permeable membranes in the presence of a growth medium or isotonic solutions. These systems can stably maintain the tissue architecture; but time availability is an essential factor. Differentiated cells are also present in these culture systems. Although, deep imaging of these issues presents a significant challenge.

One of the most common ways of 3D cell culture is the spherical cellular aggregation model known as spheroids. This model does not use any scaffold model and is easily visualised by using imaging techniques such as light or fluorescence microscopy. Henceforth, this model has been found to be very crucial for experimental research on cancer and therapeutics studies. Some spheroids such as multicellular tumour spheroids (MCTS), mammospheres, neurospheres, hepatospheres and embryoid bodies are recognised for studying the different pathological conditions. These models have been ideally used for the study of solid tumour models. The best characterised is the MCTS. These tumour spheroid models have helped a lot in better understanding cancer mechanism and in the development of a highly sensitive cancer drug testing platform. These models are highly beneficial as they mimic the in vivo tumours very well in terms of their morphology, cell proliferation, aeration, nutrient uptake and drug intake.

14.4   Methods of 3D Cell Culture Techniques

Currently, different types of 3D cell culture techniques are available at the global level. Broadly these can be categorised into two main types; scaffold-based techniques and scaffold-free techniques. These techniques are discussed in the below section:

14.4.1   Scaffold-Based Techniques

As the size and complexity of 3D culture increase, it demands a need for scaffolds or matrices. In scaffold-based 3D culturing, the cells are grown on substrates that represent the extracellular matrices making this system a lot closer to the native environment. The porosity of scaffolds promotes the transportation of oxygen, nutrients and waste, enabling cells for their easy proliferation and migration within the scaffold and eventually adherence to it. There are two scaffold categories – in vitro 3D scaffolds used for cell culturing and experimental applications such as drug testing. The second is biomedical engineering scaffolds that are exploited as supports for tissue regeneration applications. Scaffolds can be in the form of *hydrogels*, membranes and *3D matrices*. Based upon their sources, they can be divided as natural and synthetic scaffolds too (Table 14.2).

Hydrogels Scaffolds are basically polymers of hydrophilic molecules that possess a large amount of water without dissolution. Hydrogels offer several benefits over solid scaffolds, such as these provide a hydrated environment that is favourable for cell colonisation and infiltration and also high resemblance with the tissues (Rodrigues et al., 2015). The biomaterial used for designing such scaffolds plays a crucial role as these provide a 3D template for cell adhesion, proliferation and differentiation. The scaffolds could be made of different biomaterials either of natural origin (obtained from plant, animal or human tissues) such as fibrin, fibroin, glycosaminoglycans (GAGs), alginate, gelatin, chitosan, hyaluronic acid and collagen or could be synthetic such as polyethylene glycol (PEG), poly –L- lactic acid (PLLA), polycaprolactone (PCL) and polylactic acid-co–caprolactone (Afewerki et al., 2019). The cell type and the nature of the study are essential when choosing the type of scaffold.

The most commonly used matrix in 3D cultures is *collagen*, as they require low-cost maintenance, high biocompatibility and biodegradability and a very low antigenicity. The structure of collagen matrices (such as the pore size, stiffness) is also easily manipulated. This could be achieved by altering its concentration or introducing new cross-linking compounds such as glutaraldehyde and formaldehyde or physical treatment like ultraviolet or gamma irradiation (Lee et al., 2001; Ravi et al., 2015).

Table 14.2 Biomaterials used for scaffolding

Natural polymers	Synthetic polymers
Fibrin	Polyethylene glycol (PEG)
Fibroin	Poly –L- lactic acid (PLLA)
Glycosaminoglycans (GAGS)	Polycaprolactone (PCL)
Alginate	Polylactic acid- co–caprolactone
Gelatin	
Chitosan	
Hyaluronic acid	
Collagen	

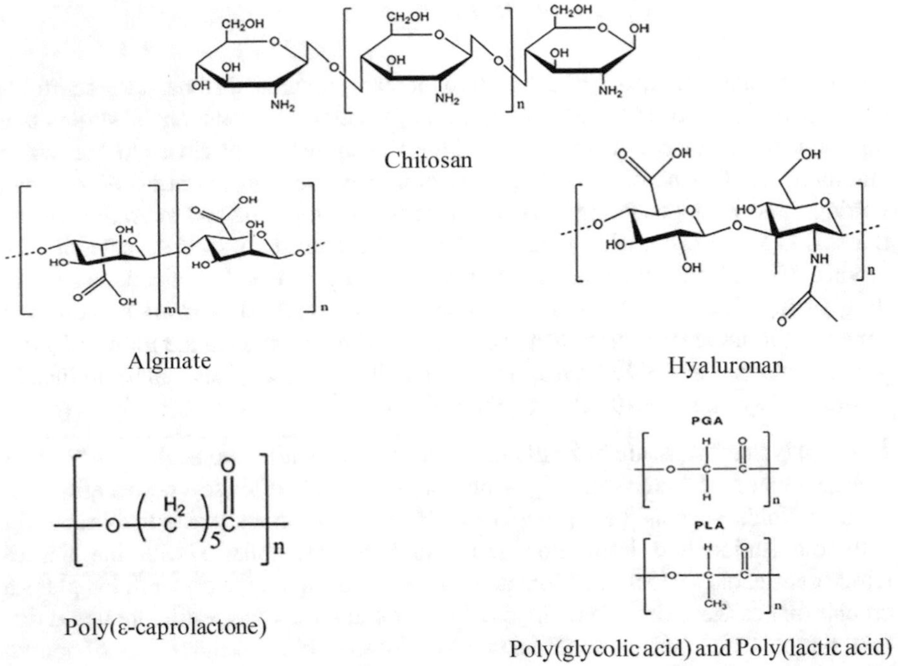

Fig. 14.2 It depicts the chemical formulas of various natural and synthetic polymers that are used for the hydrogel scaffolds

A standard method for making collagen scaffolds is freeze-drying (Carletti et al., 2011). Collagen scaffolds are used extensively for culturing osteoblast and chondrocytes and have major biomedical applications (Lee et al., 2001; Zhou et al., 2006; Negri et al., 2007) (Fig. 14.2).

Chitosanisan N-deacetylated derivative of chitin, which is found in the exoskeleton of crustaceans and insects. Chitosan is sensitive to enzymes such as chitonase and lysozyme, and the degree of degradation depends on the acetyl content. The mechanical properties of chitosan are affected by the molecular weight and degree of deacetylation (Carletti et al., 2011). Chitosan hydrogels can be prepared by ionic bonding or covalent cross-linking. Chitosan-based scaffolds have been mainly used for culturing chondrocytes for cartilage regeneration (Rogina et al., 2021). Chitosan-chondroitin sulfate scaffolds have been used to study the tumour microenvironment in prostate cancer (Xu et al., 2020). The application of chitosan-based scaffolds has also been used in improving the bioavailability of molecules. Epigallocatechin-3-gallate (EGCG) is used to differentiate mesenchymal stem cells to osteoblast, but as it is metabolised during cell culture, its bioavailability is reduced. Therefore, CS (Chitosan)/Alg (alginate)-ECN (EGCG-Chitosan nanoparticles) scaffolds have been designed for improving their bioavailability. In this type of scaffold, EGCG is loaded onto chitosan nanoparticles which are then entrapped into a chitosan alginate scaffold (Wang et al., 2021).

Glycosaminoglycans such as hyaluronic acid have also been used as a scaffold for the culture of chondrocytes, bone and skin (Carletti et al., 2011). Hyaluronic acid hydrogels have been used for delivering growth factors that are osteoinductive and angiogenic for bone tissue engineering (Rodrigues et al., 2015).

Gelatin is also used as a biomaterial for scaffolds as it shows chemical similarity to the extracellular matrix in native tissues, low cost-effectiveness, bioavailability and compatibility and low antigenicity. However, it has low solubility in concentrated aqueous media, is highly susceptible to enzymatic digestion and has a high viscosity and poor mechanical properties. Therefore, it has limited use for 3D cell culture. However, to overcome this limitation, we can synergistically use it with a wide range of polysaccharides resulting in gelatin-composite hydrogels (Afewerki et al., 2019).

Apart from the natural polymers, many synthetic polymers have also been used as scaffolds due to their high reproducibility and versatility. The processing of synthetic polymer is relatively easy in contrast to a natural polymer. The functional group of synthetic polymers can be designed due to which their structure and properties can be easily modified. For example, their degradation rate can be easily modified depending upon the molecular weight and chemical composition. This is advantageous as they can be designed according to specific applications (Donnaloja et al., 2020). However, their biocompatibility and bioactivity are less than natural polymers. So far, the most extensively used synthetic polymers for scaffolds are polyglycolic acid (PGA), polylactic acid (PLA) and polycaprolactone (PCL) (Fig. 14.2).

Among all the synthetic polymers, PCL is non-toxic in nature and a biodegradable aliphatic polyester. Its degradation rate is slow; however, it has hydrophobic properties, which create problems in cell adhesion and penetration (Donnaloja et al., 2020). To overcome these hurdles, several co-polymerisation techniques such as PCL/alginate composite scaffolds have been developed (Fig. 14.2).

14.4.2   Scaffold-Free Techniques

This category includes those techniques where cells are grown without any solid support. It includes the *forced floating method*, the *hanging drop method* and the *agitation-based methods* which are discussed below:

In the *forced floating method*, the vessel's surface is modified in a way such that the cells are not able to attach to it, which results in their forced floatation. This is a straightforward method for the formation of 3D spheroids as the floating of the cells increases the cell to cell contact promoting the sphere formation. It has been mainly used for the development of 3D spheroids in 96-well plates. In this approach, plates were coated with 0.5% poly-2-hydroxyethyl methacrylate (poly-HEMA) and dried for 3 days. This type of coating prevents the attachment of cells to the plate surface.

After this, cells were added to the plates and centrifuged, which results in adherence of cells with each other and further leads to the formation of spheroids. This has been used for the generation of spheroids of many cancerous as well as non-cancerous cells (Breslin & O'Driscoll, 2013; Kelm et al., 2003). This method is very simple and is highly reproducible. The spheroids generated in this method are highly consistent as equal numbers of cells are added to each well. This method is highly accessible for experimental studies, and the size of the spheroids can easily be modified as per the required number of cells for seeding. This has been extensively used for drug testing. Apart from the HEMA method, 1.5% agarose is also used. This also results in the formation of a thin layer on the surface of the plate, which prevents the attachment of cells. This method is also relatively simple and can be used for the long-term culture of cells. The only disadvantage of this method is that an extra step is added before culturing of cells which increases the work and time required for coating the plates. However, there are commercially pre-coated plates available, but they increase the overall cost.

In the *hanging drop method*, a small aliquot of single-cell suspension is added in 60-well microwell mini trays. After the addition of cells, the plate is kept inverted, and the cell aliquots become hanging drops and remain in place due to the surface tension. Cells are accumulated at the tip of the drop, and they continue to proliferate (Kelm et al., 2003; Timmins & Nielsen, 2007). Here, also similar to the forced floating method, the size of the spheroid can be adjusted depending upon the cells seeded. Spheroids produced by this method are tightly packed. This method is very simple for use and also highly reproducible. The only possible limitation with this method is that the volume of liquid used to generate spheroids is very low. This is because the surface tension that holds the drop does not support larger volumes (Kelm et al., 2003).

In the *agitation-based method*, cell suspensions are kept in a container that remains in continuous motion. This can be achieved by either a continuous stirring of the cells or the container is kept in rotation. Due to the continuous motion, cells tend to form cell to cell communication and adhere to each other and not to the surface of the container (Kelm et al., 2003). This method is generally of two types: spinner flask bioreactors and rotational culture systems. *Spinner flask bioreactors* possess a container for holding the cell suspension, which has a stirring element. This ensures the continuous stirring of the cell suspension. The size of the spheroid correlates with the volume of the container and can be varied. The advantage of using spinners is that these allow the long-term culturing of spheroids as the culture medium can be changed frequently. The continuous motion of the fluids also aids in the transport of nutrients to and wastes from the spheroids. The drawback of such a system is that the force experienced by the cells as a result of continuous stirring might affect cellular physiology. Rotating culture systems function similarly to the spinners except that there is no stirring rod, but the container itself rotates. In 1992, NASA developed a rotating wall vessel bioreactor for maintaining low shear force on cells in culture. This vessel consists of a chamber that is utilised for culturing of cells. The chamber is attached to a rotator which slowly rotates about a horizontal axis (Breslin & O'Driscoll, 2013; Barrila et al., 2010).

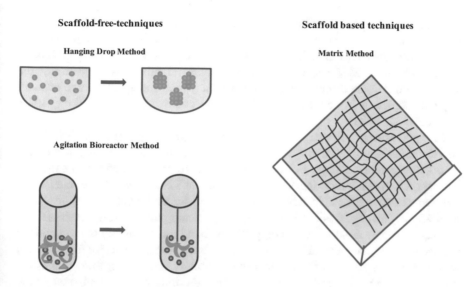

Fig. 14.3 Scaffold-based and scaffold-free techniques – 3D matrix method, hanging drop method and agitation bioreactor method

Stem cells in 3D cell culture have also been recognised in application of various biological therapeutic approaches like cell-based therapy, screening of drugs, regenerative medicine and high-throughput pharmacology. The unique properties of stem cells such as ability of self-renewal, multi-potency and clonality enable them to special use in many research areas. Human mesenchymal stem cells are mainly used due to their sectary nature for generation of spheroids culture to understand the gene expression profiles of normal physiological and diseased conditions.

All the scaffold-free and scaffold-based techniques are summarised in Fig. 14.3.

14.5   Stem Cells in 3D Spheroids and Organoids

The use of stem cells in 2D cell culture for clinical application is ineffective due to the inability of 2D culture to reflect the actual microenvironment of stem cells. Therefore, the decrease in the replicative ability of MSCs has been observed in 2D culture overtime invalidating the crucial studies. When stem cells are cultured in spheroids 3D cell culture, MSCs show a different phenotype than 2D-cultured MSCs. MSCs also have different gene expression patterns in spheroids than in 2D; e.g. MSCs show upregulation of various genes related to angiogenesis, stress response, hypoxia, inflammation, redox signalling etc. Henceforth, MSC in spheroids leads to upgradation in the development of therapeutics. Anti-inflammatory properties, reparative effect and better post-transplant survival of MSCs have also been observed in spheroid cultures. The effectiveness of MSC spheroids over MSCs in 2D culture has been observed if MSC spheroids were administrated into the kidney of mole rats with ischemia reperfusion-induced acute kidney injury. The post-administration of MSCs spheroids showed more effectiveness in guarding the kidney to curb tissue damage, bolstering vascularisation and apoptosis and enhancing renal function in contrast to MSCs of 2D cultures (Jensen & Teng, 2020; Sakalem et al., 2021).

Pluripotent stem cells are used to generally grow as organoids for their future application in analogous tissue for transplantation in humans. According to a recent report, pluripotent stem cells from renal organoids are transplanted beneath the kidney capsules of adult mice. It has been observed that these organoids are acted as the same structures like that of a kidney in vivo. Therefore, under transplantation, glomeruli are vascularised swiftly, indicating a significant path towards developing an alternative kidney replacement therapy. It was found that the ability of organoids to represent different regions of the body makes them a crucial aspect for the study the genetic diseases for better understanding. Similarly, kidney organoid for cystic fibrosis disease model has been used to study the influences of trans-membrane conductance regulator-modulating entities and more tubular organoids used for kidney disease model where microenvironment played a crucial role in cyst formation. Organoids are also recognised for their study in difficult diseased model neurodegenerative diseases like Alzheimer and Parkinson disease. The study revealed that the brain organoids prepared from the pluripotent stem cells taken from the Alzheimer patients show significant improvement when treated with β- and γ-secretase inhibitors, which displayed significant therapeutic influences (Jensen & Teng, 2020; Sakalem et al., 2021; Chaicharoenaudomrung et al., 2019).

The medical patients who are diagnosed with a chance of organ failure require a critical care and organ transplantation treatment. The critical examination of artificial tissues and organs is very necessary for successful transplantation. Therefore, 3D bioprinting emerges as a crucial technique for the fabrication of artificial tissues and organs in three-dimensional mode before transplantation for their correct manufacturing pattern of living cells in a tissue-specific manner in which cells arrange layer by layer. This advanced medical technology has made a huge impact and saves the lives of many patients who receive successful organ transplantation.

3D Bioprinting It is the most recently developed technique of 3D cell culture in which, under the computational approach, customised 3D structures are constructed, and materials will be printed out, solidified and connected together. It has wide applications like 3D art and design, prototype, industrial manufacturing and architecture, but mainly known for tissue engineering and regenerative medicine (Mazzocchi et al., 2019; Ryan et al., 2016).

14.6   Applications of 3D Cell Culture Techniques

3D Cell culture mimics the tissues and tumour structure, thus aiding in simulating the patho-physiological microenvironment of disease or outgrowth. The structural organisation like cell-cell interaction, cell-extracellular matrix (ECM) interaction, differentiation and cumulative response to drug therapy provides valuable insight into 3D cell culture models (Ravi et al., 2015; Ryan et al., 2016) (Fig. 14.4; Table 14.3). In 2D culture, the true ECM is absent. Thus, the interaction with the cells is poorly studied. ECM plays a crucial role in the adhesive properties of cells, mechano-transduction and exposure of levels of toxic compounds and soluble factors present in media. The crucial interaction of normal cells and neighbouring cells with tumour cell physiology can be studied deeply by this model. The rate of division of cells in 2D models is higher as compared to in vivo models, which show more relevance to 3Dcell culture models (Chitcholtan et al., 2013). Peela et al. have prepared a 3D tumour model by integrating the gelatin methacrylate (GeIMA) hydrogel with a two-step photolithography technique (Peela et al., 2016). They proposed that this model shows more resemblance to tumour stiffness and architecture, thus being utilised in discovering biomarkers and targeted chemotherapy. The advancement in 3D cell culture in future may lead to superior models depicting kidney, liver etc., thus enhancing the effectiveness and rapid screening of toxic compounds in an in vitro setting. The characterisation of drugs and probable drug-drug interaction and ADMET (Adsorption, distribution, metabolism, elimination, toxicity) studies in 3D cell culture might help in a vast number of successful compounds as potential drugs (Godoy et al., 2013). Jingyun Ma et al. also suggested a 3D model for glioblastoma. They found that different drugs like temozolomide and resveratrol

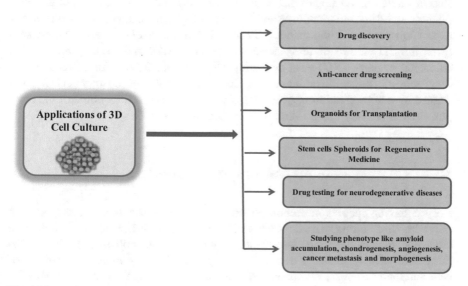

Fig. 14.4 Applications of a 3D cell culture system

Table 14.3 Application of 3D cell culture models in the study of human cancers and other pathological conditions

3D cell culture models	Applications in study of human cancers	References
3D-bioprinted agarose-alginate scaffold based, Hanging drop	Non-small cell carcinoma model	Ziółkowska-Suchanek (2021) and Amann et al. (2014)
Fibrous scaffold based	Head and neck cancer model	Young et al. (2018)
Nested hydrogel based	Osteosarcoma model	Pavlou et al. (2019)
Collagen matrix based	Glioblastoma model	Ma et al. (2018)
Fibrous scaffold based	Breast cancer model	Young et al. (2018)
Matrigel based	Lung cancer model	Pavlou et al. (2019)
Tumoroids	Colorectal cancer	Nörz et al. (2021)
3D spheroid formation with methylcellulose	Pancreatic ductal adenocarcinoma (PDAC) model	Longati et al. (2013)
3D Hanging drop	Malignant serous effusion (MSE) model	Wu et al. (2020)
Microfluidic 3D cell culture model	Vasculature study	van Duinen et al. (2015)
Silicate fibre based	Colorectal cancer	Yamaguchi et al. (2013)
3D extracellular matrix based	Antiviral drug screening	Koban et al. (2018)

in the detachable microfluidic device act similarly as in in vivo models as they inhibit the viability, proliferation and invasiveness (Ma et al., 2018). In the 2D model, the hypoxic conditions do not usually resemble the in vivo condition, but in 3D cell culture, the heterogeneous spheroidal structure allows hypoxic conditions. Recent studies on 3D models that focused on angiogenic response in hypoxic conditions observed the remarkable dysregulation in morphogenic changes and microRNA expression patterns in the model (Klimkiewicz et al., 2017). A recent report analyzed the 3D culture of drug sensitivity against AKT and mTOR inhibitor on colorectal cancer (CRC) drug screening (Nörz et al., 2021). An FDA-approved drug, gefitinib, a specific inhibitor of epidermal growth factor receptor (EGFR), was studied by a 3D cell culture system. It showed significant efficacy against the cowpox virus in primary human keratinocytes (Koban et al., 2018).

14.7 Challenges in 3D Culture Techniques

Despite having success in monolayer culture in drug discovery, it is evident that only a handful of drugs pass through the clinical phase II & III trials. Although there are many reasons, the main reason is its response and metabolism in in vivo conditions. The primary phase of drug screening is based on the monolayer culture of cells and their response. Recent advances in culture techniques enabled us a new 3D model of cell culture, which can yield valuable insight into drug screening and its metabolism. The cost of maintaining 3D cell culture is higher as compared to the monolayer cell culture system. The various 3D cell culture techniques models allow for the transfer of experimental information to fix issues that arose in their handling and upkeep. Although the importance of the 3D cell culture model is well recognised so far, it poses many challenges that make it difficult to adapt as routine culture by researchers and medical professionals. One of the major drawbacks of a 3D cell culture system is its scalability. Current formats of 3D cell culture do not provide a solution to scale up in 384- and 1536-well plates. The techniques require automated pipettes to reduce the speed of media aspiration and dispensing of media in culture plates. This step is critical as this will impact the loss of 3D cell mass and the unwanted movement of cellular mass. The user should be attentive during pipetting as viscous liquids such as collagen and Matrigel should not be pipetted out. This technique requires special training and is labour-intensive, as well as the cost of culturing is more compared to monolayer cell culture. The biggest challenges are its disease relevance and compatibility to various detection instruments.

The visualisation of 3D structure in optical microscopy is less informational as the differential light absorption and light penetration delays the image acquisition time and complex geometrical shapes. The fluorescence microscopy also depicts the structural flaw, as the structure must obtain z-stacking to process the image output. Flow cytometry also faces challenges in 3D structures, as it was principled to single-cell movement on detectors. The techniques that can sort microorganisms, different cell populations and biomarkers are not valuable for 3D structures. If the user wants to use 3D structure in FACS, then the structure should be dissociated into single-cell suspension, and the 3D structure integrity will be lost. The dissociation of a 3D structure requires trypsinization, collagenase and mechanical disruptions. This process is time-consuming and also it limits the viability of cells even after sorting.

14.8   Conclusion

Three-dimensional cell culture emerges as a link between 2D cell culture in vitro system and in vivo animal models. Among scientific workers, the 3D cell culture techniques are increasingly becoming famous due to the ease of handling cells in vitro while obtaining data that mimic in vivo conditions and prevent concern of ethical clearance in case of animal usage. The 3D cell culture techniques are increasingly becoming crucial for in vitro modelling of basic developmental processes and human diseases. 3D culture enhances the knowledge about understanding mammalian organogenesis and carcinogenesis. Their application might also be helpful for drug testing and personalised medicine regularly. Several obstacles remain to be unravelling in the use of 3D cell culture systems for physiological studies like the inclusion of all different cell types of complex tissue, realistic modelling of regional differences in ECM composition, rigidness and arrangements of the sturdy vascular system that facilitates the production of larger tissue constructs. This chapter provides information on 3D culture versus 2D culture, types of 3D cell culture techniques and their uses in various applications.

14.9 Future Prospects

We hope this detailed information on protocols of 3D cell culture will be helpful for researchers and graduate students in their cell culture studies. This chapter might be an inspiration and a starting point in developing new protocols and experimental designs as per the requirement for the specific scientific needs for advanced research in 3D cell culture.

References

Afewerki, S., Sheikhi, A., Kannan, S., Ahadian, S., & Khademhosseini, A. (2019). Gelatin-polysaccharide composite scaffolds for 3D cell culture and tissue engineering: Towards natural therapeutics. *Bioengineering & Translational Medicine, 4*(1), 96–115.

Amann, A., Zwierzina, M., Gamerith, G., Bitsche, M., Huber, J. M., Vogel, G. F., Blumer, M., Koeck, S., Pechriggl, E. J., Kelm, J. M., Hilbe, W., & Zwierzina, H. (2014). Development of an innovative 3D cell culture system to study tumour--stroma interactions in non-small cell lung cancer cells. *PloS one, 9*(3), e92511.

Barrila, J., Radtke, A. L., Crabbé, A., Sarker, S. F., Herbst-Kralovetz, M. M., Ott, C. M., & Nickerson, C. A. (2010). Organotypic 3D cell culture models: Using the rotating wall vessel to study host-pathogen interactions. *Nature Reviews Microbiology, 8*(11), 791–801.

Breslin, S., & O'Driscoll, L. (2013). Three-dimensional cell culture: The missing link in drug discovery. *Drug Discovery Today, 18*(5–6), 240–249.

Carletti, E., Motta, A., & Migliaresi, C. (2011). Scaffolds for tissue engineering and 3D cell culture. *Methods in Molecular Biology (Clifton, N.J.), 695*, 17–39.

Chaicharoenaudomrung, N., Kunhorm, P., & Noisa, P. (2019). Three-dimensional cell culture systems as an in vitro platform for cancer and stem cell modeling. *World Journal of Stem Cells, 11*(12), 1065–1083.

Chitcholtan, K., Asselin, E., Parent, S., Sykes, P. H., & Evans, J. J. (2013). Differences in growth properties of endometrial cancer in three dimensional (3D) culture and 2D cell monolayer. *Experimental Cell Research, 319*(1), 75–87.

Corrò, C., Novellasdemunt, L., & Li, V. (2020). A brief history of organoids. *American Journal of Physiology – Cell Physiology, 319*(1), C151–C165.

Donnaloja, F., Jacchetti, E., Soncini, M., & Raimondi, M. T. (2020). Natural and synthetic polymers for bone scaffolds optimization. *Polymers, 12*(4), 905.

Edmondson, R., Broglie, J. J., Adcock, A. F., & Yang, L. (2014). Three-dimensional cell culture systems and their applications in drug discovery and cell-based biosensors. *Assay and Drug Development Technologies, 12*(4), 207–218.

Freshney, I. R. (2015). *Culture of animal cells* (A manual of basic technique and specialized applications) (7th ed.). Wiley-Blackwell. ISBN: 9781118873373.

Godoy, P., Hewitt, N. J., Albrecht, U., Andersen, M. E., Ansari, N., Bhattacharya, S., Bode, J. G., Bolleyn, J., Borner, C., Böttger, J., Braeuning, A., Budinsky, R. A., Burkhardt, B., Cameron, N. R., Camussi, G., Cho, C. S., Choi, Y. J., Craig Rowlands, J., Dahmen, U., Damm, G., &

Hengstler, J. G. (2013). Recent advances in 2D and 3D in vitro systems using primary hepato-cytes, alternative hepatocyte sources and non-parenchymal liver cells and their use in investigating mechanisms of hepatotoxicity, cell signaling and ADME. *Archives of Toxicology, 87*(8), 1315–1530.

Jensen, C., & Teng, Y. (2020). Is it time to start transitioning from 2D to 3D cell culture? *Frontiers in Molecular Biosciences, 7*, 33.

Justice, B. A., Badr, N. A., & Felder, R. A. (2009). 3D cell culture opens new dimensions in cell-based assays. *Drug Discovery Today, 14*(1–2), 102–107.

Kapałczyńska, M., Kolenda, T., Przybyła, W., Zajączkowska, M., Teresiak, A., Filas, V., Ibbs, M., Bliźniak, R., Łuczewski, Ł., & Lamperska, K. (2018). 2D and 3D cell cultures – A comparison of different types of cancer cell cultures. *Archives of Medical Science: AMS, 14*(4), 910–919.

Kelm, J. M., Timmins, N. E., Brown, C. J., Fussenegger, M., & Nielsen, L. K. (2003). Method for generation of homogeneous multicellular tumor spheroids applicable to a wide variety of cell types. *Biotechnology and Bioengineering, 83*, 173–180.

Klimkiewicz, K., Weglarczyk, K., Collet, G., Paprocka, M., Guichard, A., Sarna, M., Jozkowicz, A., Dulak, J., Sarna, T., Grillon, C., & Kieda, C. (2017). A 3D model of tumour angiogenic microenvironment to monitor hypoxia effects on cell interactions and cancer stem cell selection. *Cancer Letters, 396*, 10–20.

Koban, R., Neumann, M., Daugs, A., Bloch, O., Nitsche, A., Langhammer, S., & Ellerbrok, H. (2018). A novel three-dimensional cell culture method enhances antiviral drug screening in primary human cells. *Antiviral Research, 150*, 20–29.

Koledova, Z. (2017). 3D cell culture: An introduction. *Methods in Molecular Biology (Clifton, N.J.), 1612*, 1–11. https://doi.org/10.1007/978-1-4939-7021-6_1

Lee, C. H., Singla, A., & Lee, Y. (2001). Biomedical applications of collagen. *International Journal of Pharmaceutics, 221*(1–2), 1–22.

Longati, P., Jia, X., Eimer, J., Wagman, A., Witt, M. R., Rehnmark, S., Verbeke, C., Toftgård, R., Löhr, M., & Heuchel, R. L. (2013). 3D pancreatic carcinoma spheroids induce a matrix-rich, chemoresistant phenotype offering a better model for drug testing. *BMC Cancer, 13*, 95.

Ma, J., Li, N., Wang, Y., Wang, L., Wei, W., Shen, L., Sun, Y., Jiao, Y., Chen, W., & Liu, J. (2018). Engineered 3D tumour model for study of glioblastoma aggressiveness and drug evaluation on a detachably assembled microfluidic device. *Biomedical Microdevices, 20*(3), 80.

Mazzocchi, A., Soker, S., & Skardal, A. (2019). 3D bioprinting for high-throughput screening: Drug screening, disease modeling, and precision medicine applications. *Applied Physics Reviews, 6*(1), 011302.

Negri, S., Fila, C., Farinato, S., Bellomi, A., & Pagliaro, P. P. (2007). Tissue engineering: Chondrocyte culture on type 1 collagen support. Cytohistological and immunohistochemical study. *Journal of Tissue Engineering and Regenerative Medicine, 1*(2), 158–159.

Nörz, D., Mullins, C. S., Smit, D. J., Linnebacher, M., Hagel, G., Mirdogan, A., Siekiera, J., Ehm, P., Izbicki, J. R., Block, A., Thastrup, O., & Jücker, M. (2021). Combined targeting of AKT and mTOR synergistically inhibits formation of primary colorectal carcinoma tumouroids in vitro: A 3D tumour model for pre-therapeutic drug screening. *Anticancer Research, 41*(5), 2257–2275.

Pavlou, M., Shah, M., Gikas, P., Briggs, T., Roberts, S. J., & Cheema, U. (2019). Osteomimetic matrix components alter cell migration and drug response in a 3D tumour-engineered osteosarcoma model. *Acta Biomaterialia, 96*, 247–257.

Peela, N., Sam, F. S., Christenson, W., Truong, D., Watson, A. W., Mouneimne, G., Ros, R., & Nikkhah, M. (2016). A three dimensional micropatterned tumor model for breast cancer cell migration studies. *Biomaterials, 81*, 72–83.

Ravi, M., Paramesh, V., Kaviya, S. R., Anuradha, E., & Paul Solomon, F. D. (2015). 3D cell culture systems: Advantages and applications. *Journal of Cellular Physiology, 230*(1), 16–26.

Rodrigues, M. T., Carvalho, P. P., Gomes, M. E., & Reis, R. L. (2015). Biomaterials in preclinical approaches for engineering skeletal tissues. In *Translational Regenerative Medicine*. Elsevier Inc.

Rogina, A., Pušić, M., Štefan, L., Ivković, A., Urlić, I., Ivanković, M., & Ivanković, H. (2021). Characterization of chitosan-based scaffolds seeded with sheep nasal chondrocytes for cartilage tissue engineering. *Annals of Biomedical Engineering, 49*(6), 1572–1586.

Ryan, S. L., Baird, A. M., Vaz, G., Urquhart, A. J., Senge, M., Richard, D. J., O'Byrne, K. J., & Davies, A. M. (2016). Drug discovery approaches utilizing three-dimensional cell culture. *Assay and Drug Development Technologies, 14*(1), 19–28.

Sakalem, M. E., De Sibio, M. T., da Costa, F., & de Oliveira, M. (2021). Historical evolution of spheroids and organoids, and possibilities of use in life sciences and medicine. *Biotechnology Journal, 16*(5), e2000463.

Timmins, N. E., & Nielsen, L. K. (2007). Generation of multicellular tumor spheroids by the hanging-drop method. *Methods in Molecular Medicine, 140*, 141–151.

van Duinen, V., Trietsch, S. J., Joore, J., Vulto, P., & Hankemeier, T. (2015). Microfluidic 3D cell culture: From tools to tissue models. *Current Opinion in Biotechnology, 35*, 118–126.

Wang, J., Sun, Q., Wei, Y., Hao, M., Tan, W. S., & Cai, H. (2021). Sustained release of epigallocatechin-3-gallate from chitosan-based scaffolds to promote osteogenesis of mesenchymal stem cells. *International Journal of Biological Macromolecules, 176*, 96–105.

Wu, C. G., Chiovaro, F., Curioni-Fontecedro, A., Casanova, R., & Soltermann, A. (2020). In vitro cell culture of patient derived malignant pleural and peritoneal effusions for personalised drug screening. *Journal of Translational Medicine, 18*(1), 163.

Xu, K., Wang, Z., Copland, J. A., Chakrabarti, R., & Florczyk, S. J. (2020). 3D porous chitosan-chondroitin sulfate scaffolds promote epithelial to mesenchymal transition in prostate cancer cells. *Biomaterials, 254*, 120126.

Yamada, K. M., & Cukierman, E. (2007). Modeling tissue morphogenesis and cancer in 3D. *Cell, 130*(4), 601–610.

Yamaguchi, Y., Deng, D., Sato, Y., Hou, Y. T., Watanabe, R., Sasaki, K., Kawabe, M., Hirano, E., & Morinaga, T. (2013). Silicate fiber-based 3D cell culture system for anticancer drug screening. *Anticancer Research, 33*(12), 5301–5309.

Young, M., Rodenhizer, D., Dean, T., D'Arcangelo, E., Xu, B., Ailles, L., & McGuigan, A. P. (2018). A TRACER 3D co-culture tumour model for head and neck cancer. *Biomaterials, 164*, 54–69.

Zhou, Y., Hutmacher, D. W., Varawan, S.-L., Lim, T. M., Zhou, Y., Hutmacher, D. W., Varawan, S.-L., & Lim, T. M. (2006). Effect of collagen-I modified composites on proliferation and differentiation of human alveolar osteoblasts. *Australian Journal of Chemistry, 59*(8), 571–578.

Ziółkowska-Suchanek, I. (2021). Mimicking tumor hypoxia in non-small cell lung cancer employing three-dimensional in vitro models. *Cells, 10*(1), 141.

Chapter 15
Stem Cell Culture Techniques

Rashmi Kumari, Madhu Rani, Amrita Nigam, and Anil Kumar

15.1   Introduction

Stem cells are a group of cells, which are unique and unspecialised cells that are present in the human and animal body. There are some unbelievable capabilities in stem cells like self-renewing, which means potency to regenerate into newer cells, clonogenic, having potency to form similar types of stem cells and differentiation, potency to differentiate into numerous cell lineages. All multicellular organisms have basic stem cells, which exhibit the property to differentiate into large number of adult cells. Stem cells are originated in both embryonic cells and in adult cells, which can create route for stem cell research. In humans, stem cells are found everywhere and in the whole life span means from primary stages of development to till death. A stem cell also has the ability to produce replacement cells for an extensive range of tissue and organs such as the heart, liver, pancreas and nervous system (Luke Law et al., 2019; Sachin & Singh, 2006). Stem cell gives therapeutic promises in incurable diseases, organ transplantation and early diagnosis of diseases.

R. Kumari
Genome Biology Laboratory, Department of Biosciences, Jamia Millia Islamia,
New Delhi, India

Indira Gandhi National Open University, New Delhi, India

M. Rani
Genome Biology Laboratory, Department of Biosciences, Jamia Millia Islamia,
New Delhi, India

A. Nigam
Indira Gandhi National Open University, New Delhi, India

A. Kumar (✉)
Rani Lakshmi Bai Central Agricultural University, New Delhi, India

However, for the use of stem cells culture in various applications, scientists need to have access to standardised protocols for the development, maintenance and differentiation of stem cells culture. Such "best practices" will provide comparisons of numerous studies and accelerate the improvement of these techniques. In order to standardise the stem cell culture protocol Stem Book and stem cell banks are the best resources. The current advancement in stem cell culture technique enables the development of varied stem cells such as embryonic stem cell, induced pluripotent stem cell and foetal stem cell from human origin and also the culturing of them aseptically under in vitro settings.

15.2 Historical Background

Mammalian eggs were initially fertilised outside the body in 1878, but it wasn't until the 1960s that Ernest A. McCulloch and James E., two Canadian scientists, invented the stem cell for the study field (Becker et al., 1963; Siminovitch et al., 1963). Schretzenmyr found the great achievement who were successfully done the first stem cell transplant by use of bone marrow from adult cells in 1937 (Nair, 2004). In 1959, stem cell transplants went one step ahead with the first in vitro fertilisation (IVF) in an animal (Johnson, 2019). Two different types of stem cells were recognised in the late 1960s, namely teratocarcinoma and embryonal carcinoma (EC), from the embryonic germ cells. In 1968, the first human egg was in vitro fertilised and enhanced the chance of misuse of stem cell totipotency. In an experiment on mice in 1970 it was revealed that cultured embryonic carcinoma cells might be used as models for the development of an embryo in mice. In 1981, for the first time mice embryonic germ cells were isolated from the inner cell mass of the blastocyst stage. Additionally, it was analysed that teratomas are generated in mice when embryonic stem cells were injected into it, which was grown in cultured medium under in vitro conditions. Pluripotent clonal cells were developed from 1984 to 1988, which is also known as embryonal carcinoma. It was observed that when these cells are exposed under retinoic acid they become differentiated and appear as neuron like cells and also other sorts of cells. In 1989, three primary germ lines; ectoderm, mesoderm and endoderm were formed from human embryonal carcinoma clonal cells that had minimal abilities of differentiation and replication. Another fascinating work has been done in 1994 by producing human blastocyst from inner cell mass that is centrally located in embryonic cells and maintained under aseptic culture condition which retained morphology like a stem cell. In non-human primates like monkeys, it was successfully accomplished to maintain embryonic stem cells that were extracted from the inner cell mass in 1995–1996 in an aseptic in vitro environment. In this experiment, cells were found to have pluripotent abilities that differentiate into three primary germ layers such as ectoderm, mesoderm and ectoderm. Embryonic stem cells that derived from the inner cell mass of normal human blastocyst were cultured, which were maintained by using several passaging techniques in 1998. Healthy human ES cells were successfully isolated from the inner cell mass of blastocyst stage by scientists in 2000. Further, they cultured the cells and proliferated them for a number of times under in vitro conditions, which resulted in the generation of all three primary germ layers: ectoderm (outer layer), mesoderm (middle layer) and endoderm (inner

Table 15.1 Summary of the history of stem cell research

Year	Discovery	Experimental model	References
1878	Fertilisation of mammalian egg done first time outside the body of an organism	Mammals	Becker et al. (1963)
1937	First bone marrow transplantation done		Nair (2004)
1959	First in vitro fertilisation (IVF) technique successfully done to produce animal	Animal	Friedenstein et al. (1976)
Late 1960	Teratocarcinoma is known to develop from embryonic germ cells, however it was also shown that embryonal carcinoma cells may be other types of stem cells	Mice	–
1968	Fertilisation of the first human egg was successfully done in vitro and the risk of exploitation of totipotency in stem cells was observed	Human	–
1970s	Embryonic cells cultured under in vitro conditions might be models for embryonic development	Mice	–
1981	Embryonic stem (ES) cells can be isolated from the inner cell mass of blastocyst in mammals	Mouse	–
1984–1988	Developed embryonal carcinoma (EC) cells		–
1989	Three primary germ layers were derived from a clonal line of embryonal carcinoma cells	Human	–
1994	Generated blastocyst and cultured the inner cell mass	Human	–
1995–1996	ES cells were aseptically sustained under in vitro conditions, which were isolated from the inner cell mass of the blastocyst in a non-human primate	Monkey	Nair et al. (2004)
1998	Passaging done for maintenance of ESCs culture	Human	–
2000	Successfully derived embryonic stem cells from the inner cell mass of the blastocyst by scientists	Human	–
2001	Every stem cell carries the capacity to divide and replenish itself, which is known as self-renewing qualities	Human	–
2004	Proved hematopoietic stem cells are pluripotent stem cells	Human	Ballen et al.
2012	Revealed that mature cells can be converted into pluripotent stem cells	Human	Colman (2013)
2013	Produced embryonic stem cells from foetal cells	Human	Rajpoot and Tewari (2018)
2014	Produced embryonic stem cells from adult cells	Human	Rajpoot and Tewari (2018)

layer). In addition to this, it was also noted that ES cells were transferred into immune-deficient animals and formed when injected.

Gesine Koegler et al. discovered in 2004 that umbilical cord blood contains pluripotent stem cells, according to a 2001 research study on human umbilical cord tissue, the source of stem cells used in adult transplants (Ballen et al., 2015). In 2012, it was discovered that mature cells in an organism can be converted into stem cells in vivo; later the cells were successfully cultured to become pluripotent. For this

Fig. 15.1 Various sources
of stem cells

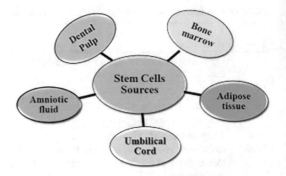

great achievement, scientists Shinya Yamanaka and John Gurdon have been awarded
the Nobel Prize in the areas of Physiology or Medicine (Colman, 2013). As a result
of the ongoing work on human ES cell lines and the aseptic cultivation of several
novel cell types, stem cell research has been advancing day by day. Various scientist
groups are working on it so that the cell could be differentiated in vitro. Shoukhart
Mitalipov and his colleagues by using therapeutic cloning produced embryonic stem
cells from foetal cells at the Oregon National Primate Research Centre, Beaverton in
2013 (Rajpoot and Tewari 2018). Dieter Egli and Young Gie Chung successfully
developed human embryonic stem cells from human adult stem cells followed by
therapeutic cloning in 2014 (Rajpoot and Tewari 2018) (Table 15.1).

15.3   Properties of Stem Cells

Stem cells exist in a whole life span of all living organisms and exhibits three important properties which are discussed below:

Self-Renewing Every stem cell carries the capacity to divide and replenish itself,
which is known as self-renewing qualities. Stem cells proliferate mitotically and
maintain themselves in an undifferentiated state (Avasthi et al., 2008).

Unspecialised or Undifferentiated All types of stem cells are unspecialized or
undifferentiated in behavior. These are basic cells found as a mass of cells that differentiate later in the course of cell division.

Specialized Each and every stem cell is likely to differentiate into specialised cells
and play a variety of specific functions in an organism. Due to these features stem
cells can generate either pluripotent or multipotent types of stem cells.

15.4   Sources of Stem Cells

There are various sources of stem cell that can be isolated from human origin, which
are discussed below (Fig. 15.1).

15.4.1 Human Umbilical Cord

In the human umbilical cord, the blood and tissues are the purest, youngest and richest source of stem cells. They are multipotent stem cells. After baby birth mesenchymal and hematopoietic stem cells can be easily collected from this source. These cells can be easily collected from the umbilical cord after baby birth, which contains hematopoietic and mesenchymal stem cells. The hematopoietic stem cells can generate blood cells and cells of the immune system, while on the other hand mesenchymal stem cells are involved in the formation of cartilage, bone and other types of tissues. As cord blood is a source of stem cells, it can be stored or banked for the future for applications like stem cell therapy.

15.4.2 Bone Marrow

It is one of the sources of stem cells which possesses a soft and spongy tissue located at the center of bones such as in hips and thigh bones. Mesenchymal stem cells originate in the bone marrow, hence it is common source of stem cells. In addition, bone marrow also generates blood cells such as RBC, WBC and platelets or blood stem cells that make it unique and attractive for possible use in regenerative medicine and disease therapy.

15.4.3 Adipose Tissue

The adipose tissue is a type of loose and soft connective tissue found in humans, which is one of the chief sources of stem cell. These cells are mesenchymal cells, which have the capacity of self-renewal and multipotency.

15.4.4 Amniotic Fluid

It is one of the high accumulated embryonic stem cells source in humans, which can differentiate into numerous forms of cells, for example, nerve, muscle, skin, cardiac and bone. Amniotic fluid has defensive properties and it is present around the amnion or sac. The best source of amniotic fluid is cord tissue and amniotic fluid. Nowadays after baby birth amniotic fluids are being cryopreserved or frozen for therapeutic purposes (Larson & Gallicchio, 2017).

15.4.5 Dental Pulp

In recent years a new source of stem cells has been revealed, which is the dental pulp. Stem cells can be extracted from the tissue of human dental pulp from permanent teeth and exfoliated deciduous teeth (Shi et al., 2020). Dental pulp tissue is where mesenchymal stem cells are produced and are widely distributed.

15.5 Classification of Stem Cell Types Based on Origin

Stem cells are divided into three main groups according to the origin and distribution of cells such as embryonic stem cells, foetal stem cells and adult stem cells (Fig. 15.2).

15.5.1 Embryonic Stem Cells

As per the name, embryonic stem cells have only one origin in animal and human embryo, which have the tendency of self- replicating pluripotent stem cells. These cells are derived at the developmental stage from the embryo just before the time of implantation in the uterus (Avasthi et al., 2008). A blastocyst is a small hollow sphere of cells, which develops after 4–5 days of fertilisation of an egg or sperm. At the centre of the blastocyst inner cell mass exists comprising 30–34 cells that further differentiate into variety of cells.

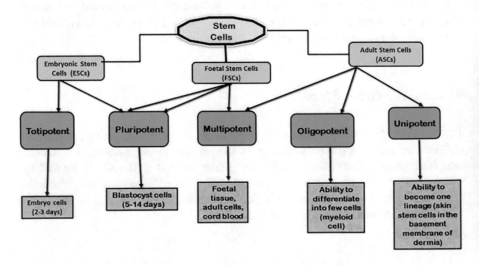

Fig. 15.2 The classification of stem cells based on their origin and potency

15.5.2 Adult Stem Cells

These cells are found all over the body after embryonic development, which multiply by mitotic cell division to regenerate the damaged tissue and replace dead cells. Adult stem cells are multipotent in nature and get differentiated into several distinct cell types. Numerous adult stem cells are found in an organism for example haematopoietic stem cells, spermatogonial stem cells, mesenchymal stem cells from adipose tissue etc.

15.5.3 Pluripotent Stem Cells

These types of stem cells possess very similar properties to embryonic stem cells. In recent years, due to their similarities with embryonic stem cells various scientists have focused on the use of pluripotent stem cells, such as adding some specific gene through engineered induced pluripotent stem cells (iPSCs) that results into reprogramming of somatic cells into a pluripotent cell.

15.6 Classification of Stem Cells Based on Potency

On the basis of differentiation capabilities stem cells can be characterised into five types: totipotent, pluripotent, multipotent, oligopotent and unipotent (Fig. 15.2).

15.6.1 Totipotent

These cells likely differentiate into any variety of cells in an organism. For example, an egg or zygote is formed within the first few cells divisions after fertilisation.

15.6.2 Pluripotent

The category of this stem cell has capabilities to differentiate into all kinds of cells in an organism. For example embryonic stem cells and induced pluripotent stem cells (iPSCs) that divide and develop into three primary germ layers i.e. endoderm, mesoderm and ectoderm.

15.6.3 Multipotent

The cells belonging to this class have the tendency to differentiate into genetically determined cells. For example adult stem cells (hematopoietic, mesenchymal and neural stem cells). RBCs, WBCs and platelets are hematopoietic stem cells. Adipose tissues are also multipotent stem cells.

15.6.4 Oligopotent

These types of cells have the tendency to differentiate into quite a limited number of cells. For example, lymphoid or myeloid stem cells (adult).

15.6.5 Unipotent

The cell has the potential to develop its own type along a single lineage, but it has the self-renewal property. For example muscle stem cells (adult) (Kalra & Tomar, 2014).

15.7 Stem Cell Culture Techniques

Cells mainly from blastocyst and inner cell mass and developed in an artificial or in vitro setting are known as stem cell culture. Stem cells propagation in the laboratory entails specialised, high-quality media and expert culture techniques involving thawing frozen stocks, plating cells in culture vessels, changing media, passaging and cryopreservation. Stem cell culture conditions should not be suboptimum because it can generate uninvited differentiation or cellular senescence in stem cell. Stem cell differentiation in vivo was induced by a variety of factors some of which might be mimicked in stem cell cultures. In research application, it is imperative to select correct stem cell lines because some cell lines are immortal and can be cultured indeterminately. Before initiating the culture of stem cells all cell manipulations, preparations of the tissue culture vessel and preparations of the culture medium must be executed under aseptic settings in a Microbiology Safety Cabinet (MSC) level II.

The MSC should be well cleaned before and after processing each cell line by applying 70% ethanol on the base surfaces. During work in an MSC, a single cell line should be handled at a time so that the probabilities of wrong labelling or cross-contamination between different stem cell lines are reduced. For cryopreservation of stem cell lines, it is advised a minor stock of cryovials is kept as a backup for use in the future. To perform hESCs culture, it is important to routinely check that the equipment used to culture or to make reagents are stored at their appropriate temperature and conditions and also check they are working within their specifications.

15.7.1   Stem Cell Lines

A stem cell line is a diverse group of certain stem cells that can be cultured in in vitro conditions. These can be obtained from either humans or animals tissues and can replicate from long durations where these are either embryonic stem cells, adult stem cells or induced stem cells. Stem cell lines entail high care in their maintenance, handling and preservation so that their stem cell properties and potency for differentiation are retained. Stem cell lines are widely used for various purposes such as research, regenerative medicine, screening of drugs, toxicity testing and cellular transplantation therapy.

15.7.2   Standardisations for Specific Uses

Standardisation of stem cell lines is very important because it can be used for studies in a specialised area. Production of a stem cell line is also one of the great considerations but it involves attention regarding definite characteristics along with various fundamental issues which relate to all types of human or animal cell culture. Moreover, it is also a very essential task to check, or recheck, cell lines properly so that they could be maintained and also for any changes that could be observed such as in functional expression and marker-like characteristic morphology, structure, mechanism of biotransformation of the xenobiotic and cytoskeletal marker.

15.7.3   Maintenance of Aseptic Condition

Safety measures should be followed to protect the laboratory personnel from any type of health risk while the aseptic condition is to be maintained to keep cultured cells free from any contamination, as even a minor amount of contamination can hamper scientific data and result in unusual results. The aseptic condition can be hampered by microorganisms, which can invade cell culture lab via infected media, equipment, opened or contaminated culture vessels, cell culture components, etc. hence, maintain aseptic environment inside the lab.

The gloves, PPE kit and laminar cabinets should be disinfected with 70% ethanol before starting the work. The materials and equipment that are to be used in cell culture practices should be sanitised and kept in a laminar flow hood before starting the work. The UV light should be kept on for 15–30 min before commencing the work. Put off the contaminated hand gloves by touching anything outside after the work under laminar flow hood. Talking, sneezing and coughing should be done in the direction away from the hood and do not disturb the flow of air. It is advised not to enter the lab if personnel feel sick of cough, cold and sneeze as it can contaminate the cultures. Disinfection should be done of all equipment and materials between the consecutive works.

All these procedures are important to maintain the aseptic condition and can also minimise the use of antibiotics in culture medium. Antibiotics are added to the culture medium if aseptic conditions are somehow compromised. However, the use of antibiotics should be avoided, as it is not an essential component of the culture medium formulation, of no nutritive value and has also been reported to have harmful effects on certain cells in culture. It can also trigger the evolution of antibiotic-resistant strains of microorganisms. That is why the use of antibiotics in the culture medium is to be avoided. Although, if the aseptic condition is somehow compromised, the following criteria are followed for selecting antibiotics: Antibiotics can be used either alone or in combination. It is recommended that when antibiotics are used in the combination, the cocktail should be capable of eliminating a wider spectrum of microbial contaminants like gram positive and gram negative bacteria, fungi, mycoplasma and yeast.

A combination of *Penicillin, Streptomycin* and *Gentamicin* is used in the proper amount to take care of most of the microbial contaminants except fungi, yeast and mycoplasma.

Amphotericin B or *Nystatin* can be used to control fungal and yeast infections.

Ciprofloxacin, Erythromycin or *Tylosin* is used to control the infection of Mycoplasma. However, it is again emphasised that the best alternative to antibiotics is a practice of good aseptic conditions and avoid their use in cell culture.

15.7.4   Safety Features for Stem Cell Culturing

- Cultured cells (like HIV) release infectious agent that may be hazardous, and also there are reagents used that may be corrosive, toxic or mutagenic. These can be toxic to both laboratory personnel and the environment. So, these should be handled properly.
- Hazard groups should be determined for pathogenic or disease-causing cell types and safety features should be implemented according to that.

Hazard group 1 (e.g. *Escherichia coli* K12) and *Hazard group 2* (e.g. *Staphylococcus aureus*) serve low to moderate health risks to handling personal, and treatment options are available on infection from both of these groups.

Hazard group 3 (e.g. Severe Acute Respiratory Syndrome-CoV) and *Hazard group 4* (e.g. Ebola) serve high risk to laboratory personnel, and treatment option may also be not available.

- Every lab member should have their protective equipment (PPE) kit, which must be worn during entering the lab process.
- Handling personnel should minimise the exposure of skin by wearing full sleeves shirts, long trousers, fully covered shoes etc.
- Eating, drinking and storage of other consumable items in the lab is not recommended.
- There should be immediate disposal of sharp items via biomedical waste designated boxes.
- The hand should be washed properly before entering and after leaving the cell culture lab.
- The whole lab along with the equipment (like laminar flow hood, incubators) should be cleaned regularly with the help of disinfectants.

15.7.5   Technique of Embryonic Stem Cell Culture (Fig. 15.3)

Immuno-Surgery or Dissection In 1998, Thomson et al for the first time generated hESC line using inner cell mass (ICM) from fertilised embryos at the blastocyst stage. ICM cells have properties OF pluripotency, with the capacity to develop into any cell type. Traditionally, immuno-surgery or dissection techniques were employed for this. ICMs are plated onto a mouse fibroblast feeder (MEF) layer culture system that has undergone radiation treatment, and cells were cultivated in a mixture rich in growth factors and serum (Laowtammathron et al., 2018).

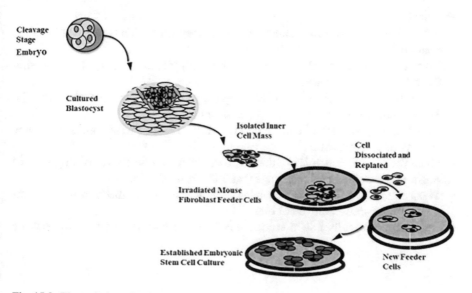

Fig. 15.3 The technique for the generation of embryonic stem cells

This method's primary drawback is that hESCs are cultivated in the presence of animal products that are harmful to hESCs, such as non-human sialic acid and N-glycol neuromeric acid (NeuGc).

Reagents Required for Human Embryo Stem Cell Culture
1. hESC Culture Medium; Dulbecco's modified Eagle medium,
2. Phosphate buffered saline (PBS).
3. Foetal bovine serum (FBS).
4. Inactivated MEF medium.
5. 0.1% Gelatin solution
6. Human basic fibroblast growth factor (bFGF) solution.
7. Collagenase solution.
8. Trypsin solution etc.

Their preparation strategies, amount and maintenance protocol vary a lot depending upon the type of cells to be cultured.

15.7.6 Thawing of Frozen Stocks

Cells that are stored or that arrive at the cell culture laboratory from any external source should be in a frozen state (temperature < 130 °C) because all the biological processes are paused at these temperatures. It is essential that the cells should be thawed to reuse them for culturing. Thawing is somehow the process to reactivate the cells and make them suitable for culturing. Precise thawing is very important to protect the viability of the culture and for the recovery of cells.

Protocol
- Take 10 ml of pre-warmed complete culture medium which is warmed by using water bath.
- Place the frozen cryovials into the water bath (30 °C immediately after removing them from liquid nitrogen.
- When two-third of the content is completely thawed, place the vials into the laminar hood after wiping it with 70% ethanol.
- Add 1 ml of pre-warmed culture into the vials in a drop-wise manner to reduce the risk of osmotic stress.
- Once the vials are completely thawed, transfer them into the remaining 9 ml of pre-warmed culture and centrifuge it at $200 \times g$ for 2 min.
- Wash again the pellet which contains cells with a culture medium to get rid of any extra cryopreservative left out.
- Resuspend the cells into a culture medium and transfer them to the culture vessels.

15.7.7   Gelatin Coating of Culture Plates

In the case of culturing stem cells, gelatin coating is done on culture plates to enhance the attachment or differentiation of embryonic stem cells (Fig. 15.4) and (Table 15.2).

Protocol
- Warm 0.1% of gelatin solution to 37 °C.
- Add 2 ml of gelatin solution in each well.
- Place the culture plates in the laminar flow hood after wiping the external surface with 70% ethanol.
- Put 2 ml of gelatin solution in each well.
- Slowly rotate the culture plates in every direction for uniform distribution of solution in the well.
- Place the culture plate in the CO_2 incubator at 37 °C for 4 h.

15.7.8   Media Change

Culture media needs to be changed at regular intervals, as the previously contained media may be deprived of nutrients or there may be the accumulation of waste materials secreted by cultured cells themselves, which may cause toxicity and lead to the destruction of cell lines.

Protocol
- Warm the culture media to 37 °C and place it in a laminar flow hood after applying the 70% ethanol on the external surface. Also, transfer the Human Embryo Stem Cell (hESC) culture plates from incubator to laminar airflow hood. (It is advised to use 2.5 ml of culture medium for each well).

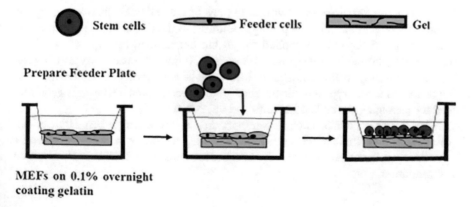

Fig. 15.4 Schematic representation of Feeder-dependent stem cell culture

Table 15.2 It shows the stem cell category, its types and sources and also suitable culture conditions

Stem cells category	Types of stem cell		Sources of stem cell	Stem cell culture system	Serum-free culture media
Pluripotent stem cells	Induced pluripotent stem cells (iPSCs)	iPSCs	Post-natal tissues i.e. immediate after baby birth	Feeder-dependent system with MEF or HFF; or and extracellular matrix	Yes
	Embryonic stem cells (ESCs)	Human embryonic stem cells (hESCs)	Embryonic cells from blastocyst stage		Yes
		mESCs	Blastocyst stage embryos	Feeder dependent culture system or gelatin	
Multipotent stem cells	Adult stem cells	Mesenchymal stem cells (MSCs)	Post-natal tissues i.e. immediate after baby birth	Feeder-free culture system	No
		Cord stem cell (CSCs)	Tumors	Feeder-free culture system	No
					Yes

- Remove the already present or old media completely with the help of the Pasteur pipette.
- Add 2.5 ml of fresh complete culture medium to each well of the culture plate and return the plates immediately to the CO_2 incubator.

15.7.9 Passaging

To maintain the log phase of growth or to prevent the cell colonies from being too dense, cultured cells need to be passaged within 4–5 days of culture. However, this schedule of 4–5 days is not applied for all but depends on the type of cell lines because growth kinetics are different for every cell line. The use of trypsin is good for the automated, healthy passage for hESC. Make sure good manufacturing practice is mentioned in trypsin bottle because gradient recombinant trypsin is generally used for passaging the cells (Zakrzewski et al., 2019).

In hESC, the growth is much slower in the first 1–2 weeks after being thawed but grows faster after this period is over. Hence, hESC passaging is done if the feeder layer becomes 2 weeks old or colonies become too dense or colonies show increased differentiation.

Protocol
- Place the hESC plates from incubator to laminar hood.
- Strain out the already present media and add 1.5 ml fresh culture medium to each cell containing well.
- Mark the undifferentiated area well at bottom of the plate.
- Remove the undifferentiated cells and break them into aggregated form by pipetting them for few times.
- Smoothly transfer the 1.5 ml of culture medium with swimming cell colony to the freshly prepared media in a 1:1 ratio.
- Keep the seeded culture plates in the incubator again and allow the cells to grow.
- In the older plates, add 2.5 ml of culture medium and keep it also into the incubator and observe the culture for 1–2 days.

15.7.10   Cryopreservation

Cryopreservation process is done to keep the cells for future use. Cryopreservation is done when the cells are in their log growth period. To keep the cells in frozen condition, a cryoprotectant is used (like dimethyl sulfoxide-DMSO or glycerol); this also helps in prohibiting the formation of any catastrophic intracellular or extracellular crystals. However, the issue with DMSO is that it is toxic at higher concnetrations, so it is always important to keep this in mind while adding.

Protocol
- Discard the culture plates from the CO_2 incubator and put them in a cleaned laminar airflow hood.
- Remove the already present medium from wells and wash it twice with wash buffer (0.5 mM EDTA or PBS depending upon the cell cultured).
- Add 0.5 mM EDTA to loosen the attachment of cells to the material.
- Once the cells get detached from the surface, suspend it in cryoprotectant and centrifuge at $200 \times g$ for 2 min.
- Discard the supernatant and take pellet in an appropriate amount of cryoprotectant or freezing medium and preserve it in cryovials tube.
- Place the cryovials in 2–8 °C, then transfer them to −80 °C freezer immediately and keep them there overnight.
- The next day early morning, transfer the cryovials into a (−196 °C) liquid nitrogen freezer on dry ice.

15.7.11   Precautions and Troubleshooting Tips (Table 15.3)

15.8   Limtations and Challenges in Stem Cell Research

Due to various ethical and other problems associated with stem cell research, here we are describing some limitations or challenges of stem cell research:

Ethical Issues It is one of the major considerable issues linked with stem cells usage due to the involvement of human or animal origin cell lines which create a serious ethical question. The stem cell is one of the advanced research areas in the field of biological research. But due to the use of human or animal origin cell lines various countries have made very strict rules and controls and also some other countries are working on it so that human or animal stem cells could be used less. They are currently observing regulations to determine which laws should be upheld and how that might serve the greater good. Therefore, it is crucial that ethical concerns are approved by the appropriate authority before any human material for a new cell line is considered for establishment.

Use of Human/Animal Embryonic Stem Cells It is a critical challenge in stem cell based research and are responsible for raising the ethical issues as well. Hence, worked on stem cells ethical approval obtained from relevant authorities is very important because it is a multistep process that takes a long time.

Religious and Political Impediments These issues also create obstruction in the field of stem cell research.

Transplant Difficulties in the production of stem cells for transplant in the laboratories to the target cells in the humans.

Self-Renewal Self-renewal properties in stem cells can be attained in vitro under artificial conditions in the laboratories which might impede stem cell differentiation. It is also one of the limitations or challenges that in vivo embryonic stem cells do not permanently renew themselves, but instead, sometimes stem cells rapidly differentiate into diverse cell lineages of the three primary embryonic germ layers i.e. ectoderm, mesoderm and endoderm.

Mutation Occurrence and risk of mutations in the source of stem cell lines might cause major problems in the transplantation.

Tendency to differentiate into numerous cell types is one of the foremost problems regarding stem cell research. Unsurprisingly, the involvement of spontaneous differentiation can cause adult cells and embryonic stem cells, which develop as a result of the association of differentiation factor having some certain amount, to differentiate into a variety of cell types. Moreover, differentiation of the embryonic stem cells is not synchronisable yet so maybe chances of a mixture of cells present in several development phases.

Table 15.3 Problems and troubleshooting tips during stem cell culture

Problem	Observation	Possible Solution/Precaution
Low viability of hESCs after the thaw	Little to no colonies visible within 4 days after recovery	Confirm that cryovials are thawed quickly and that medium is added to the cells very slowly by swirling the tube At thaw 10 μm ROCK inhibitor add but do not usage it in routinely manner. 10 μM ROCK inhibitor after thawing, but do not use it frequently Check that cells were banked or preserved at log phase stage of the growth having low levels of differentiation Grow small and healthy colonies and passage done with low split ratio (1:1 or 1:2)
Fewer cells are viable after passaging	Cells do not attach properly Non-typical morphology High levels of cell death Proliferation does not occur in cell	Retain more confluent culture by using a lower split ratio Make sure during passaging of cells they Are in the log growth phase Start work fastly or minimize EDTA incubation duration because high exposure of EDTA may form clump size If cells do not come off easily, upsurge the incubation period of EDTA. Always keep in mind to rinse cells gently because harsh rinsing of cells results in aggregation of very small / single cell suspension Check the matrix expiry date, the plates' right coating, and the batch with the manufacturer if the problem persists despite other potential causes having been eliminated.

(continued)

Table 15.3 (continued)

Problem	Observation	Possible Solution/Precaution
Spontaneous differentiation occurs	No defined edges in the colonies Less compact cells appear within the colonies Flattened and bigger or fibroblastic size cells appear	Ensure cells are being cultured as per protocol or recommendations (i.e. daily feeding of cells) Always check that all reagents are freshly prepared (i.e. used within 2 weeks) Avoid the culture plates put outside of the incubator that may cause temperature decrease and fluctuation and also cause exposure to light Colony density of the cells decreases after passage, so plating only collects a few cells per cm^2 Differentiated cells remove by scrapping and use pipette tip to differentiated away do not touch or disturb the intact colonies of hESCs In this process avoid or do not disrupt the colonies of hESC and also not to scrape forcefully of the matrix layer If good hESCs colonies are exist or present between differentiated areas pick colonies manually having good hESCs morphology using a pipette tip can be considered. It is suggested to pick several colonies of cells and with the help of pipette tips cut it into small pieces, cut them into pieces with a pipette tip, then it lifts, remove and pass them to a new and aseptic 1:6 well
Non-uniform distribution of colonies within a plate	There are certian areas which have very high density of hES cells and from here the cells start to differentiate from the middle. Moreover to areas having hardly any colonies of the cells	Confirm matrix as it should be accurately coated in whole surface area of the tissue culture vessel Make sure that the cell aggregates are uniformly distributed through gently rocking the plate back and forth and side to side Be careful when plate putting into the incubator and leave uninterrupted for 24 h
Proper scraping is needed to dislodge cells	Colonies do not easily come off the petri plate with 2–3 rinses by using a 1 ml pipette	Make sure that incubation duration of EDTA and temperature are according with matrix use Increase incubation period of EDTA Observe cells should not be confluent more than 70% Avoid overgrown cells colonies centrally because sometimes requirement of passage the cells with a very minimum confluent plate having fewer but healthy and strong colonies, lower split ratio
Poor attachment and significant upsurge in cell death post passage or after passage	Cells start to lift off even if they appear to be attached after passage	Rather than changing culture medium from uppermost wells of the culture plates with new medium to assure desire amount of nutrients and leave cell uninterrupted for extra 24 h to allow aggregates to fully attach Exchange medium slowly and carefully and avoid fast medium addition into the plate

15.9 Applications of Stem Cells

In recent times, stem cell therapy has played a very promising role for various pathological conditions and in scientific research (Zakrzewski et al., 2019). Below we have enlightened the applications of stem cells in various areas (Fig. 15.5).

Medicinal Uses Stem cells play a very crucial role in the areas of personalised medicine development. The study on stem cells is well explained and involves numerous mechanisms that happen during human body development that aid in the early detection of diseases that are linked to developmental problems. Stem cell therapy is also used for treating many diseases like spinal cord injury, tendon ruptures and diabetes mellitus.

Hematopoietic Stem Cell Transplantation Since more than 50 years, it is the most studied and recognised model for the study of tissue-specific stem cells. Currently, multipotent HSC transplantation is well-known for stem cell therapy in humans; umbilical cord or peripheral blood and bone marrow are the stem cell sources where target cells or tissues are generally derived (Rocha et al., 2006). Diseases like leukemia and anemia, which are caused by dysfunction of the hematopoietic system, can be successfully treated by HSC transplantation.

Drug Testing Testing and development of new type of drug in the field of biological research is one of the best applications of stem cell. Testing of drug can be done in precise differentiated stem cells from pluripotent stem cells, and if any sudden or undesirable effects develop the alterations in drug formulas can be done to achieve the desired amount of effectiveness.

Stem Cell Substitute for Arthroplasty Osteoarthritis (a problem associated with joints) and tendon damage are general problems these days (Tempfer et al., 2020). However, arthroplasty is the most common treatment procedure to follow, but it is not recommended to young personnel, as it requires various surgical procedures in

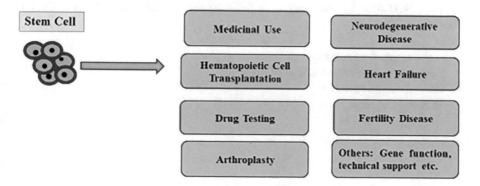

Fig. 15.5 The possible applications of stem cells

the future terms. So, stem cell therapy will be the best treatment option in case of osteoarthritis and tendon damage.

Neurodegenerative Disease Treatment Studies have proved that stem cell therapy is not only capable of delaying the development of neurodegenerative diseases like Parkinson's disease (PD) and Alzheimer's disease (AD) but also can also strike out the cause of the problem. Induced-PSCs-based therapy is commonly used to treat both PD and AD. In the case of AD, it was examined that the transplantation of derived human iPSC-cholinergic NPCs into a transgenic mouse model of AD alleviated performance of spatial memory by producing cholinergic neurons (Fouad, 2019). Therefore, treatment of deadly AD via stem cell therapy might show a promising therapeutic approach.

Heart Failure Chronic heart failure in an individual is one of the foremost reasons for cardiovascular incidence and death rate as compare to other diseases in worldwide. It occurs due to the loss of functional heart muscle, which is either caused by ischemic heart disease, hypertension or idiopathic factors. It is found that the use of patients' own bone marrow-derived stem cells and peripheral blood-derived stem cells can be used to treat heart failure disease (Mummery et al., 2003; Vanderlaan et al., 2003).

Fertility Diseases Katsuhiko Hayashi et al. (Hayashi et al., 2011) worked on mice and observed that sperm can be formed by using iPSCs in 2011. They have been able to successfully deliver the robust pups in sterile mice, by injecting the iPSC collected from fertile mice. In another experiment, they also worked on female mice and showed fully functional eggs can be formed from iPSCs.

Cell-Based Therapy Cell-based therapies are used to treat some common diseases like osteoarthritis, neurodegenerative diseases, macular degeneration (Sun et al., 2017), strokes (Liu, 2013) and diabetes (Shahjalal et al., 2018). Since stem cells have the potency to differentiate into specific cells, they can be used to restore impaired and destroyed tissues. At present, diabetes mellitus, which is a metabolic disease, is treated by inducing stem cells to differentiate and develop insulin-forming cells.

Miscellaneous Applications
- Reparation and reconstruction of organs.
- Gene function research.
- Technical support.

15.10   Conclusion

Stem cell culture technique is useful for generating a large number of cells in a Petri dish within a few days. Embryo stem cells have the capacity to develop into three primordial germ layers and also exhibit pluripotency. Embryonic cells and their derivatives can provide an admirable in vitro model systems, to understand the progression and pathogenic mechanism in various diseases.

Moreover, stem cell culture techniques will serve as an exceptional platform for various applications such as in drug discovery, neurodegenerative disease, medicinal uses and HSCs transplantation. In addition, the use of human cells carries a higher risk of human pathogens and requires strict adherence to the respective safety measures and assurance of ethical provenance. Therefore, it is essential to receive sufficient training to ensure the quality of the work, protect the workforce, and prevent expert researchers and technicians from wasting time and resources. Hence, technical improvements should be well assessed and criticised in order to govern which of the available culture techniques are more reliable, influence, cost effective, and companionable that provides better impact with the optimum for stem cell culture under artificial conditions.

15.11   Future Prospective

We believe that advancement in stem cell techniques like isolation, maintenance and differentiation might provide an opportunity to run into the great expectations of regenerative medicine. Stem cell culture is very crucial because it possesses potential for both diagnosis and treatment of various fatal diseases. Thus, an understanding of the use of stem cell techniques will be essential for students, scientists and clinicians in the future.

References

Avasthi, S., Srivastava, R. N., Singh, A., & Srivastava, M. (2008). Stem cell: Past, present and future – A review article. *Internet Journal of Medical Update, 3*(1), 22–30.

Ballen, K. K., Logan, B. R., Laughlin, M. J., et al. (2015). Effect of cord blood processing on transplantation outcomes after single myeloablative umbilical cord blood transplantation. *Biology of Blood and Marrow Transplantation, 21*(4), 688–695.

Becker, A. J., McCulloch, E. A., & Till, J. E. (1963). Cytological demonstration of the clonal nature of spleen colonies derived from transplanted mouse marrow cells. *Nature, 197*, 452–454.

Colman, A. (2013). Profile of John Gurdon and Shinya Yamanaka, 2012 Nobel laureates in medicine or physiology. *Proceedings of the National Academy of Sciences, 110*(15), 5740–5741.

Fouad, G. I. (2019). Stem cells as a promising therapeutic approach for Alzheimer's disease: A review. *Bulletin of the National Research Centre, 43*, 52.

Friedenstein, A. J., Gorskaja, J. F., & Kulagina, N. N. (1976). Fibroblast precursors in normal and irradiated mouse hematopoietic organs. *Experimental Hematology, 4*(5), 267–274.

Hayashi, K., Ohta, H., Kurimoto, K., Aramaki, S., & Saitou, M. (2011). Reconstitution of the mouse germ cell specification pathway in culture by pluripotent stem cells. *Cell, 146*(4), 519–532.

Johnson, M. (2019). Human in vitro fertilisation and developmental biology: a mutually influential history. *Development, 146*(17), dev183145.

Kalra, K., & Tomar, P. C. (2014). Stem cell: Basics, classification and applications. *American Journal of Phytomedicine and Clinical Therapeutics, 2*(7), 919–930.

Laowtammathron, C., Chingsuwanrote, P., Choavaratana, R., et al. (2018). High-efficiency derivation of human embryonic stem cell lines using a culture system with minimized trophoblast cell proliferation. *Stem Cell Research & Therapy, 9*, 138.

Larson, A., & Gallicchio, V. S. (2017). Amniotic derived stem cells: Role and function in regenerative medicine. *Journal of Cell Science & Therapy, 8*, 3.

Law, L. A., Hunt, C. L., & Wenchun, Q. (2019). *Scientific basis for stem cell therapy* (pp. 715–726). Springer.

Liu, J. (2013). Induced pluripotent stem cell-derived neural stem cells: New hope for stroke? *Stem Cell Research & Therapy, 4*, 115.

Mummery, C., Ward-van Oostwaard, D., Doevendans, P., et al. (2003). Differentiation of human embryonic stem cells to cardiomyocytes: Role of coculture with visceral endoderm-like cells. *Circulation, 107*, 2733–2740.

Nair, V. (2004). Stem cell transplantation. *API Medical Update, 14*, 366–377.

Ovsianikov, A., Yoo, J., & Mironov, V. (2018). *3D Printing and Biofabrication (Reference series in biomedical Engineering)*. Springer.

Rajpoot, S., & Tewari, G. (2018). Review on stem cells: Basics classification and applications. *International Journal of Pharmaceutical Sciences Review and Research, 49*(2), 48.

Rocha, V., et al. (2006). Clinical use of umbilical cord blood hematopoietic stem cells. *Biology of Blood and Marrow Transplantation, 12*(1), 34–34.

Sachin, K., & Singh, N. P. (2006). *Indian Journal of Human Genetics, 12*(1), 4.

Shahjalal, H. M., Dayem, A. A., Lim, K. M., Jeon, T. I., & Cho, S. G. (2018). Generation of pancreatic β cells for treatment of diabetes: Advances and challenges. *Stem Cell Research & Therapy, 9*, 355.

Shi, X., Mao, J., & Liu, Y. (2020). Pulp stem cells derived from human permanent and deciduous teeth: Biological characteristics and therapeutic applications. *Stem Cells Translational Medicine, 9*(4), 445–446.

Siminovitch, L., McCulloch, E. A., & Till, J. E. (1963). The distribution of colony-forming cells among spleen colonies. *Journal of Cellular Physiology, 62*, 327–336.

Sun, S., Li, Z. Q., Glencer, P., Cai, B. C., Zhang, X. M., Yang, J., & Li, X. R. (2017). Bringing the agerelated macular degeneration high-risk allele age-related maculopathy susceptibility 2 into focus with stem cell technology. *Stem Cell Research & Therapy, 8*, 135.

Tempfer, H., Lehner, C., Grütz, M., Gehwolf, R., & Traweger, A. (2020). Biological augmentation for tendon repair: Lessons to be learned from development, disease, and tendon stem cell research. In J. Gimble, D. Marolt, R. Oreffo, H. Redl, & S. Wolbank (Eds.), *Cell engineering and regeneration*.

Vanderlaan, R. D., Oudit, G. Y., & Backx, P. H. (2003). Electrophysiological profiling of cardiomyocytes in embryonic bodies derived from human embryonic stem cells. *Circulation Research, 93*(1), 1–3.

Zakrzewski, W., Dobrzyński, M., Szymonowicz, M., et al. (2019). Stem cells: Past, present, and future. *Stem Cell Research & Therapy, 10*, 68.

Chapter 16
Identification and Removal of Biological Contamination in the Media and Cell Suspensions

Vaishnavi Shishodia, Divya Jindal, Sarthak Sinha, and Manisha Singh

16.1 Introduction

Contamination holds back the use of the cell culture technique as an effective analytical tool in defining the experimental set-ups, and it is one of the commonest concerns shared by researchers across the globe. Since each cell culture system is different, the effects of contamination inside the system are also different. It is a known fact that the contamination is usually persistent since it is invisible, unnoticed, hidden and generally innocuous within a cell culture system. We also need to understand that the cell cultures are living organisms and that they respond to their surroundings and, as a result, to contamination too. Further, contamination can infiltrate the cell culture system mostly through the environment in the form of either physical, chemical or biological constituents. Although living organisms metabolise and reproduce, biological contamination is the most significant risk to the cell culture system. As a result, the chance of disease propagating to certain other cell cultures increases due to cross-contamination. Metabolism broadens the variety of potential influences on the cell culture system by reducing key nutrients and adding new by-products such as distinct enzymes, antigens and toxins. Biological contamination is the in vitro counterpart of an emergent disease in a cell culture system. As a result, the risk of infection propagating to certain other cell cultures also increases due to cross-contamination. Metabolism broadens the variety of potential influences on the cell culture system by reducing key nutrients and adding new by-products such as distinct enzymes, antigens and toxins. Biological contamination is the in vitro counterpart of an emergent disease in a cell culture system. The implementation of standardised, effective cell culture techniques and quality control processes

V. Shishodia · D. Jindal · S. Sinha · M. Singh (✉)
Department of Biotechnology, Jaypee Institute of Information Technology (JIIT), Noida, Uttar Pradesh, India

© The Author(s), under exclusive license to Springer Nature Switzerland AG 2023
S. Mani et al., *Animal Cell Culture: Principles and Practice*, Techniques in Life Science and Biomedicine for the Non-Expert,
https://doi.org/10.1007/978-3-031-19485-6_16

is the solution to resolve the concern of cell contamination (Corning Inc., 1987; Ryan, 2012). However, the chemical contaminants, including impurities in media, serum samples, detergents, endotoxins, plasticisers, along with biological ones (moulds, bacteria, yeasts, viruses and mycoplasma), and cross-contamination by different cell lines, are considered the major sources of cell lines contamination. Although it is difficult to entirely discriminate against contamination and propagate them in a contamination-free environment, it is quite possible to minimise its occurrence and severity by learning about its origins and using effective sterile techniques (Freshney, 1981). The convenience with which a pollutant can be discovered might sometimes be the key to determining it. Fungal, bacterial and yeast contamination are frequently seen by naked eyes and can destroy cells present in culture media quickly, although subtle morphology can make major microbial intrusions like mycoplasma more difficult to detect. Viruses, on the other hand, cannot be spotted by normal light microscopy and are thus frequently discovered as a result of unexplained separation or another indicator of poor cell integrity or perhaps even death of cells (Freshney, 1981). Contamination and even wilful mistaken identity of research cell lines have been one of the most serious issues plaguing the biology industry in the previous years.

16.2 Cell Culture Contaminants

16.2.1 Biological Contaminants

Cultured cell with contamination of microbial organisms necessitates careful monitoring and constant vigilance.

16.2.1.1 Mycoplasma

Mycoplasma, bacteria, yeast, fungi and viruses are the most common microbial contaminants in cell culture. Mycoplasmas are the smallest prokaryotes (0.3–0.8 m in diameter) and their appearance in cultures is usually not visible. Bacteria are yet another common cell culture contaminant (Wang, 1976). They typically came to know via not-so-good aseptic technique and are frequently identified in cultural media. With aerobic bacteria, there will be a decrease in acidity or alkalinity (yellow with phenol red as an indicator) and an increase in obscurity.

16.2.1.2 Bacteria

Bacteria, moulds and yeasts may be found almost anywhere and can instantly colonise and thrive in the rich and largely non-defended environment that cell cultures provide. These microorganisms are the highly typically encountered cell culture contaminants due to their size and rapid growth rates. Microbes may generally be recognised in culture after some days of getting contaminated in the absence of antibiotics, either by direct microscopic inspection or by indirect microscopic observation (MacMichael, 1986).

16.2.1.3 Fungi

Fungi (moulds) are the commonly found contaminants present as tiny filamentous mycelia or clumps of spores. In their advanced stages, they could take over culture as furry growths ranging in hues from white to black. Fungi are abundant in nature and often infiltrate cultures through the air. Seasonal variations, climate control and warming can have a big impact on the atmosphere.

16.2.1.4 Virus

The three primary concerns for virus contamination in cultured cells for therapeutic production are cell resources, cell culture supplies and the introduction of the cultured cells procedure flow to the operators or environment. Viruses are the most challenging pollutants to detect, yet, unless they are cytopathic, they may have a minor impact on their host cells. Bovine serum intended for use in tissue culture and cell lines has been reported to include viral contamination including bovine viral diarrhoea virus (BVDV), bovine herpesvirus (BHV), bovine parainfluenza-3 and epizootic haemorrhagic illness (Rottem & Barile, 1993).

As a result of their very compact size, it is very difficult to detect viruses in culture, demanding techniques that are impracticable for most research investigating institutions (Case-Gould, 1984). They are also difficult to eliminate from the complex sample, sera, as well as other solutions because of their small size. Most viruses, on the other side, have unique needs for their real host species' cellular machinery, limiting their capability to disseminate cultures of cells from all various species. Viruses are thus more prevalent in cultured cells than many researchers, while they are rarely a severe problem until they cause cytopathic or other negative effects on the cultures. Yeasts are typically small oval-shaped organisms that are significantly smaller than mammalian cells. In short chains, they can be seen as "budding" from other yeast particles. These chains will eventually form multiple branches.

16.2.1.5 Yeast

Yeasts typically infect cell culture materials via airborne routes and are easily transmitted through contact (Fogh et al., 1971; Fogh, 1973). Free-living and parasitic single-celled protozoa, such as amoebas, have been discovered as cell culture contaminants. Amoebas, which are mostly found in soil, may produce spores and can be easily extracted from the air, tissues and even laboratory staff's throat and nose samples. They exhibit cytopathic effects comparable to viral damage and can destroy a culture in as little as 10 days. Amoebas are difficult to recognise in culture because of their slow growth and morphological resemblance to grown cells unless they have already been identified as contaminants. Fortunately, incidences of this kind of contamination are rare, but it is important to be aware of the risks (Barkley, 1979; Robinson et al., 1956).

16.2.2   Chemical Contaminants

A cultured cell contaminant is an ingredient in the culture system that is unwanted due to its potential for contamination and unfavourable impact on the system or even its components usage. The impurities in sera, media and water, detergents, plasticisers and endotoxins are the prime sources. The contaminant components can be classified into two primary categories – chemical and biological contaminants. The presence of any non-living material is best defined as chemical contaminants, and this has unfavourable consequences for the cellular environment (Gibco). Even basic nutrients become poisoned as a result of this cult if the concentrations are high enough. The sole issue has been majorly related to serum toxicity along with the other growth factor alterations that might result in modifications that are not always desirable. Furthermore, most of the bacterial or fungal contamination can be visually detected in cell culture media and prevented with antibiotic treatments. Other biological contaminants, such as mycoplasma or undesired/other cell lines, are more difficult to detect but can still be monitored using commercially available testing kits. Moreover, other than this all endotoxins contamination can also occur after opening reagents or from glassware/plasticware; therefore, regular endotoxin testing is always needed. Chemical contamination, in contrast to biological contamination, receives less attention and is more difficult to detect and avoid. Endotoxins, a type of lipopolysaccharide that forms the outer cell wall of Gram-negative bacteria, are among the most dangerous chemical contaminants, and unlike live bacteria they cannot be seen visually in cell culture media. Though they can be easily detected by using the Limulus Amebocyte Lysate assay, which can be seen in the water, serum and various culture additives (especially those created through microbial fermentation). Endotoxins, on the other hand, cannot be depleted with antibiotics, necessitating the use of specialised endotoxin removal solutions. Also, they have been shown to negatively affect the growth or performance of cell cultures. Water is employed in many steps of the tissue and cell culture process in in vitro research because it is a main component of media and buffer, it dissolves additives and medications, and it is used to rinse glassware, plasticware, and bioreactors. Therefore, the quality of the water may have a substantial influence on the results of cell culture experiments. Other types of contaminants, such as plasticisers eluted from plastic instruments or trace elements, can be found in water too. These substances can also affect cells in a culture medium. It has also been reported that some toxic substances are even eluted out from sterilised microfilters. Invasion of cell culture by bacteria, yeast or mould is always a problem, and researchers go to considerable measures to avoid it. Typically, these pollutants may be seen with the unaided eye or optical microscope. Chemical contamination, on the other hand, can alter the cultured cell's growth, shape and behaviour, while staying invisible to the human eye (Wolf & Quimby, 1973; Wolf, 1979).

The serum has been the main source of inducing endotoxins in cell cultures, and most manufacturers have greatly reduced this concern by handling raw materials too under aseptic settings. It has been also noticed that when the endotoxin issues are anticipated or detected in the cultures, its cause is mainly related to not-well-maintained water systems, particularly those utilising ion exchange resins, which might harbour substantial numbers of endotoxin-producing bacteria and would need to be checked. Bacterial contamination can produce rapid variations in media pH and contaminate clean cultures. Furthermore, endotoxin, a complicated lipo-polysaccharide produced by Gram-negative bacteria, is produced by the majority of Gram-negative bacteria (LPS). Macrophages and mononuclear phagocytes generate several pro-inflammatory cytokines in response to endotoxin activation. Endotoxins harm a wide range of cells, as well as those that lack CD14 endotoxin receptors (Corning Inc., 1987). Changes in cell development and function, as well as cloning efficiency and recombinant protein production, are all possibilities.

16.2.3 Inorganic Ions

Heavy metals such as arsenic, copper, chromium, nickel and cadmium have indeed been observed to be harmful to glial and neuronal cell structures. Further, many research studies have exhibited that the DNA nuclease II and glucose-6-phosphate dehydrogenase are both inhibited by magnesium. Thus, manganese enables DNA polymerase to integrate ribose rather than deoxyribose into nascent DNA strands of the cells.

16.2.4 Organic Compounds

Smaller molecules (organic) like humic acids, tannins, pesticides and endocrine disruptors are commonly found in raw water supplies and might sustain in water coming from the tap. These substances are keen to affect cell development and should thus be removed during the production of laboratory-grade water using appropriate processes. Furthermore, water is used in many cell culture-related instruments, such as autoclaves and incubators (Ryan, 2012; Freshney, 1981).

16.3   Role of Antibiotics and Antimycotics

Antibiotics are classified based on whether they are bactericidal or bacteriostatic in their action on bacterial cellular components. Antibiotics that are bactericidal kill germs, whereas bacteriostatic antibiotics prevent them from growing. In mammalian culture systems, there are seven essential needs for antibiotic supplements: (i) remove microbiological pollutants (bactericidal over bacteriostatic is desirable), (ii) do not impair mammalian cell growth and metabolism, (iii) offer protection for the duration of the experiment and (iv) do not influence any eventual usage of mammalian cells (Lincoln & Gabridge, 1998).

Antibiotics should not be used in cell culture regularly even though they foster the growth of antibacterial drug strains and allow low-level pollutants to persist, which can advance to full-scale pollution once the antibiotic is deleted from the media, as well as concealing mycoplasma illnesses as well as other cryptic contaminants. Furthermore, certain antibiotics may well have a cross-reaction involving cells, causing problems with the cellular methods under investigation. Antibiotics should only be used as a last resort and for a brief duration, but they should be removed from the culture as soon as feasible. If antibiotic-free cultures are used for a long time, they must be preserved in parallel to screen for cryptic infections (Luzak et al., 1993).

16.4   Sources of Contamination

16.4.1   Biological Sources

By sharing media and reagents, microbial species can produce physical, biological, and chemical pollutants (being able to share media and reagents, using unplugged pipettes, improper handling and use of non-sterile reagents, unplanned spilling or abnormal contact with inanimate objects), and biological (being able to share media and reagents, using unplugged pipettes, poor handling or use of non-sterile reagents, unplanned spilling or abnormal contact with inanimate (direct or indirect contact on hands). Biological agents can contaminate any type of microbial laboratory. A unicellular organism capable of reproducing and may creep, crawl or fall into cultured cells might induce biological contamination in the system. Mycoplasmas, which include other cell cultures, insects, arthropods, protozoa, moulds, yeasts, viruses, walled bacteria and/or wall-less bacteria, can all infiltrate the system. When a cell culture is treated in an open system, it is vulnerable to microbial contamination. The surrounding world, approach, gear and supplies of the cell culturist define the level of hazard. Biological contamination can set off a chain of metabolic processes unique to each culturing system and the contaminant (McGarrity, 1988).

Contaminated cell culture is the most common source of mycoplasmal contamination in a cell culture facility. From development on the surface of its former host to the symbiotic proliferation of a host of organisms in a cell culture system, a commensal and/or pathogenic organism has evolved (Barile, 1977). Aside from that, human isolates account for a significant portion of the mycoplasmal contamination identified in cell culture. *Mycoplasma orale*, *Mycoplasma fermentans* and *Mycoplasma salivarium* are all common isolates. *M. buccale*, *M. faucium*, *M. genitalium*, *M. hominis* and *M. pirum* are other mycoplasma species linked with humans that are infrequently isolated from cell cultures (Freshney, 1981). Human activities produce polluted airborne particles and aerosols, which are accidentally introduced into the cell culture system.

16.4.2   Chemical Sources

The major components include toxins present in the medium, sera, water, metal ions, enterotoxins, free radicals, cleansers and miticide or pesticide residues. Furthermore, unwanted impurities in the gases utilised in the atmospheric CO_2 incubator are added.

16.4.2.1   Water

Water is a one-of-a-kind substance that expands when frozen. This aspect must be taken into account when choosing storage tanks for frozen aqueous solutions. A reagent purity breach is caused by a cracked storage vessel. Water is a solvent that includes metal ions, chemical molecules, bacterial endotoxins and other substances (Lincoln & Lundin, 1990). With more purification, this becomes a better solvent. Metal ions, chemical molecules, endotoxin and other contaminants will leak from glassware, tubes and pipes in highly filtered water. One element of the chemical pollution of water is the production of steam for steam lines and autoclaves, as well as the purification of water for in-house use which is distilled or through injection systems. Routine maintenance conducted on steam turbines, water purification equipment and pipelines for preventing scale accumulation may introduce toxic chemical pollutants.

16.4.2.2   Sera

Animal sera used in cell culture media remain a potential source of biological and chemical contamination; sera are collected from herds of individual animals and processed and efficiently controlled using different procedures from vendor to vendor; there is an unknown and uncontrollable variation in the concentrations of hormones, growth regulators, enzymes, fatty acids and other substances between lots. Sera should be thoroughly evaluated for development support, toxicity and contamination before being purchased. Serum growth support and toxicity will range from cell line to cell line based on the expression of differentiated function, species of origin, cell culture medium composition and other factors. If a variety of cell culture systems are being employed, growth support and toxicity testing must take this into account (Gabridge & Lundin, 1989; McGarrity et al., 1985).

16.5   Testing and Detection of Different Contaminants in Media

Animal sera used in cultured cell media continue to be potential sources of chemical or biological contaminants; sera are collected from herds of organisms and efficiently controlled using various techniques from vendor to vendor; concentration levels of hormones, growth regulators, enzymes, fatty acids, and other substances differ between lots in an unknowable and uncontrollable manner. Before purchasing sera, it should be extensively assessed for development funding, toxicity and contamination. Serum development support and toxicity will range from cell line to cell line depending on the expression of distinct function, kind of source, cell growth medium composition and other factors. If a range of cell culture techniques are utilised, development assistance and drug screening must take this into account (Upchurch & Gabridge, 1983).

Table for different sources of contaminants

Chemical contamination	Biological contamination
Impurities such as metal ions, endotoxins and other toxins present in media, sera and water	Bacteria, moulds, yeasts, viruses, protozoa, mycoplasmas
Plasticisers in plastic tubing and storage bottles	Contact with non-sterile supplies, media or solutions
Exposure to fluorescent light causes photoactivation of tryptophan, riboflavinor HEPES. This leads to the generation of free radicals in the media	During culture manipulation, transfer or incubation, particulate or aerosol discharge occurs
Generation of residues from germicides or pesticides which are used to sterilise incubators, equipment, and laboratories	Growth into culture vessels

16.6   Testing and Detection of Different Contaminants in Cell Cultures

16.6.1   Bacteria

The culture medium's turbidity has risen, and it now seems hazy. If the mixture contains phenol red as a pH indicator, a quick colour shift to yellow signals an abrupt reduction in pH. Bacteria are a fraction of the size of eukaryotic cells. They appear as black rod-like formations, spheres or spiral structures under the microscope, and they can be found as cells, pairs, chains or clusters. Bacteria may be observed at total magnification ranging from 100x to 400x utilising phase contrast (Upchurch & Gabridge,1983)

The new standard for the identification of bacteria is DNA sequencing. Other approaches, including culture-based and molecular approaches, are also employed. These are time-consuming and require the assistance of experienced individuals. To identify microorganisms, many analytical techniques based on VOC monitoring have been applied. Gas chromatography-mass spectrometry (GC-MS) is the gold standard for VOC analysis, while Proton-Transfer Reaction, Secondary Electrospray Ionization and Selected Ion Flow Pipe are also used in conjunction with MS (Hay et al., 2008).

16.6.2   Fungi

Fungi and mould might appear as tiny, isolated colonies that float on the medium's top and are grey, white or greenish. Visual assessment of the culture vessel for colonies that are lying on the surface of the particles and will be overlooked during the microscopic inspection is crucial (Veilleux et al., 1996). The media will become turbid and hazy if there is a large quantity of pollution, and patches on the vessel top may form. Fungal infection can occasionally raise the pH of the medium, turning phenol-red-containing media pink. The characteristic fibre-like hyphae are formed when fungal spores reach a culture container and begin to sprout. Fungi that develop here on the bottom of the dish/flask are simpler to see under a microscope than those which float on top (Wise et al., 1995).

16.6.3   Virus

Viruses are one of the most challenging cell culture toxins to detect, needing microscopy methods that are out of reach for most research labs. Endogenous retroviruses can be produced either from patients or the host animal's cell sources, and endogenous retroviruses have been discovered in numerous biotechnologically important cell lines. Viruses discovered in animal-derived materials used to cultivate cells infect cells often. Viruses are exceedingly harder to eliminate from biological medium, sera and other solutions due to their tiny size. Nevertheless, since most viruses were host- and even tissue-specific, their capacity to attack other species or tissues may be limited (Yogev & Razin, 1996).

Aside from the possible harm to cultured cells, utilising virally contaminated cell cultures puts the health of laboratory personnel at risk. When working with tissues or cells from humans or other primates, extra caution should be taken to prevent viral infection (HIV, hepatitis B, Epstein-Barr, and mammalian herpes B virus, among others) from spreading to laboratory workers via cell cultures. Institutional environmental safety authorities should be consulted about procedures for dealing with potentially dangerous tissues, cultures or viruses (Ryan, 2020; Weiss, 1978).

16.6.4 Mycoplasma

Mycoplasma Contamination: Prevention, Detection, and Elimination. Mycoplasma can easily spread to other areas of the lab after contaminating a culture. Controlling the sources of mycoplasma contamination in the labs is extremely difficult. Because certain mycoplasma species can be found on human skin, they can be spread in a poor aseptic way (Adams, 1980). They can also arise from contaminated supplements like a foetal bovine serum and, more crucially, infected cell cultures. Once it has infected a culture, it can quickly travel around the lab, contaminating other places. For successful mycoplasma contamination management, strict adherence to excellent lab procedures such as good aseptic technique is required, and routine mycoplasma testing is highly recommended (Mycoplasma culture, DNA staining and PCR-based detection are the three most popular methods for detection). Mycoplasmas affect the cell culture's metabolism by creating metabolites and absorbing resources, such as nucleic acid precursors and critical amino acids. Individual species in this diverse group may ferment sugars, hydrolyse arginine, oxidise pyruvate or metabolise urea, glutamine, putrescine and other compounds by unknown methods. This could set off a chain reaction of metabolic events that affect protein, RNA and DNA synthesis, as well as purine synthesis, purine salvage and nucleotide pools. So, even if you do not suspect the existence of this contamination in the lab for Mycoplasma Identification and Mitigation, it is necessary to set a process for mycoplasma checking of cultures (Lundin & Lincoln, 1994).

Medications, either alone or in combo with antifungals, are a standard technique in tissue culture. Continued or improper antibiotic usage, like the use of therapeutic antibiotics recommended by doctors, can lead to the formation of resistant strains that are difficult to remove and may demand the use of subsequent antibiotics that are hazardous to cell cultures. In the presence of an antibiotic, a fresh study has raised new concerns regarding the likelihood of the changed expression of genes in cultured cells.

16.6.5 Yeast

Yeasts can be found almost anywhere, and they can swiftly colonise and thrive in the rich, largely undefended environment that cell cultures provide (Del Giudice & Gardella, 1984). These microorganisms are the most typically encountered cell culture contaminants due to their size and rapid growth rates. Yeasts may generally be spotted in a culture within a few days of getting infected in the absence of antibiotics, either by direct microscopic observation.

The medium becomes foggy as the salinity of the medium grows. The earliest phases of yeast infection are impossible to discern macroscopically because the pH varies very slightly, leading to little or no change in colour in phenol red-based media. At magnifications varying from 100x to 400x, yeasts may be observed using phase contrast (Barkley, 1979). Phase-contrast enhances detection, especially at low levels of contamination. Yeasts emerge between the cells as brilliant ovoid particles. They can take the form of single cells, chains or branches (Del Giudice & Gardella, 1984).

16.7   Problem of Cross-Contamination by Other Species

Cross-contamination of many cell lines with HeLa and other fast-growing cell lines is a serious issue with possibly lethal consequences. Cross-contamination can be avoided by purchasing cell lines from trustworthy cell banks, regularly checking the characteristics of the cell lines and using the correct aseptic technique. Cross-contamination in cell cultures can be confirmed using DNA fingerprinting, karyotype analysis and isotype assessment.

16.7.1   Methods for the Identification and Elimination of Cross-Contamination

Table for the source of contamination/method of containment

Type of contamination	Identification	Elimination	References
Biological			
Bacterial	Physical appearance, phenolred, microscope, DNA sequencing methods	Antibiotics, either alone or in cocktails.	Ricardo Franco-Duarte (2019)
Fungi/ Yeast	Physical appearance, phenolred, under a microscope, the yeast appear as individual ovoid or spherical particles, which may bud off smaller particles.	Fungizone	Guillaume-Gentil (2022)
Virus	Microscopy methods	Filtration through 0.2 μm membranes	Sinclair (2018)
Mycoplasma	Mycoplasma culture, DNA staining method and PCR-based detection.	Antibiotics, either alone or in cocktails with antifungals	Nikfarjam and Farzaneh (2012)
Chemical			
Detergent, Endotoxin and Sera	Microscope, 10% sodium hypochlorite and 70% isopropanol	Using good aseptic techniques, routinely monitoring for contamination, using frozen cell repository and antibiotics	Saher Taher (2010)

16.8 Effects of Contamination on Media and Cell Suspensions

Contamination Due to Bacteria Bacteria are difficult to discern under low power microscopy; however, there are certain telltale signs of contamination. The area between cells will appear evenly granular, and the bacteria's movement may cause it to shimmer. Flasks and cell culture reagents may also get foggy, forming a thin layer on the growth vessel or reagent bottle's surface that dissipates when the vessel is moved. An abrupt shift in pH is frequently caused by bacterial infection. The media will become acidic and yellow if the aerobic infection is present in the culture. In cell culture labs, aerobes are the most prevalent source of bacterial contamination. The medium will become basic and pink as an outcome of the contamination if the bacteria are anaerobic. Bacterial toxins commonly inhibit cell activity and, as a result, destroy cell cultures (Del Giudice & Tully, 1996).

Mycoplasmas Mycoplasmas did not often produce pH which changes in the medium or has major toxic effects on mammalian cells in the early stages of contamination. They can modify cell growth properties, cell metabolism, nucleic acid synthesis, chromosomal abnormalities, cell membrane antigenicity and virus susceptibility, as well as alter transfection rates and virus susceptibility.

Yeasts In most cases, yeast contamination in cell culture occurs through airborne transmission; however, yeasts can easily "colonise" an incubator and spread to other cultures through contact with contaminated flask or dish surfaces during cell culture manipulation. The easiest of the major cell culture contaminants to cure is yeast.

16.9 Prevention of Possible Contamination and Safe Handling of Cell Cultures – Maintaining Aseptic Conditions

Cell cultures can be kept to reduce the frequency and severity of culture-related problems, such as contamination. Lack of basic culture management practices, especially in larger labs, frequently leads to long-term concerns, putting everyone in danger of contamination (Del Giudice & Tully, 1996).

- *Use good antimicrobial techniques* – The goal of the antimicrobial technique is to form a barrier between ambient microbes and your cultured and sterile items while yet allowing you to work with them. There have been various thriving ways of creating and maintaining aseptic cell cultures.
- *Reduce accidents* – Accidents almost always include people, thus preventing them requires taking stress and human behaviour into account. When labelling treatments, cultures and other goods, be very careful, use standardised record-

keeping forms wherever feasible and make media and alternatives using documented processes and formulation sheets, to name a few ideas for preventing mishaps.

- *Keep the laboratory clean* – The amount of contamination in the laboratory can be reduced by reducing the number of airborne particles and aerosols, particularly around the incubator and laminar flow hood.
- *Routinely monitor for contamination* – Testing all new lots of chemicals, medium and notably sera, as well as testing the water cleanliness at least once a year with the help of the most sensitive culture assay available is the best strategy to avoid contamination.
- *Use frozen cell repository strategically* – In labs, a cryogenic cell repository is used to avoid the need for large-scale culture transfers and to provide replacements for cultures that have been eliminated due to contamination or accidents. Freezing cultures also slow down biological time, inhibiting the emergence of new characteristics that can emerge in actively growing cells as a result of environmental or age-related changes.
- *Use antibiotics sparingly if at all* – Antibiotics should not be used instead of appropriate aseptic procedures, although they might be used by planning to avoid the loss of culture and important studies (Gabridge, 1985; Barile, 1973; Perlman, 1979).

16.10   Conclusion

Even as the successful use of cell culture technology expands in research institutions and biopharmaceutical manufacturing sites around the world, cell culture contamination remains a potential hazard. Losses because of cell culture contamination, particularly through mycoplasma, cost cell culture users millions of dollars each year in the United States alone; money that could otherwise be spent on more research. For most cell culture workers, contamination is an issue. At least 23% of those polled said their cultures had been contaminated with mycoplasma. The overuse of antibiotics is one of the major causes of widespread contamination (Wang, 1976). Cryptogenic pollutants, notably mycoplasmas, are predicted to grow more prevalent as percent of respondents frequently use antibiotics. Because of the seriousness of mycoplasma contamination and its global dispersion, it is vital to review the key sources of mycoplasma contamination as well as the basic procedures for preventing it in your lab (Weiss, 1978). Common sense and the implementation of some basic cell culture handling methods should help to mitigate the threat and increase the use and application of cell culture technology in research and industry. To build a competent cell culture management program, you must first have a basic grasp of cell culture contamination. Incoming cell line quarantine and routine mycoplasma contamination monitoring will significantly reduce the occurrence of biological contamination in the laboratory. Antibiotics are no longer used, which means there will be fewer lapses in aseptic practice. Cross-contamination will be reduced and the implications of any contamination occurrence will be minimised if contamination is detected more quickly.

Although it can never be completely eradicated, it can be better regulated and the harm it does is substantially lessened by obtaining a good insight into the nature of contamination and using certain fundamental ideas.

References

Adams, R. L. P. (1980). *Cell culture for biochemists* (Vol. 172, pp. 1–11). Elsevier/Plenum Press.

Barile, M. F. (1973). *Mycoplasmal contamination of cell cultures: Mycoplasma-virus-cell culture interactions in contamination in tissue culture* (pp. 131–171). Academic.

Barile, M. (1977) Mycoplasma contamination of cell cultures: A status report. In *Cell culture and its applications* (pp. 291–334). Academic.

Barkley, W. E. (1979). Safety considerations in the cell culture laboratory. *Methods in Enzymology: Cell Culture, 58*, 36–44.

Case-Gould, M. J. (1984) Endotoxin in vertebrate cell culture: Its measurement and significance. In *Uses and standardization of vertebrate cell lines* (pp. 125–136).

Corning Inc General procedures for the cell culture laboratory (1987)

Del Giudice, R. A., & Gardella, R. S. (1984). *Mycoplasma infection of cell culture: Effects, incidence, and detection* (In vitro monograph No. 5) (pp. 104–115).

Del Giudice, R. A., & Tully, J. G. (1996). Isolation of mycoplasmas from cell cultures by axenic cultivation techniques. *Academic, 11*, 411–418.

Fogh, J. (Ed.). (1973). *Contaminants in tissue culture*. Academic.

Fogh, J., Holmgren, N. B., & Ludovici, P. P. (1971). A review of cell culture contaminations. *In Vitro, 7*(1), 26–41.

Franco-Duarte, R. (2019). Advances in chemical and biological methods to identify microorganisms- from past to present. *Microorganisms, 7*(130).

Freshney, R. I. (1981) Culture of animal cells: A manual of basic technique.

Gabridge, M. G. (1985). Vessels for cell and tissue culture. In *Setting up and maintenance of tissue and cell cultures* (Vol. C102, pp. 1–19).

Gabridge, M. G., & Lundin, D. J. (1989). *Cell culture user's guide to mycoplasma detection and control*. Bionique Testing Laboratories.

Guillaume-Gentil, O., Gäbelein, C. G., Schmieder, S., Martinez, V., Zambelli, T., Künzler, M., & Vorholt, J. A. (2022). Injection into and extraction from single fungal cells. *Communications Biology, 5*(1), 180.

Hay, R. J., Reid, Y. A., Miranda, M. G., & Chen, T. C. (2008). Current techniques for cell line authentication. *Biotechnology International*, 143–147.

Lincoln, K. L., & Gabridge, M. G. (1998) Cell culture contamination: Sources, consequences, prevention and elimination. In *Methods in cell biology* (Vol. 57, pp. 49–65). Academic.

Lincoln, C. K., & Lundin, D. J. (1990). Mycoplasma detection and control. *United States Federation for Culture Collection Newsletter, 20*(4), 1–3.

Lundin, D. J., & Lincoln, C. K. (1994). *Mycoplasmal contamination of cell cultures within the clinical diagnostic laboratory*. The American Clinical Laboratories.

Luzak, J., Pawar, S. A., Knower, S. A., Cox, M. S., Dubose Jr., J., & Harbell, J. W. (1993). *Trends in the incidence and distribution of mycoplasma contamination detected in cell lines and their products*.

MacMichael, G. J. (1986). *The adverse effects of UV and short wavelength visible radiation on tissue culture*. American Biotechnology Laboratory.

McGarrity, G. J. (1988) Working group on cell culture mycoplasmas. Annual report to the international research program in comparative mycoplasmology.

McGarrity, G. J., Sarama, J., & Vanaman, V. (1985). Cell culture techniques. *ASM News, 51*, 170–183.

Nikfarjam, L., & Farzaneh, P. (2012). Prevention and detection of mycoplasma contamination in cell culture. *Cell Journal, 13*(4), 203–212.

Perlman, D. (1979). Use of antibiotics in cell culture media. *Methods in Enzymology: Cell Culture, 58*, 110–116.

Robinson, L. B., Wichelhausen, R. H., & Roizman, B. (1956). *Science, 124*, 1147–1148.

Rottem, S., & Barile, M. F. (1993). Beware of mycoplasmas. *Trends in Biotechnology, 11*, 143–150.

Ryan, J. A. (2012). *General guide for identifying and correcting common cell culture growth and attachment problems*. Corning, Inc.

Ryan, J. A. (2020). *Endotoxins and cell culture*. Corning, Inc. Technical Bulletin.

Saher Taher. (2010). Cell culture contamination.

Sinclair, T. R., Robles, D., Raza, B., van den Hengel, S., Rutjes, S. A., de RodaHusman, A. M., de Grooth, J., de Vos, W. M., & Roesink, H. D. W. (2018). Virus reduction through microfiltration membranes modified with a cationic polymer for drinking water applications. *Colloids and Surfaces A: Physicochemical and Engineering Aspects, 551*, 33–41.

Upchurch, S., & Gabridge, M. G. (1983). De novo purine synthesis, purine salvage, and DNA synthesis in normal and Lesch-Nyhan fibroblasts infected with Mycoplasma pneumoniae. *Infection and Immunity, 39*, 164–171.

Veilleux, C., Razin, S., & May, L. H. (1996). *Detection of mycoplasma infection by PCR* (Vol. 11, pp. 431–438). Academic.

Wang, R. J. (1976). Effect of room fluorescent light on the deterioration of tissue culture medium. *In Vitro, 12*, 19–22.

Weiss, R. A. (1978). Why cell biologists should be aware of genetically transmitted viruses. *National Cancer Institute Monograph, 48*, 183–190.

Wise, K. S., Kim, M. F., & Watson-McKown, R. (1995). *Variant membrane proteins* (Vol. I, pp. 227–241). Academic.

Wolf, K., (1979). Laboratory management of cell cultures. In *Methods in enzymology: Cell culture* (Vol. 58, pp. 116–119). Academic.

Wolf, K., & Quimby, M. (1973). Towards a practical fail-safe system of managing poikilothermic vertebrate cell lines in culture. *In Vitro, 8*(4), 316–321.

Yogev, D., & Razin, S. (1996). *Cloned genomic DNA fragments as probes* (Vol. 11, pp. 47–51). Academoc Press.

Chapter 17
Analysis of Cell Growth Kinetics in Suspension and Adherent Types of Cell Lines

Vaishnavi Shishodia, Divya Jindal, Sarthak Sinha, and Manisha Singh

17.1 Introduction

Cell culture is a multidimensional process that involves isolating cells from animals or plants and growing them under regulated artificial circumstances, usually outside of their native habitat (Philippeos, 2012). As discussed in the previous chapters, the majority of cell lines start as primary cultures derived from diced or enzyme-dispersed tissue and are made up of a variety of cells, which preserve the features of the tissue from which they were derived. Later, after a certain duration, primary cultures attain confluency, and due to the cellular expansion, the tissue culture flasks get occupied and need to be distributed equally at this time, typically with proteolytic enzymes like trypsin and then subcultured into individual cells. Now, this culture is commonly referred to as a cell line after this first passage with each passing day (Kaur & Dufour, 2012). Further, as the faster-growing cells proliferate in consecutive subcultures, the cellular population becomes more homogeneous. For a wide range of purposes, in vitro cell propagation has become a routine procedure in many laboratories and the variety of cell types that are generated in enormous amount (Verma et al. 2020). Since the nutrient content concentrations that influence the physicochemical and physiological aspects of cell development are completely understood, the current cell-culture methods allow for effective control of the cellular environment (Segeritz & Vallier, 2017). Cell line culturing allows researchers to analyse the immunological markers including cytokines and other proteins, assess drug dose efficacy and run cytotoxic experiments. A cell line is a long-lived cell culture that will continue to proliferate in the presence of a proper complete medium and sufficient space. The fact that cell lines have transcended the upper

V. Shishodia · D. Jindal · S. Sinha · M. Singh (✉)
Department of Biotechnology, Jaypee Institute of Information Technology (JIIT), Noida, Uttar Pradesh, India

251

specification limit and have become everlasting distinguishes them from cell strains. The Hayflick limit (or Hayflick Effect) is the maximum number of times a typical cell population can divide before it stops, most likely because the telomere length reaches a critical length (Hayflick, 1979).

Moreover, it has been also observed that the cell lines must be kept in the exponential phase to ensure survival, genetic stability and phenotypic stability (Verma et al. 2020). This implies that they must be subcultured frequently before approaching the stationary growth phase, well before the monolayer becomes completely confluent, or before a suspension that reaches the maximum suggested cell density. Creating a growth curve for each cell line is essential for evaluating the cell line's growth parameters (Geraghty, 2014).

In general, primary cultures are subcultured in a 1:2 ratio and often these continuous cell lines multiply at a faster speed and have a substantially greater split ratio when subcultured (Geraghty, 2014). The number of times the cells have been subcultured into a new vessel is known as the passage number. The number of populations doubling level (PDL) since the culture began is roughly equal to the passage number for diploid cultures (Torsvik et al., 2014). Continuous cell lines, on the other hand, are passaged at larger split ratios, therefore this is not the case. As a result, the PDL for continuous cell lines is unknown. Since the nutritional availability influences mammalian cell growth, a threshold cell size is required for DNA synthesis and carrying on the mitosis process. As a result of their cell cycle and cell division, each class of organisms has a distinct growth expansion pattern. Above all recognising the growth kinetics of different types of cell lines is the foundation for developing an efficacious evaluation system that further results in the highest possible product yield. Growth kinetics studies in cell lines happen to be an autocatalytic reaction where the rate of cellular growth is proportional to the cell concentration (Sakthiselvan et al., 2019). Moreover, there are direct and indirect approaches that are used to determine cell concentration. Direct methods include dry weight measurements of cell mass concentration and cell number density, turbidity (optical density) and plate counts, whereas, indirect techniques used for determining cell density include estimating the concentration of proteins, ATP, or DNA.

17.2   Kinetic Characterisation of Cell Culture

Cell growth is defined as a rise in the stack and the overall area caused by chemical, biological and physical processes (Tyson & Novak, 2014). Changes in the cell's macromolecular and chemical contents are used to track animal development, and each organism's growth pattern is unique. The development of cell lines during the log phase is critical for cell analysis owing to the respective division. During mitosis and DNA synthesis, a cut-off point in the cell area is required because nutrition availability determines the development of cells in mammals. Growth kinetics is indeed a reaction in which cell growth rate is proportional to cell concentration (Kovárová-Kovar & Egli, 1998). Cell concentration is determined using both direct and indirect methods (Wilson et al., 2017).

A batch growth kinetics obeys a growth curve, with the lag phase serving as the first period in which cells adjust to a new environment. If the media is supplied with much more than one sugar, several lag periods occur, and this form of development is known as diauxic growth (Chu & Barnes, 2016). The log phase follows the lag phase, in which cell mass and number expand exponentially before depletion of nutrients occurs, signalling the start of the deceleration phase (Rolfe, 2012). The link between a particular growth rate as well as its substrate concentration is explained by animal growth kinetics. The kinetics of animal growth is heavily influenced by the experimental culture conditions (Kovárová-Kovar & Egli, 1998). In batch culture, the composition and state of animal cells change over time, allowing the rate of growth in biomass concentration to be tracked (Vrabl et al., 2019).

Statistical methods are kinetic models that describe the link between rates and reactant/product concentrations and allow the pace of translation of reactions into products to be predicted (Goutsias, 2007). As a result of this simulated model, the design approach of the operating circumstances and process operating design for optimal product production became possible. Because the quantitative prediction of process variables is difficult, researchers mostly employed qualitative models instead of statistical models for gene expression systems. Complex mathematical models for the design of diverse bioprocesses have been produced as a result of developments in life science experimental methodologies and the use of sophisticated computer technologies (Abt, 2018). Industrial Biotechnology makes extensive use of such mathematical models, saving time and money through a thorough grasp of product yield optimisation tactics (Almquist, 2014). Other possible applications for mathematical models include expanding the variety of substrates, reducing undesired product production and optimising fermentation operations as a whole.

17.3   Cell Kinetics in Batch Culture

The rate of cell growth is proportional to the cell concentration in growth kinetics, which is an autocatalytic reaction. Cell concentration is determined using both direct and indirect methods. The batch growth kinetics follows a growth curve, with the lag phase serving as the interval during which cells acclimate to a new environment. When the media is provided with a large number of sugars, many lag periods occur, and this type of development is referred to as diauxic growth (Chu & Barnes, 2016). The log phase comes after the lag phase, during which cell mass and number rise significantly before nutrients are depleted, signalling the start of the deceleration phase. A deceleration phase occurs as harmful substances accumulate, followed by a standstill phase in which the rate of increase equals the rate of mortality. The kinetics of continuous growth are obtained through a perpetual treatment procedure in which the rate-limiting food concentration governs growth.

17.3.1 Lag Phase

The cells are adjusting to their new surroundings throughout this time. Enzymes are created from scratch. There may be a minor rise in cell mass and volume, but still no increase in overall number. Insufficient inoculum size, poor inoculum quality (high percent of dead cells), inoculum age and nutrient-deficient medium all prolong the process. It is possible that cell lysis will occur, resulting in a decrease in viable cell mass (Bertrand, 2019).

17.3.2 Log Phase

The cells inside the medium have acclimated to their new surroundings and then they develop at the end of the lag phase. During the log phase, the growth of the cell is at its peak (max). During the growth phase, the cells grow faster and the cell mass (or the number of live cells) doubles regularly (Liu, 2017).

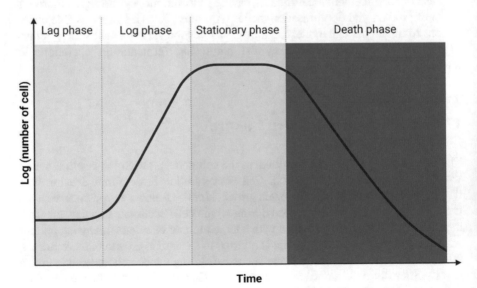

17.3.3 Time Profile of the Concentration of Cells, Nutrients and Metabolites

Whether cells coordinate their development with nutrition availability is a major topic in cell biology. Cells have evolved sophisticated ways to detect nutritional status and modify their activity to sustain growth or deal with stress. The AMP-activated kinase (AMPK) operates as a master energy sensor, modulating cellular behaviour about energy stress, whereas the target of rapamycin (TOR) controls cell development by monitoring amino acid levels and growth-boosting signals (Yuan et al., 2013). The dioxygenase family, which includes prolylhydroxylase, lysine demethylase and DNA demethylase, has emerged as potential metabolic sensors for controlling gene expression and physiological processes (Iyer, 2009; Loenarz & Schofield, 2008). As a result, nutrients and metabolites play an active role in cellular regulation via several methods.

The medium is chosen appropriately depending on the nutritional needs as well as the function for which the cultivated cells are required. The most widely used mediums are Eagle's basal medium and minimum essential media. Serum can be used to complement the media (Philippeos, 2012).

Because particular nutrients are quickly used up and depleted, further feeding of those nutrients is frequently necessary. Glucose, glutamine and cystine are among them. Suspension cultures are grown on media that are devoid of calcium and magnesium, as their absence reduces surface attachment (Bhatia et al., 2019a).

Non-nutrient chemicals are frequently added to the mixture to improve cell growth results. The addition of sodium carboxymethyl cellulose to the medium helps to reduce mechanical damage caused by forced aeration or the forces created by the agitating impeller. In stirred and aerated cultures, polyglycol (trade name Pluronic F-68) in the media minimises foaming (Gigout et al., 2008).

17.3.4 Variation of Cell Morphology

Cell morphological characteristics have been demonstrated to vary in response to disturbance with small-molecule chemicals, including form, size, intensity and roughness of cellular compartments. Changes in cell morphological characteristics have been linked to changes in cellular function, and molecular mechanisms of action have been inferred using image-based cell profiled or cell morphological profiling. The LINCS Project recently analysed gene expression and performed graphics cellular profiling on 9515 distinct chemicals in cell lines (Cheng & Li, 2016).

These findings allow the researchers to investigate the relationship between transcription and cell shape. Previous approaches to studying cell phenotypes have focused on identifying potential genes as parts of known pathways, RNAi morphological profiling and documenting morphological abnormalities; however, these approaches lack a clear model to relate transcriptome changes to morphological changes. To address this, we suggest using a cell morphology enrichment analysis to determine the relationship between transcriptome changes and cell morphology changes (Nassiri & McCall, 2018). The method may also be utilised to anticipate related effects on cellular morphology for a new transcriptome query. They have used a human bone osteosarcoma cell line to illustrate the efficacy of our technique by examining cell morphological changes (Mohseny, 2011).

17.3.5 Determination of Cell Population Heterogeneity for Kinetic Studies

The combined activity of cells from huge populations inside a tissue or animal colony characterises organisms and animal systems. Individual cells often display a significant level of phenotypic variability within every genetically homogenous population (Lu, 2007). Cellular heterogeneity, on the other hand, might be undesirable from a bioprocess or medical standpoint, since it can lower biotechnological process productivity or lead to the establishment of tumour resistance in cancer therapy (Altschuler, 2010; Mansoori, 2017). The importance of cellular heterogeneity for both basic understanding and applications of living systems has prompted the development of experimental techniques for elucidating heterogeneity through high-throughput measurement techniques on single cells, as well as the creation of computational models to visualise the emergence, dynamics and effects of cellular variability.

Mathematical models that represent the dynamics of a huge number of live cells are known as cell population models (Charlebois & Balázsi, 2019). Unlike traditional population models, which solely represent the population size, the models discussed here account for heterogeneity by enabling individual cells inside a potentially vast population to adopt distinct states. Heterogeneity in cellular populations is fundamentally a statistical property. From observations of a tiny cell count overtime or a big cell count in a single moment, a wide range of cellular activity can be inferred. In practice, statistics for a cellular community are frequently calculated from snapshots of large numbers of cells. Flow cytometry, electrophysiology, microscopy and single-cell PCR or sequencing are just a few of the technologies that can be used to acquire single-cell properties (Valihrach et al., 2018). For some technologies, the cellular characteristics possibilities are well defined, while for others, they are less obvious. These models may incorporate both population dynamics, which vary the number of cells, and cellular dynamics, which influence the course of individual cells. Intracellular activities in metabolism, signalling and gene control are the most common sources of cellular dynamics.

17.3.6   Cell Kinetics in Continuous Culture

A continuous cell culture, sometimes known as immortal cell culture, appears to be capable of an infinite number of population doublings. In vitro, neoplastic or malignant transformation traits may or may not be expressed in these cells. Continuous cell lines are often aneuploid, with chromosomal numbers ranging from diploid to tetraploid. There is also a lot of variety in chromosomal number and structure among the population's cells (heteroploidy).

Stirred culture and continuous culture are very similar. However, in this technique, the old liquid medium is replaced with the fresh liquid medium regularly to maintain the physiological states of the growing cells. Normally, the liquid medium is not changed until some nutrients in the media have been depleted, and the cells are kept in the same medium for an extended period. As a result, the cell's active development phase reduces nutrition depletion. Due to the constant flow of nutrient medium in a continuous culture system, nutritional depletion does not occur, and the cells always remain in an active growth phase (Sakthiselvan et al., 2019).

Some chief properties of continuous cell lines:

• Reduction of density limitation of growth
• Aneuploidy
• Growing in the semisolid media
• Reduced serum requirement

17.3.7   Advantages of Continuous Culture

• Cell cycle mechanics and specialised cell function are studied.
• Toxicology testing is used to investigate the effects of new medications.
• Replacement of non-functional genes with functional gene-carrying cells in gene therapy.
• Cancer cell characterisation, as well as the effects of different drugs, and radiation on cancer cells.
• Vaccines, mABs and pharmaceutical medicines are manufactured.
• Virus manufacturing for use in vaccine production.
• In culture, continuous cell lines demonstrate cell faster development and higher cell densities.
• There may be protein-free and serum-free media for commonly used cell lines on the market.
• Large-scale bioreactors might be used to cultivate the cell lines in suspension (Segeritz & Vallier, 2017).

17.3.8   Applications of Continuous Culture

As they are frequently obtained from cancerous cells, continuous cell lines are sub-cultured indefinitely in glass, plastic surfaces or solutions. These cell lines are frequently employed due to their ability to develop quicker and for longer periods, as well as their low nutritional requirements and excellent plating efficiency. It is also employed in drug discovery and large-scale production of biological substances such as vaccines and therapeutic proteins. Some common examples are HeLa, HEp2 and BHK 2 (Bhatia et al., 2019b; Righelato & Elsworth, 1970).

17.3.8.1   Estimation of Specific Rates of Cell Metabolism in Continuous Culture

Specific rates (glucose, glutamine, amino acids) or metabolite production rates (ammonia, lactate, monoclonal antibody) are significant criteria for cell culture reactor performance and control (Li, 2010).

The specific rate is denoted by μ and it is estimated by using the following formula:

$$\frac{dX(t)}{dt} = \mu Xv(t) - DX(t)$$

Where $X(t)$ is the total cell concentration, $Xv(t)$ is the viable cell concentration and D is the dilution rate.

In continual chemostat cultures, data were only collected after at least 3 days of steady-state. In the spinner reactor, which was used as a chemostat, several continuous cultivations were carried out. At various cell densities, steady-state values were obtained. With a growing number of cells and lowering the concentration of glucose, a decreasing glucose absorption rate can be seen in the chemostat data (Lee et al., 2011).

17.4   Influences of Physiological Conditions and Rate Equations

17.4.1   Effect of Nutrients

Micronutrients, including minerals and vitamins, are required for DNA metabolic pathways and are thus just as necessary as macronutrients for survival. Genomic instability affects homeostasis without the right nutrition, leading to chronic illnesses and certain types of cancer (Ferguson, 2015). Cell-culture media attempt to replicate the in vivo environment, resulting in vitro models that can be used to predict cell responses to various stimuli (Arigony, 2013).

17.4.2   Effect of Temperature

The ideal temperature for cell growth is mostly determined by the host's body temperature, with anatomical temperature variations influencing it to a lesser amount (for example, inversion of skin may be lower than the temperature of skeletal muscle). Overheating is a more significant concern to cell cultures than underheating, hence the incubator temperature is normally adjusted slightly lower than the ideal temperature (González-Alonso, 2000).

17.4.3   Effect of pH

pH changes in the cell culture environment can affect almost every cellular process, including metabolism, cell proliferation and membrane potential. Many organelles, such as mitochondria, have a little different pH than the cytoplasmic pH, but if the variances are too large, the organelles' functionality suffers. At pH 7.4, most typical mammalian cell lines grow effectively, and there is little variation among cell strains. Some altered cell lines, on the other hand, have been demonstrated to grow better in slightly more acidic settings (pH 7.0–7.3), whereas some normal fibroblast cell lines prefer slightly more basic surroundings (pH 7.4–7.3) (Michl et al., 2019).

17.4.4   Dissolved Oxygen and Oxygen Uptake Rate

Since oxygen is required for cell respiration and division, it is vital to include it in cell culture for maximal energy, protein production and cell growth. The suggested operating range for dissolved oxygen is 30–40%. Protein degradation due to oxidation can occur if dissolved oxygen levels become too high. Carbon dioxide levels that are too high can also impair cell development and protein production. While dissolved oxygen is an important process characteristic, proper process performance requires an appropriate oxygen supply. The rate of oxygen transport must be in harmony with the rate of cell consumption to promote optimal growth and/or product creation (Fleischaker & Sinskey, 1981; Palomares & Ramirez, 1996).

17.4.5 Dissolved Carbon Dioxide

The pH of the culture is regulated by the growth media, which also shields the cells from pH changes. An organic or CO_2-bicarbonate-based buffer is typically used to achieve this. Because the pH of the medium is based on a delicate balance of dissolved carbon dioxide (CO_2) and bicarbonate, changes in atmospheric CO_2 can have an impact (HCO_3). Exogenous CO_2 is required when using a CO_2-bicarbonate-based medium, particularly if the cells are cultivated in open plates or changed cell lines are cultured at high concentrations (Palomares & Ramirez, 1996).

17.4.6 Osmolality and Salts

Both cell volume regulation and recombinant protein output have been connected to osmolality. Hyperosmotic circumstances usually increase in cell size. Both cell volume regulation and recombinant protein output have been connected to osmolality. Hyperosmotic circumstances usually increase in cell size (Alhuthali, 2021).

While inorganic salts such as Ca^{2+}, Mg^{2+}, Na^+ and K^+ are of the essential components of culture media, they aid to maintain adequate water and nutrient balance.

17.4.7 The Rate Law for Cell Growth, Death and Productivity

Cell Growth Physical, biological and chemical variables all influence the increase in mass and physical size and this refers to cell growth. Changes in the cell's macromolecular and chemical contents are used to measure animal growth, and each growth pattern is unique. Because nutrition influences cell development in mammals, the cell's size should be kept to a minimum so that DNA synthesis and mitosis may occur. In a growth kinetics reaction, the result alone acts as a catalyst, and the cell's growth is precisely proportional to its concentration. Cell concentration is determined using both direct and indirect methods. The population doubling period, time lag and saturate density of the cell line may all be calculated and characterised for a specific cell type. A normal culture serves as the foundation for a development curve with three parts: log, lag and plateau (Bhatia et al., 2019b).

Lag Phase The lag phase is the early growth period of a subculture and re-seeding when the cell population requires time to recover. This is a period during which the population's number of cells develops exponentially as just a result of continual division. The original seeding densities, cell rate of growth as well as the density at which cell proliferation is inhibited determine the length of a log phase. Before rapid expansion, the cell number remains steady. The cell replaces glycocalyx elements lost during trypsinisation, adheres to the substrate and expands out during this phase. The cytoskeleton reappears during the spreading phase, and its recurrence is most likely an important aspect of this process (Rolfe, 2012).

Log Phase This is a period during which the population's number of cells develops exponentially as just a result of continual division. The original seeding densities, cell rate of growth, as well as the density at which cell proliferation is inhibited determine the length of a log phase. Because the growth fraction and viability are high (typically 90–100%), and the population is at its most uniform, this phase represents the most repeatable form of the culture. The cell culture, on the other hand, may not be synchronised, and the cells may be distributed randomly throughout the cell cycle (Sakthiselvan et al., 2019).

Stationary Phase At the end of the log phase, the culture becomes collinear as growth rates slow and cell multiplication slows owing to exhaustion in some cases. The growing surface has been occupied, and the cells are in contact with each other. The culture reaches the stationary phase at this point, with the growth fraction dropping to between 0% and 10%. Furthermore, the composition and voltage of a cell surface may alter, as well as the percentage of specialised vs. structural proteins generated (Watanabe & Okada, 1967).

Decline Phase In this phase, cell death predominates, and the number of viable cells decreases. Cell death is caused by the natural course of the biological cycle and not by a lack of nutrition.

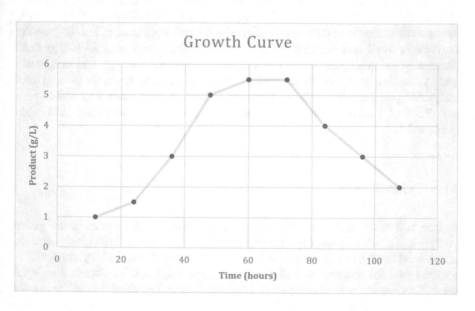

Cell Death Since its discovery in 1972, cell death, particularly apoptosis, has been a major focus of scientific research. Due to a scarcity of resources including glycogen, glutamate, growth regulators and air, or the existence of hazardous metabolites such as ammonia and lactate, cells quickly suffer apoptosis at the end of batch culture. The situation is more complicated in fed-batch and continuous cultures since nutrients are constantly provided and metabolites are eliminated in some circumstances. When nutrients are at critical small doses, or even in response to various environmental stress advantages such as rising osmotic pressure, ionic strength large swings, oxygen contours and deprivation due to inadequate blending at high cell densities as well as the prospective generation of reactive oxygen species, cells can undergo apoptosis. Apoptosis can also be triggered by increased hydrodynamic forces caused by bubble bursting just at the surface of the liquid, gas sparging, liquid flow and gas entrainment, including impeller stream energy dissipation. As a result of the rise in apoptosis, productivity, product quality, expansion and viable cell amount may suffer (Krampe & Al-Rubeai, 2010).

The use of siRNA technology to manipulate pro-apoptotic genes has aided in the establishment of apoptosis-resistant cell lines. A study used siRNA technology to knockdown Bax and Bak to 10% of their usual levels, resulting in 30–50% high viability of fed-batch and CHO batch cultures and a 35% rise in interferon product titre. This research also found that CHO with genetically inactivated targets was resistant to cytotoxic lectins, UV irradiation, nutritional deprivation and hyperosmotic conditions. Another study used zinc-finger nuclease technology to cleave Bax and Bak in the CHO cell line, which focused on the same pro-apoptotic genes. Double-blow cells produced 2 to 5 times more IgG than wild-type cells and were even more resistant to apoptosis brought on by starvation, staurosporine, and sodium butyrate in small-scale culture systems. Another study employed siRNA to mute those pro-apoptotic genes alg-2 and requiem, whose microarrays had previously been identified as up-regulated during fed-batch culture. Because of a greater sialylation pattern, the transformed cells had higher overall number of viable cells and 2.5-fold greater recombinant interferon gamma release (Lim, 2006; Xu et al., 2019).

17.5 Conclusion

Animal cell culture is now widely used, having evolved from a strictly investigative process to a standard technological component of many fields of biological research. It is crucial in all disciplines of bioscience, including medicine and agriculture. It is a useful tool for researching cell processes, physiology and biochemical components. The main benefit of cell culture is the stability and repeatability of the results produced by employing clonal cells. They are used in bioscience research and biological repair as building blocks and stem cells. In mammalian cells, a lot of progress has been achieved recently in terms of growth and cell division, but the circuits that control these activities have yet to be examined. The development of good cell culture laboratories should be prioritised, especially in developing nations where they are not yet fully established. Over the last decade, researchers have focused on the regulation of molecules of death of the cells and the interchange of specific anti- and pro-death proteins. Furthermore, we had a strong grasp of two of them, type I and type II programmed death of cells (apoptosis and autophagy), allowing us to influence their pathways via genetic and environmental techniques. Nutrient feeding and metabolic modification are currently the most effective methods for controlling cell death. Through intense process development, this technique based on fed batch culture has resulted in significantly higher productivity. The integration of cell death with metabolic deception and feed intake procedures is likely to provide a strong impetus for future development in order to increase the productivity of bioreactors and improve cell robustness, provided that the metabolic burden caused by the over-expression of the survival gene is kept to a minimum.

References

Abt, V. B. (2018). Model-based tools for optimal experiments in bioprocess engineering. *Current Opinion in Chemical Engineering, 22*, 244–252.

Alhuthali, S. (2021). Osmolality effects on CHO cell growth, cell volume, antibody productivity and glycosylation. *International Journal of Molecular, 22*(7), 3290.

Almquist, J. (2014). Kinetic models in industrial biotechnology – Improving cell factory performance. *Metabolic Engineering, 24*, 38.

Altschuler. (2010). Cellular heterogeneity: do differences make a difference? *Cell, 141*(4), 559–563.

Arigony. (2013). The influence of micronutrients in cell culture: A reflection on viability and genomic stability. *BioMed Research International, 2013*, 597282–597282.

Bertrand, R. L. (2019). Lag phase is a dynamic, organized, adaptive, and evolvable period that prepares bacteria for cell division. *Journal of Bacteriology, 201*(7), e00697–18.

Bhatia, S., Naved, T., & Sardana, S. (2019a). Culture media for animal cells. In *Introduction to pharmaceutical biotechnology* (Vol. 3, pp. 4-1–4-33). IOP Publishing.

Bhatia, S., Naved, T., & Sardana, S. (2019b). Introduction to animal tissue culture science. In *Introduction to pharmaceutical biotechnology* (Vol. 3, pp. 1-1–1-30).

Charlebois, D. A., & Balázsi, G. (2019). Modeling cell population dynamics. *In Silico Biology, 13*(1-2), 21–39.

Cheng, L., & Li, L. (2016). Systematic quality control analysis of LINCS data. *CPT: Pharmacometrics & Systems Pharmacology, 5*(11), 588.

Chu, D., & Barnes, D. J. (2016). The lag-phase during diauxic growth is a trade-off between fast adaptation and high growth rate. *Scientific Reports, 6*(1), 25191.

Ferguson. (2015). Genomic instability in human cancer: Molecular insights and opportunities for therapeutic attack and prevention through diet and nutrition. *Seminars in cancer biology, 35*(Suppl), S5–S24.

Fleischaker, R. J., & Sinskey, A. J. (1981). Oxygen demand and supply in cell culture. *European Journal of Applied Microbiology and Biotechnology, 12*(4), 193–197.

Geraghty. (2014). Guidelines for the use of cell lines in biomedical research. *British Journal of Cancer, 111*(6), 1021–1046.

Gigout, A., Buschmann, M., & Jolicoeur, M. (2008). The fate of Pluronic F-68 in chondrocytes and CHO cells. *Biotechnology and Bioengineering, 100*, 975–987.

González-Alonso. (2000). Heat production in human skeletal muscle at the onset of intense dynamic exercise. *The Journal of Physiology, 524*(Pt 2), 603–615.

Goutsias, J. (2007). Classical versus stochastic kinetics modeling of biochemical reaction systems. *Biophysical Journal, 92*(7), 2350–2365. https://doi.org/10.1529/biophysj.106.093781

Hayflick, L. (1979). The cell biology of aging. *Journal of Investigative Dermatology, 73*(1), 8–14.

Iyer, L. (2009). Prediction of novel families of enzymes involved in oxidative and other complex modifications of bases in nucleic acids. *Cell Cycle, 8*(11), 1698–1710.

Kaur, G., & Dufour, J. M. (2012). Cell lines: Valuable tools or useless artifacts. *Spermatogenesis, 2*(1), 1–5.

Kovárová-Kovar, K., & Egli, T. (1998). Growth kinetics of suspended microbial cells: From single-substrate-controlled growth to mixed-substrate kinetics. *Microbiology and Molecular Biology Reviews, 62*(3), 646–666.

Krampe, B., & Al-Rubeai, M. (2010). Cell death in mammalian cell culture: Molecular mechanisms and cell line engineering strategies. *Cytotechnology, 62*(3), 175–188.

Lee, K., Boccazzi, P., Sinskey, A., & Ram, R. (2011). Microfluidic chemostat and turbidostat with flow rate, oxygen, and temperature control for dynamic continuous culture. *Lab on a Chip, 11*, 1730–1739.

Li, F. (2010). Cell culture processes for monoclonal antibody production. *mAbs, 2*(5), 466–479.

Lim, S. F. (2006). RNAi suppression of Bax and Bak enhances viability in fed-batch cultures of CHO cells. *Metabolic Engineering, 8*(6), 509–522.

Liu, S. (2017). Chapter 11 – How cells grow. In S. Liu (Ed.), *Bioprocess engineering* (2nd ed., pp. 629–697). Elsevier.

Loenarz, C., & Schofield, C. J. (2008). Expanding chemical biology of 2-oxoglutarate oxygenases. *Nature Chemical Biology, 4*(3), 152–156.

Lu, T. (2007). Phenotypic variability of growing cellular populations. *Proceedings of the National Academy of Sciences of the United States of America, 104*(48), 18982–18987.

Mansoori, B. (2017). The different mechanisms of cancer drug resistance: A brief review. *Advanced Pharmaceutical Bulletin, 7*(3), 339–348.

Michl, J., Park, K. C., & Swietach, P. (2019). Evidence-based guidelines for controlling pH in mammalian live-cell culture systems. *Communications Biology, 2*(1), 144.

Mohseny. (2011). Functional characterization of osteosarcoma cell lines provides representative models to study the human disease. *Laboratory Investigation, 91*(8), 1195–1205.

Nassiri, I., & McCall, M. N. (2018). Systematic exploration of cell morphological phenotypes associated with a transcriptomic query. *Nucleic Acids Res, 46*(19), e116–e116.

Palomares, L., & Ramirez, O. (1996). The effect of dissolved oxygen tension and the utility of oxygen uptake rate in insect cell culture. *Cytotechnology, 22*, 225–237.

Philippeos. (2012). Introduction to cell culture. *Methods in Molecular Biology, 806*, 1–13.

Righelato, R. C., & Elsworth, R. (1970). Industrial applications of continuous culture: Pharmaceutical products and other products and processes**Dr. R. C. Righelato contributed the section on pharmaceutical products to this article. In D. Perlman (Ed.), *Advances in applied microbiology* (Vol. 13, pp. 399–417). Academic.

Rolfe. (2012). Lag phase is a distinct growth phase that prepares bacteria for exponential growth and involves transient metal accumulation. *Journal of Bacteriology, 194*(3), 686–701.

Sakthiselvan, P., Meenambiga, S., & Madhumathi, R. (2019). Kinetic studies on cell growth. *Cell Growth*, 1–9.

Segeritz, C. P., & Vallier, L. (2017). Cell culture: Growing cells as model systems in vitro. In *Basic science methods for clinical researchers* (pp. 151–172). Academic.

Torsvik, A., Stieber, D., Enger, P., Golebiewska, A., Molven, A., Svendsen, A., et al. (2014). U-251 revisited: Genetic drift and phenotypic consequences of long-term cultures of glioblastoma cells. *Cancer Medicine, 3*(4), 812–824.

Tyson, J. J., & Novak, B. (2014). Control of cell growth, division and death: Information processing in living cells. *Interface Focus, 4*(3), 20130070.

Valihrach, L., Androvic, P., & Kubista, M. (2018). Platforms for single-cell collection and analysis. *International Journal of Molecular Sciences, 19*(3), 807.

Verma, A., Verma, M., & Singh, A. (2020). Animal tissue culture principles and applications. In *Animal Biotechnology* (pp. 269–293). Academic.

Vrabl, P., Schinagl, C. W., Artmann, D. J., Heiss, B., & Burgstaller, W. (2019). Fungal growth in batch culture – What we could benefit if we start looking closer. *Frontiers in Microbiology, 10*, 2391.

Watanabe, I., & Okada, S. (1967). Stationary phase of cultured mammalian cells (L5178Y). *The Journal of Cell Biology, 35*(2), 285–294.

Wilson, C., Lukowicz, R., Merchant, S., Valquier-Flynn, H., Caballero, J., Sandoval, J., et al. (2017). Quantitative and qualitative assessment methods for biofilm growth: A mini-review. *Research & Reviews: Journal of Engineering and Technology, 6*(4), 201712.

Xu, W., Jiang, X., & Huang, L. (2019). RNA interference technology. *Comprehensive Biotechnology*, 560–575.

Yuan, H.-X., Xiong, Y., & Guan, K.-L. (2013). Nutrient sensing, metabolism, and cell growth control. *Molecular Cell, 49*(3), 379–387.

Chapter 18
In Vitro Cytotoxicity Analysis: MTT/XTT, Trypan Blue Exclusion

Shalini Mani and Geeta Swargiary

18.1 Introduction

Cytotoxicity analysis is the measure of a substance or compound's ability to cause damages or even death to the cells. Cells exposed to cytotoxic substances may experience different forms of cellular damages by various modality of actions including the cell membranes disruption, unalterable binding to receptors, inhibition of protein biosynthesis, polydeoxynucleotide elongation and enzymatic reactions that may prevent the proliferation of actively growing and dividing cells (Ishiyama et al., 1996). In addition to that, these cells may also undergo different types of death processes such as apoptosis (programmed cell death), necrosis (uncontrolled cell death) and autophagy. Therefore, there are different types of in vitro methods or assays to determine the cytotoxicity or cell death via these mechanisms. Moreover, such methods of cytotoxicity analysis need to be cheaper, reliable and reproducible. Because the in vitro cytotoxicity or cell viability analysis is the very first step to check whether a compound or a substance saves or kills the cells. In vitro cytotoxicity analysis, as well as cell viability assays with cultured cells, are broadly used in fundamental research and drug discovery to screen libraries for toxic compounds. Therefore, cytotoxicity in drug discovery is a critical endpoint for evaluating the fate and effects of a compound in the cells. Thus, the application of different cytotoxicity and cell viability assays has been gaining massive interest over the recent past years. At present, these assays are most commonly implemented in the fields of oncological researches to assess both the toxicity of the compound and their inhibiting potential towards the development of tumour cells during the drug discovery.

S. Mani · G. Swargiary (✉)
Centre for Emerging Diseases, Department of Biotechnology, Jaypee Institute of Information Technology, Noida, India

© The Author(s), under exclusive license to Springer Nature Switzerland AG 2023
S. Mani et al., *Animal Cell Culture: Principles and Practice*, Techniques in Life Science and Biomedicine for the Non-Expert,
https://doi.org/10.1007/978-3-031-19485-6_18

It is always necessary to keep the note of remaining viable and dead cells after the treatment with the cytotoxic compounds. So, the main reason behind this booming interest is due to their rapidity, inexpensiveness and likelihood for automation, as well experiments on the human cell lines might be relatively appropriate than testing on in vivo model systems i.e., no requirement of animal use. Also, it adds as an advantage by allowing tests on a larger amount of samples. Nevertheless, in vitro analyses are technologically inadequate to substitute in vivo experiments, (Chrzanowska et al., 1990). A wide range of cytotoxicity and cell viability assays are currently used in the fields of toxicology and pharmacology. The choice of assay method is crucial in the assessment of the interaction type (Sliwka et al., 2016). Moreover, depending on the type of analysis methods, the assays can be categorised in different ways.

18.2 Categories of In Vitro Cytotoxicity and Cell Viability Assays

Even though there are diverse categories of cytotoxicity and cell viability analyses, they can be majorly categorised based on their assessment forms, resultant colour variations, fluorescence, luminescence, etc. So, it can be categorised as dye exclusion methods, colorimetric assays, fluorometric assays and luminometric assays. Figure 18.1 is the schematic diagram for various assays that are included in these categories.

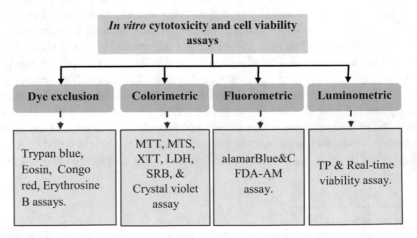

Fig. 18.1 Categories of the in vitro methods for analysing the degree of damaged, dead or viable cells

18.2.1   Dye Exclusion Methods

The dye exclusion method is the simplest and extensively used method to estimate the quantity of live and/or healthy cells in a cell population. In this technique, the dye is excluded by the living and/or healthy cells, whereas the dead or non-viable cells take up the dye. The dye exclusion method also allows determining the membrane integrity in the cells. There are many dyes such as eosin, Congo red, erythrosine B and trypan blue utilised for in vitro dye exclusion method (Bhuyan et al., 1976; Krause et al., 1984). Amongst the exemplified dyes, trypan blue is the most widely used (Aslantürk and Çelik, 2013a, b, c; Eisenbrand et al., 2002). However, the dye exclusion method is difficult for the experimental practice of bulky sample sizes and time taking (Yip et al., 1972). So, a few factors must be taken care of in the dye exclusion method. For example, cells that are critically damaged by the cytotoxic agents could take several days to lose their membrane integrity, but the viable cells may remain proliferating throughout this period. Moreover, at the end of the experiment, most of the severely injured cells may not be dyed due to the early breakdown of the cells/cellular components. Therefore, these factors can reason an erroneous estimation of the dead cells in case of the assays, grounded on the percentage viability of the cells (Weisenthal et al., 1983; Son et al., 2003; Lee et al., 2005).

Dye exclusion methods have got distinctive pros to analyse chemosensitivity. Moreover, they are relatively simpler, rapid, involve small numbers of cells and can detect cell killing in non-dividing cell populations. More studies are required to explore the possible role of dye exclusion methods in chemosensitivity analysis (Weisenthal et al., 1983). Most importantly, these dyes are intended for suspended cell lines and are not recommended for direct use on the monolayer cells, rather may be used after trypsinisation of the monolayer cells (Krause et al., 1984).

18.2.2   Colorimetric Methods

Colorimetric methods are the most widely used methods in the estimation of cell viability and cytotoxicity. The colorimetric methods are designed based on the quantification of a biochemical indicator to estimate cells' metabolic activity. In colorimetric methods, the reagents change colour in reaction to the viable cells, thereby letting the colorimetric quantification of cell viability by spectrophotometer. Unlike the dye exclusion method, colorimetric methods are appropriate for the adherent as well as suspended cell lines. Colorimetric assays are simple and easy to perform, as well as reasonably inexpensive (Präbst et al., 2017; Mosmann, 1983). Several companies also offer commercial colorimetric assay kits where the experimental processes of these methods are presented in kit packages.

18.2.3   Fluorometric and Luminometric Assays

Similar to the colorimetric methods, fluorometric and luminometric methods are also simple methods to estimate the percentage cell viability and cytotoxicity. The fluorometric assays require a fluorescence microscope, fluorometer, fluorescence microplate reader or flow cytometer. Fluorometric techniques extend several advantages over conventional dye exclusion and colorimetric methods. For example, fluorometric assays are easy to perform and appropriate for both adherents and suspended cell lines. Moreover, fluorometric and luminometric assays are extra subtle in comparison to the colorimetric assays (O'brien et al., 2000; Page et al., 1993; Riss et al., 2016).

On the other side, luminometric assays could be conveniently executed in 96-well and 384-well microtitre plates and detected by a luminometric microplate reader (Duellman et al., 2015; Mueller et al., 2004; Riss et al., 2016). In luminometric assays, the addition of reagents develops stable glow-type signals (light signals) that are detected by a luminometer. So, this characteristic of luminance is utilised to generate both viability and cytotoxicity values from the same well (Niles et al., 2009).

Commercial kits for fluorometric assays and luminometric assays are also available. However, the underlying disadvantage is that the fluorometric assays or fluorescence-based techniques, as well as luminometric assays, often require expensive as well as complex instrumentation.

Overall, the choice of the assays depends on the type as well as the aim of the study. So, the zest of the assay types is explained above and since the detailed protocols for all these categories will be very vast, therefore, in this chapter we discuss the selective and the most commonly used assays such as the Trypan blue exclusion method, MTT assay, XTT assay, LDH assay and SRB assay for cell cytotoxicity analysis.

18.3   Commonly Used In Vitro Cytotoxicity Analysis Methods

18.3.1   Trypan Blue Exclusion (TBE)

TBE method is utilised to estimate the quantity of viable as well as dead cells in a cell suspension. Trypan blue is a diazo dye frequently employed to selectively stain the dead cells in blue colour. The characteristic nature of trypan blue is that they are membrane-impermeable and hence enter the cells with compromised membrane. Due to the highly negatively charged nature of trypan blue, it is excluded from the intact cell membrane of the living cells. Moreover, trypan blue binds to the intracellular proteins and stains the cells a blue colour. Therefore, the principle of TBE method is that living cells have undamaged cell membranes that reject some dyes,

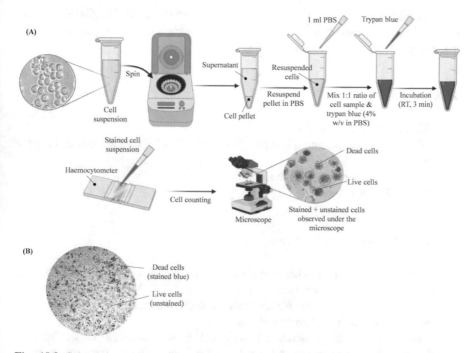

Fig. 18.2 Schematic representation of trypan blue-stained cells, where the clear cytoplasm denotes the live cells, and the dark blue-coloured cytoplasm denotes the dead cells

such as trypan blue, eosin or propidium, but the dead cells uptake these dyes (Strober, 2015). In TBE method, the cell suspension and trypan blue are mixed, then visually assessed to measure the uptake or exclusion of the dye by the viable and non-viable cells. The schematic representation of the TBE method is shown in Fig. 18.2.

Advantages
Simple and easy to perform, inexpensive, less time-consuming and a good indicator of membrane integrity (Ruben, 1988; Stone et al., 2009).

Disadvantages
Cell counting is usually performed on a haemocytometer, which is prone to the possibility of counting inaccuracies due to inadequate dilution and distribution of the cells, inadequate filling or incidence of air bubbles in the haemocytometer compartment (Ruben, 1988).

Difficulty in parallel processing of larger sample sizes especially when the precise timing of progressive cytotoxic effects is needed (Yip & Auersperg, 1972).

Difficult to distinguish the difference between healthy cells and weak cells that are alive, but slowly losing their functions.

Trypan blue is toxic and shows adverse effects on mammalian cells (Kim et al., 2016). Therefore, exposing the cells to trypan blue for a long time can cause cyto-toxicity and reduce the number of viable cells.

18.3.1.1 Materials

(a) PBS or serum-free media.
(b) 0.4% trypan blue prepared in PBS (Recommended to store in dark and filter if stored for longer period).
(c) Haemocytometer for cell counting.
(d) Trypsin EDTA solution.

18.3.1.2 Protocol

(a) Firstly, the adequate number of cells (adherent or non-adherent) are cultured in the 24-well plate and are incubated at 37 °C and 5% CO_2 for various times (e.g. 18–24 h). This incubation time allows the cells to adhere to the bottom of the plate (in the case of adherent cell lines).
(b) Once the cells have adhered, administer them with different concentrations/doses of the desired compound(s) followed by incubation for the required treatment period (e.g. 24, 48 and 72 h).
(c) As the incubation period is over, wash the treated and the untreated cells with PBS and prepare the cell suspension from each of the plates separately.
(d) Centrifuge the treated cell suspension for 5 min at 100 Å ~ g.
(e) Discard the supernatant and then re-suspend the cell pellet in 1 ml PBS or incomplete media (i.e., media without serum).
     Since the serum proteins get stained with trypan blue, it may develop ambiguous results. So, it is recommended to do the determinations in serum-free solution.
(f) Mix 1:1 ratio of the cell suspension and trypan blue (4% w/v in PBS) and incubate the mixture for ~3 min at room temperature (RT).
     Make sure that the cells are counted within 3–5 min of mixing with trypan blue. Because longer incubation periods may lead to cell death and reduce the counts of viable cells as well as lead to false-positive results.
(g) To determine whether cells take up or exclude the dye, smear a drop of the trypan blue/cell mixture to a haemocytometer and count the live (viable) and dead (nonviable) cells separately. Live cells will appear transparent and dead ones are stained dark blue.
(h) The total number of viable cells per ml of aliquot is attained by multiplying the total number of viable cells by 2 (the dilution factor for trypan blue).
(i) Calculate the percentage (%)of viable cells by using the following equation:

Fig. 18.3 Sample graph for trypan blue exclusion results

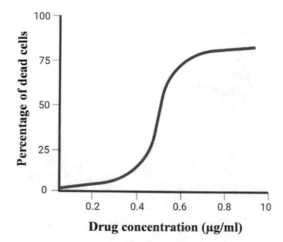

$$Viable\ cells(\%) = \frac{Total\ number\ of\ viable\ cells\,per\,ml\,of\,aliquot}{Total\ number\ of\ cells\,per\,ml\,of\,aliquot} \times 100$$

Here, 1 aliquot represents the amount of cell suspension added in each well of the plate.

Total number of cells per ml of aliquot is the total number of cells cultured on each well and the total number of viable cells per ml of aliquot is obtained from (h).

Results are represented in the form of graphs showing the concentration of the drug on the x-axis and the percentage of viable cells on the y-axis. The sample graph is shown in Fig. 18.3.

18.3.2   MTT Assay

MTT: 2-(4, 5-dimethyl-2-thiazolyl)-3, 5-diphenyl2H tetrazolium bromide.

MTT assay is the most frequently used simple colorimetric approach to estimate the cytotoxicity or the cell viability in vitro (Mosmann, 1983).The basic theory behind this method is that the cell viability is estimated by determining the mito-chondrial function of the cells, which is done by assessing the activity of mitochon-drial reductases such as succinate dehydrogenase (Stone et al., 2009). MTT is a yellow-coloured tetrazolium salt (Berridge et al., 2005). In MTT assay, NADH of the viable cells reduces the MTT to purple-coloured insoluble formazan crystal Therefore, the presence of more viable cells will produce more formazan crystals that have to be suspended by using detergents such as DMSO or isopropanol. Only the suspended form of the formazan is possible to quantify in a 96-well plate by using a spectrophotometer. So, the darker purple colour and high absorbance repre-sent the presence of more viable cells, whereas the fading purple colour and decreas-ing absorbance indicate the reducing number of viable cells, which also suggest the

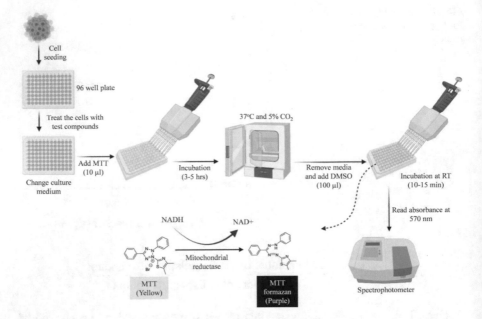

Fig. 18.4 Schematic representation of the MTT assay procedure

cytotoxic effect of the treated substance. As the MTT assay is performed in multi-well plates, it provides the ease of setting up experiments for a large number of samples in very little time. The process of MTT assay is presented in Fig. 18.4.

18.3.2.1 Materials

(a) MTT
(b) PBS
(c) Isopropanol or DMSO

18.3.2.2 Protocol

(a) Prepare the MTT solution by dissolving 5 mg of MTT powder in 1 ml PBS solution.
(b) Add 10^4 cells and 100 µl of culture medium to each well of the 96-well plate. Remember to consider three replicates for each concentration. Then incubate the cells in 5% CO_2 and 37 °C for ~12 h to allow the cells to adhere to the plate bottom.

(c) Once the cells adhere, treat the cells with various concentrations of the desired compound, followed by incubation. Here, the incubation period is your required treatment time (e.g. 24, 48 and 72 h)).

(d) After the treatment period is over, discard the media and then pour 100 μl of culture medium into each of the wells followed by the addition of 10 μl of MTT solution per well. Then keep the plates for incubation in 5% CO_2 and 37 °C for 3–4 h.

(e) Discard the supernatant, followed by the addition of 100 μl of the suitable detergent isopropanol or DMSO for dissolving the formazan crystals formed by the MTT. Preferably, slowly shake the plates for 10–15 mins to obtain a uniformly dissolved solution.

(f) Now, read at plates at wavelength 570 nm using a spectrophotometer.

(g) Compute the percentage of viable cells as follows:

$$\text{Viable cells}(\%) = \frac{\text{Total number of viable cells per ml of aliquot}}{\text{Total number of cells per ml of aliquot}} \times 100$$

(h) The data are reported in the form of a cell survival diagram (Fig. 18.5).

Advantages
- Simple, safe, high reproducibility and easy to use.
- Widely used to determine both cytotoxicities as well as cell viabilities.

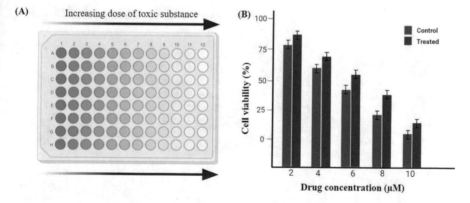

Fig. 18.5 Representation of the MTT assay results. (**a**) Sample of the 96 well-plate after MTT assay plate where the purple colour indicates MTT formazan. (**b**) Cell survival data are presented in the form of a bar graph

Disadvantages
- MTT formazan is not soluble in water, and it forms purple-coloured needle-shaped crystals in the cells. So, before the measurement of absorbance, it has to be solubilised by using organic solvents such as DMSO or isopropanol.
- Additional sets of control experiments are required to minimise the risk of false-positive or false-negative results produced by background interference due to the presence of serum and/or phenol red in the culture medium.

18.3.3   XTT Assay

XTT (Sodium 3'-[1- (phenylaminocarbonyl)- 3,4- tetrazolium]-bis (4-methoxy6-nitro) benzene sulfonic acid hydrate) assay was first introduced by Scudiero in 1988 (Scudiero et al., 1988). Similar to MTT, XTT is also a tetrazolium salt that was used as an alternative for MTT (Roehm et al., 1991). The major reason behind this alternative is that the MTT assay produces a water-insoluble formazan that necessitates an additional step for solubilising the dye to quantity its absorbance. In this assay, XTT reduces to orange-coloured formazan in metabolically active cells. The underlying mechanism of XTT reduction is the mitochondrial succinoxidase and cytochrome P450 systems as well as flavoprotein oxidases (Altman, 1976). XTT is water-soluble which simplifies the measurement of its intensity with a

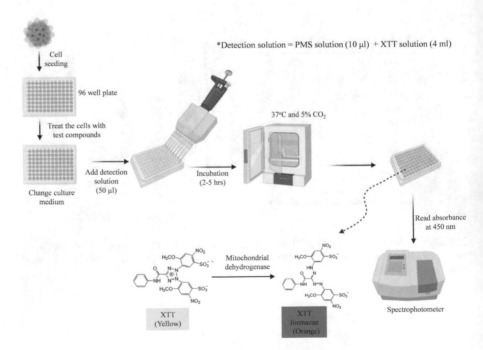

Fig. 18.6 Schematic representation of the XTT assay procedure

spectrophotometer. Moreover, the technique of XTT is simple for the measurement of cell proliferation, and hence it is an excellent solution to measure cells, determine their viability and cytotoxicity as well. Figure 18.6 illustrates the procedure of XTT assay.

18.3.3.1   Materials

(a) XTT
(b) PBS
(c) PMS (Phenazine methosulfate)

18.3.3.2   Protocol

(a) 10^4 cells with 150 µl culture medium are added to each well of 96-well plate and are incubated in 5% CO_2 and 37 °C for 12–24 h. For each concentration, three replicates are considered.
(b) After the incubation period is over, discard the media, wash with PBS and add 150 µl fresh culture medium to each of the wells.
(c) Add various concentrations of the desired compound/substance to each well and incubate for the required treatment time (e.g. 24, 48 and 72 h).
(d) After the treatment period is over, discard the media from each well and then pour 100 µl of culture media into each of the wells.
(e) Prepare PMS solution by dissolving PMS (3 mg) in 1X PBS (1 ml).
(f) Prepare XTT solution by dissolving XTT (4 mg) in cell culture media (4 ml).
(g) Add 10 µl of the PMS solution to the 4 ml of XTT solution to create the detection solution.
(h) Immediately, add 50 µl of detection solution to each well.
(i) Incubate the plates for 2–5 h at 37 °C and 5% CO_2.
(j) Place the reaction on a shaker for a short period to mix the dye in the solution.
(k) Measure the absorbance in a spectrophotometer at 450 nm wavelength immediately.

(l) The percentage of cell survival is calculated by using the following formula:

$$\text{Cell survival}(\%) = \frac{\text{Ab}(\text{Test}) - \text{Ab}(\text{Blank})}{\text{Ab}(\text{Control}) - \text{Ab}(\text{Blank})} \times 100$$

*Ab indicates absorbance
Cell inhibition (%) = 100 − Cell survival (%)
(m) The XTT assay results are presented in the form of a cell inhibition graph as shown in Fig. 18.7.

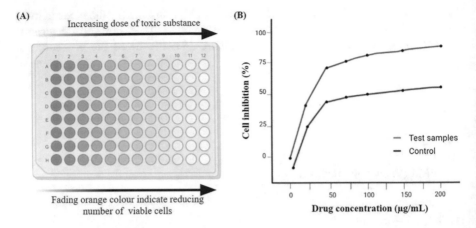

Fig. 18.7 Representation of the XTT assay results. (**a**) Sample of 96-well plate after XTT assay where the orange colour indicates the water-solubilised formazan. (**b**) Cell survival data are presented in the form of a graph

Tip
XTT solution and PMS solution can each be stored at −20 °C in the dark and should be stable for 9 months. If XTT solution changes colour or forms crystals, dispose of it properly.

Advantage
XTT technique is a fast, sensitive, convenient and safer technique with higher accuracy and sensitivity.

Disadvantage
XTT method is based on the reducing tendency of viable cells having activated mitochondrial dehydrogenase. So, alterations in the reducing tendency of viable cells due to enzymatic regulatory actions, pH, ion concentrations in the cells, cell cycle variations or added environmental influences might disturb the reading of absorbances.

18.3.4   Sulforhodamine B (SRB) Assay

SRB assay is a fast and sensitive colorimetric technique used for determining cytotoxicity in adherent as well as suspension cell cultures. This method was first described by Skehan and colleagues, who developed SRB assay to utilise in the disease-orientated, large-scale anticancer drug discovery program of the National Cancer Institute (NCI), launched in 1985. SRB is a bright pink-coloured aminoxanthene dye comprising of two sulfonic groups. In lightly acidic environments, SRB interacts with the basic amino acid residues of proteins in trichloroacetic acid (TCA)

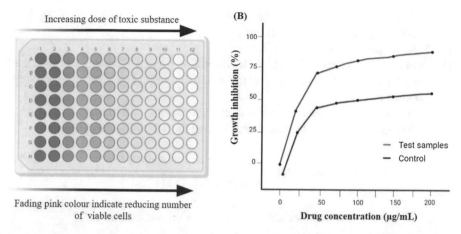

Fig. 18.8 Representation of the SRB assay results. (**a**) Sample of 96-well plate after SRB assay (**b**) Cell survival data are presented in the form of a graph

fixed cells for delivering a sensitive index of proteins in the cells. SRB method is used for assessing colony formation and extinction (Skehan et al., 1990) (Fig. 18.8).

18.3.4.1   Materials

 (i) Trichloroacetic acid (TCA)
 (ii) SRB
(iii) Acetic acid
(iv) Tris base solution

18.3.4.2   Procedure

(a) Similar to all the above-discussed assays, the process of cell seeding in a 96-well plate and treatment of the test compound is the same.
(b) Rinse the cells with PBS, followed by the addition of 100 µl culture media in each well.
(c) Gently add 25 µl cold 50% (w/v) TCA to each well. Note that the TCA should not be mixed in the plate, as it can detach some cells from the surface of the plate. So, it is recommended to gently pour the TCA into the media supernatant.
(d) Rinse the plate approximately four times in slow-running tap water, then gently tap the plate on a blotting paper to take out the excess water. Rest the plate for air-drying at RT.
(e) Add 50 µl of 0.04% (w/v) SRB solution to each well and rest the plate at RT for 1 h.

(f) Quickly wash the plate for approximately four times with 200 μl of 1% (v/v) acetic acid, which will take off the unbound dye.
(g) Rest the plate for air-drying at RT.
(h) Add 50 μl to 100 μl of 10 mM Tris base solution (pH 10.5) to each well and then keep the plate on an orbital shaker for about 10 min for solubilising the protein-bound dye.
(i) Take the spectrophotometric reading at a wavelength of 510 nm.
(j) Calculate the cell-growth inhibition % using the below-given equation:

$$\text{Cell growth } (\%) = \frac{\text{Ab}(\text{Test}) - \text{Ab}(\text{Blank})}{\text{Ab}(\text{control}) - \text{Ab}(\text{Blank})} \times 100$$

$$\text{Growth inhibition } (\%) = 100 - \text{Cell survival } (\%)$$

Note: Control represents the untreated cells

Advantage
SRB technique is a fast, easy and sensitive method that provides excellent linearity with cell number, permits the practice of saturating dye concentrations, is not dependent on intermediary metabolism, produces a determined endpoint and high reproducibility. Moreover, this technique is not much sensitive towards variations due to environmental factors.

Disadvantage
SRB technique requires a homogeneous cell suspension. Cellular clumps or aggregates must be avoided for efficient performance.

18.3.5   LDH Assay

LDH (lactate dehydrogenase) test is a colorimetric technique for assessing cellular cytotoxicity. This method determines the quantity of the stable LDH enzymes in the cytosol, liberated by the damaged cells. LDH enzyme is normally found within the cytoplasm of the cells. Upon reduced viability of the cells, the plasma membrane becomes more porous and hence the LDH enzymes are liberated into the cell culture medium. So, the liberated LDH is quantitatively determined with a coupled enzymatic reaction, where diaphorase (a catalyst) converts tetrazolium salt (iodonitrotetrazolium (INT)) into a red colour formazan. Firstly, LDH catalyses the transformation of lactate to pyruvate and thus NAD is reduced to NADH/H+. Secondly, diaphorase transfers H/H+ from NADH/H+ to the INT, which is reduced to red formazan (Decker & Lohmann-Matthes, 1988; Lappalainen et al., 1994). The activity of LDH is determined as the oxidation of NADH or reduction of INT over a definite time. The maximal absorbance of red formazan is at a wavelength of 492 nm. Therefore, the optical density (OD) of the samples in the LDH assay is measured at 490 nm. Triton X-100 is a detergent frequently used as a positive control in this assay for assessing the maximal LDH liberated from the cells. (Schins et al., 2002). The protocol is adapted from (Kumar et al., 2018).

18.3.5.1   Materials/Reagents

(a) 2-p-iodophenyl-3-p-nitrophenyl tetrazolium chloride (INT) solution
   Dissolve INT in PBS to a final concentration of 100 mM.
(b) 1-methoxyphenazine methosulfate (MPMS) solution
   Dissolve MPMS in PBS to a concentration of 100 mM. It can be stored for up to 1 month at 4 °C.
(c) LDH assay substrate solution is prepared as follows:

- L-(+)-lactic acid (0.054 M)
- β-NAD+ (1.3 mM)
- INT solution (0.66 mM)
- MPMS solution (0.28 mM)

- Dissolve the above-mentioned components in 0.2 M Tris–HCl buffer (pH 8.2). 1 ml of this substrate solution is needed for 20 reactions.

- Use freshly prepared LDH assay substrate solution (from stock solution).

(d) Lysis solution (9% [v/v] Triton X-100, for measuring total cellular LDH).
(e) Stop solution (50% dimethylformamide (DMF) and 20% SDS at pH 4.7).
   Other stop solutions such as 1 N hydrochloric acid can also be used for stopping the reaction. However, DMF/SDS solution is preferable when used with culture medium comprising Phenol Red, as DMF/SDS solution counteracts the background absorption.

18.3.5.2   Procedure

(a) Similar to all the above-discussed assays, the process of cell seeding in a 96-well plate and treatment of the test compound is the same.
(b) Rinse the cells with PBS, followed by the addition of 100 μl culture medium in each well.
(c) Add 15 μL of lysis solution to all the wells, followed by centrifugation of the plate at 250 g for 4 min.
(d) Collect 50 μL of each supernatant and put it to the wells of fresh 96-wellplate respectively.
(e) Now add LDH assay substrate (50 μL) to the medium, cover the plate with foil or a small opaque box for avoiding direct contact with light and then followed by incubation for approx. 15–30 min at 37 °C and 5% CO_2.

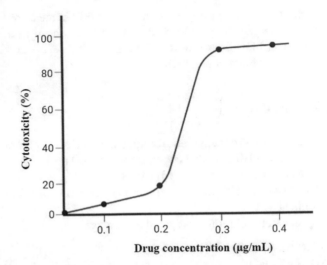

Fig. 18.9 Representation of the LDH assay result presented in the form of a graph

(f) After incubation, stop the reaction by adding the stop solution (100 μL).
(g) Take the OD at a wavelength of 490 nm. Remember that the reading should be taken within 1 h of adding the stop solution.
(h) Determine the concentration of target cells yielding absorbance values at least two times the background absorbance of the medium control.
(i) Calculate the percentage cytotoxicity by using the below-given formula:

$$\% \text{ Cytotoxicity} = \frac{\text{Experimental LDH Release} \left(OD_{490}\right)}{\text{Maximum LDH release} \left(OD_{490}\right)}$$

(j) The sample result of LDH assay is shown in Fig. 18.9.

Advantage
Due to the loss of intracellular LDH, which is released into the culture media and indicates that the cell membrane is broken and hence the cell death is not reversible, the LDH assay is a highly accurate, quick, and simple evaluation technique (Decker & Lohmann-Matthes, 1988; Fotakis & Timbrell, 2006).

Disadvantage
The main drawback of LDH assay is that serums have innate LDH activity. So, LDH assay is limited to the conditions of low serum or no serum that restricts the assay culture period and minimises the scope of the assay, as it could no longer allow the determination of cell death caused under normal growth conditions (Kumarasuriyar, 2007).

References

Altman, F. P. (1976). Tetrazolium salts and formazans. *Progress in Histochemistry and Cytochemistry, 9*(3), III-51.

Aslantürk, O. S., & Çelik, T. A. (2013a). Antioxidant, cytotoxic and apoptotic activities of extracts from medicinal plant Euphorbia platyphyllos L. *Journal of Medicinal Plants Research, 7*(19), 1293–1304.

Aslantürk, Ö. S., & Çelik, T. A. (2013b). Potential antioxidant activity and anticancer effect of extracts from Dracunculus vulgaris Schott. Tubers on MCF-7 breast cancer cells. *International Journal of Research in Pharmaceutical and Biomedical Sciences, 4*(2), 394–402.

Aslantürk, O. S., & Çelik, T. A. (2013c). Investigation of antioxidant, cytotoxic and apoptotic activities of the extracts from tubers of Asphodelus aestivus Brot. *African Journal of Pharmacy and Pharmacology, 7*(11), 610–621.

Berridge, M. V., Herst, P. M., & Tan, A. S. (2005). Tetrazolium dyes as tools in cell biology: New insights into their cellular reduction. *Biotechnology Annual Review, 11*, 127–152.

Bhuyan, B. K., Loughman, B., Fraser, T. J., & Day, K. J. (1976). Comparison of different methods of determining cell viability after exposure to cytotoxic compounds. *Experimental Cell Research, 97*(2), 275–280.

Chrzanowska, C., Hunt, S. M., Mohammed, R., & Tilling, P. J. (1990). The use of cytotoxicity assays for the assessment of toxicity. In *EHT 9329, final report to the Department of the Environment.*

Decker, T., & Lohmann-Matthes, M. L. (1988). A quick and simple method for the quantitation of lactate dehydrogenase release in measurements of cellular cytotoxicity and tumor necrosis factor (TNF) activity. *Journal of Immunological Methods, 115*(1), 61–69.

Duellman, S. J., Zhou, W., Meisenheimer, P., Vidugiris, G., Cali, J. J., Gautam, P., Wennerberg, K., & Vidugiriene, J. (2015). Bioluminescent, nonlytic, real-time cell viability assay and use in inhibitor screening. *Assay and Drug Development Technologies, 13*(8), 456–465.

Eisenbrand, G., Pool-Zobel, B., Baker, V., Balls, M., Blaauboer, B. J., Boobis, A., Carere, A., Kevekordes, S., Lhuguenot, J. C., Pieters, R., & Kleiner, J. (2002). Methods of in vitro toxicology. *Food and Chemical Toxicology, 40*(2–3), 193–236.

Fotakis, G., & Timbrell, J. A. (2006). In vitro cytotoxicity assays: Comparison of LDH, neutral red, MTT and protein assay in hepatoma cell lines following exposure to cadmium chloride. *Toxicology Letters, 160*(2), 171–177.

Ishiyama, M., Tominaga, H., Shiga, M., Sasamoto, K., Ohkura, Y., & Ueno, K. (1996). A combined assay of cell vability and in vitro cytotoxicity with a highly water-soluble tetrazolium salt, neutral red and crystal violet. *Biological & Pharmaceutical Bulletin, 19*(11), 1518–1520.

Kim, S. I., Kim, H. J., Lee, H. J., Lee, K., Hong, D., Lim, H., Cho, K., Jung, N., & Yi, Y. W. (2016). Application of a non-hazardous vital dye for cell counting with automated cell counters. *Analytical Biochemistry, 492*, 8–12.

Krause, A. W., Carley, W. W., & Webb, W. W. (1984). Fluorescent erythrosin B is preferable to trypan blue as a vital exclusion dye for mammalian cells in monolayer culture. *The Journal of Histochemistry and Cytochemistry, 32*(10), 1084–1090.

Kumar, P., Nagarajan, A., & Uchil, P. D. (2018). Analysis of cell viability by the lactate dehydrogenase assay. *Cold Spring Harbor Protocols, 2018*(6), pdb-prot095497.

Kumarasuriyar A (2007) Cytotoxicity detection kit (LDH) from Roche applied science

Lappalainen, K., Jääskeläinen, I., Syrjänen, K., Urtti, A., & Syrjänen, S. (1994). Comparison of cell proliferation and toxicity assays using two cationic liposomes. *Pharmaceutical Research, 11*(8), 1127–1131.

Lee, J. C., Lee, K. Y., Son, Y. O., Choi, K. C., Kim, J., Truong, T. T., & Jang, Y. S. (2005). Plant-originated glycoprotein, G-120, inhibits the growth of MCF-7 cells and induces their apoptosis. *Food and Chemical Toxicology, 43*(6), 961–968.

Mosmann, T. (1983). Rapid colourimetric assay for cellular growth and survival: Application to proliferation and cytotoxicity assays. *Journal of Immunological Methods, 65*(1–2), 55–63.

Mueller, H., Kassack, M. U., & Wiese, M. (2004). Comparison of the usefulness of the MTT, ATP, and calcein assays to predict the potency of cytotoxic agents in various human cancer cell lines. *Journal of Biomolecular Screening, 9*(6), 506–515.

Niles, A. L., Moravec, R. A., & Riss, T. L. (2009). In vitro viability and cytotoxicity testing and same-well multi-parametric combinations for high throughput screening. *Current Chemical Genomics, 3*, 33.

O'brien, J., Wilson, I., Orton, T., & Pognan, F. (2000). Investigation of the Alamar blue (resazurin) fluorescent dye for the assessment of mammalian cell cytotoxicity. *European Journal of Biochemistry, 267*(17), 5421–5426.

Page, B., Page, M., & Noel, C. (1993). A new fluorometric assay for cytotoxicity measurements in-vitro. *International Journal of Oncology, 3*(3), 473–476.

Präbst, K., Engelhardt, H., Ringgeler, S. and Hübner, H., (2017). Basic colourimetric proliferation assays: MTT, WST, and resazurin. In *Cell viability assays* (pp. 1–17). Humana Press.

Riss, T. L., Moravec, R. A., Niles, A. L., Duellman, S., Benink, H. A., Worzella, T. J., & Minor, L. (2016). Cell viability assays. *Assay Guidance Manual [Internet]*.

Roehm, N. W., Rodgers, G. H., Hatfield, S. M., & Glasebrook, A. L. (1991). An improved colourimetric assay for cell proliferation and viability utilizing the tetrazolium salt XTT. *Journal of Immunological Methods, 142*(2), 257–265.

Ruben, R. L. (1988). *Cell culture for testing anticancer compounds. In advances in cell culture* (Vol. 6, pp. 161–197). Elsevier.

Schins, R. P., Duffin, R., Höhr, D., Knaapen, A. M., Shi, T., Weishaupt, C., Stone, V., Donaldson, K., & Borm, P. J. (2002). Surface modification of quartz inhibits toxicity, particle uptake, and oxidative DNA damage in human lung epithelial cells. *Chemical Research in Toxicology, 15*(9), 1166–1173.

Scudiero, D. A., Shoemaker, R. H., Paull, K. D., Monks, A., Tierney, S., Nofziger, T. H., Currens, M. J., Seniff, D., & Boyd, M. R. (1988). Evaluation of a soluble tetrazolium/formazan assay for cell growth and drug sensitivity in culture using human and other tumor cell lines. *Cancer Research, 48*(17), 4827–4833.

Skehan, P., Storeng, R., Scudiero, D., Monks, A., McMahon, J., Vistica, D., Warren, J. T., Bokesch, H., Kenney, S., & Boyd, M. R. (1990). New colourimetric cytotoxicity assay for anticancer-drug screening. *JNCI: Journal of the National Cancer Institute, 82*(13), 1107–1112.

Sliwka, L., Wiktorska, K., Suchocki, P., Milczarek, M., et al. (2016). The comparison of MTT and CVS assays for the assessment of anticancer agent interactions. *PLoS One, 11*(5), e0155722.

Son, Y. O., Kim, J., Lim, J. C., Chung, Y., Chung, G. H., & Lee, J. C. (2003). Ripe fruits of Solanum nigrum L. inhibits cell growth and induces apoptosis in MCF-7 cells. *Food and Chemical Toxicology, 41*(10), 1421–1428.

Stone, V., Johnston, H., & Schins, R. P. (2009). Development of in vitro systems for nanotoxicology: Methodological considerations. *Critical Reviews in Toxicology, 39*(7), 613–626.

Strober, W. (2015). Trypan blue exclusion test of cell viability. *Current Protocols in Immunology, 111*(1), A3-B.

Weisenthal, L. M., Dill, P. L., Kurnick, N. B., & Lippman, M. E. (1983). Comparison of dye exclusion assays with a clonogenic assay in the determination of drug-induced cytotoxicity. *Cancer Research, 43*(1), 258–264.

Yip, D. K., & Auersperg, N. (1972). The dye-exclusion test for cell viability: Persistence of differential staining following fixation. *In Vitro, 7*(6), 323–329.

Chapter 19
Applications of Animal Cell Culture-Based Assays

Pallavi Shah, Anil Kumar, and Rajkumar James Singh

19.1 Introduction

In vivo models offer a better prospective of the activity of substances when tested in animals as compared to in vitro models (for example in cells), but their use has been banned in numerous countries because of several ethical and economic concerns. This has led to a tremendous increase in the use of in vitro cell-based models for various applications starting from biomarker identification, genetic manipulation, drug discovery programs, stem cell research to toxicity analysis. Often these in vitro tests are defined as bioassays. With the increase in the understanding of the underlying biochemical processes involved in cellular functions and different disease pathologies, there has been a tremendous increase in the development of a number of various cell-based assays.

19.2 Biomarker Identification

A biomarker can be defined as "a characteristic that is objectively measured and evaluated as an indicator of normal biological processes, pathogenic processes, or pharmacological responses to a therapeutic intervention (Strimbu & Tavel, 2010)". Classification of biomarkers is governed by several parameters, which include

P. Shah · A. Kumar (✉) R. J. Singh
Division of Veterinary Biotechnology, ICAR – Indian Veterinary Research Institute, Izatnagar, Uttar Pradesh, India

Rani Lakshmi Bai Central Agricultural University, Jhansi, Uttar Pradesh, India

© The Author(s), under exclusive license to Springer Nature Switzerland AG 2023
S. Mani et al., *Animal Cell Culture: Principles and Practice*, Techniques in Life Science and Biomedicine for the Non-Expert,
https://doi.org/10.1007/978-3-031-19485-6_19

characteristics such as biomarkers based on imaging (computed tomography, magnetic resonance imaging) or molecular biomarkers. Molecular biomarkers include non-imaging biomarkers which possess biophysical properties, allowing them to be determined in biological samples. These include biomarkers based on nucleic acids such as gene polymorphisms or mutations and gene expression analysis and other small molecules such as peptides, proteins, lipids and metabolites. Further classification of biomarkers can be based on their application, such as biomarkers used in disease diagnosis, biomarkers for assessing disease stages, biomarkers used in disease prognosis such as cancer biomarkers and biomarkers used for monitoring the clinical response to a potential novel treatment. Biomarkers are also utilized during early drug development such as pharmacodynamic biomarkers to monitor a certain pharmacological response and help in dose optimization studies (Huss, 2015). Modern day advances in the fields of genomics, proteomics and molecular pathology have helped characterize numerous candidate biomarkers with potential clinical value.

19.2.1 Cell Culture-Based Cancer Biomarker Identification

Cancer causes millions of deaths round the globe every year. Present cancer biomarkers have low diagnostic specificity and sensitivity and have not yet made any deep impact on reducing the cancer burden (Kulasingam & Diamandis, 2008). Limited understanding of cancer initiation and progression is one of the major reasons for the lack of effective diagnostic tools and therapeutics. Carcinogenesis is a complex process and involves alterations at the genomic, proteomic and metabolomic levels. Alterations at the genome level include gene mutations and alterations in the expression of protooncogenes, tumour suppressor genes and DNA repair genes. Alterations at the proteome level include protein alterations in the serum, plasma (secretome) and cell surface protein expression. This is further accompanied by the alterations in metabolite composition at the metabolome level. These alterations are often targeted as potential biomarkers for diagnostic, prognostic and therapeutic biomarkers. High-throughput genetics, genomics, proteomics, many non-invasive imaging techniques and other technologies allow measurement of several biomarkers (Bhatt et al., 2010).

Advances in studying cancer pathobiology heavily depend on different experimental model systems to decipher disease biology (van Staveren et al., 2009). Cancer cell line models are important in biomedical research and are critical gene discovery tools in human cancer research. Even though animal models aid in understanding the progression of cancer in vivo, they do not illustrate the molecular mechanism causing the initiation of carcinogenesis. Cancer cell lines act as a valuable model for replicating the various stages of initiation and progression of carcinogenesis in vitro. The cell line models are useful in the identification of biomarkers and potential therapeutic targets (Raju et al., 2017). Cell lines prove to be invaluable

Table 19.1 Details of various cell lines used for cancer biomarker identification

Cell line (s)	Cancer Studied	Marker (s)	Method Used for Study	Conclusion	References
DOK, NOK, KB, HN5, HN13, FaDu, Hep-2, CAL27, SCC-4, Tca8113	Oral cancer	Expression of miRNA-451, c-myc expression	Real-time PCR	miRNA 451 has tumour suppressor role by down-regulating c-myc expression	Wang et al. (2015)
HMEC, HCC1599, HCC1143, HCC1937, HCC202, HCC2218, MCF10a, HMT-3522-S1, MDA-MB-453, MFM223, MCF7	Breast cancer	Levels of IDH2, CRABP2 and SEC14L2 proteins	SILAC labelling, 2D-PAGE, LC -MS	High levels of IDH2 and CRABP2 and low levels of SEC14L2 have been used as prognostic markers for breast cancer	Geiger et al. (2012)
LCCLs, HepG2 and Huh6	Hepatocellular carcinoma	*FGF19, CK19, TP53, TSC-1, TSC-2, MET, NQO1* gene (s) expression	RNA sequencing, analysis of the transcriptome, miRNA profiling, reverse phase protein assay	LCCLs can act as a valuable resource for discovery of drug-biomarkers	Caruso et al. (2019)
ALST, CAOV3, DOV13, OVCA3, OVCA420, OVCA429, OVCA432, OVCA433, OVCA633, and SKOV3	Ovarian cancer	Osteopontin gene and protein expression	Real-time PCR, immunohistochemistry and ELISA	The findings suggest an association between plasma levels of osteopontin and ovarian cancer and hence its potential for use as a diagnostic marker	Kim et al. (2002)
SW480, SW620, and HT-29 and T84	Colorectal cancer	Akt, STAT3, AMPKα and bad activation	Angiogenesis-related antibody array, intracellular signalling arrays, activated receptor tyrosine kinase arrays, Western blot analyses	IL-6-induced activation of Akt, STAT3, AMPKα and bad- and down-regulation of EGFR, HER2 receptor, insulin R and IGF-1R can be utilized as biomarkers for CRC	Chung et al. (2015)

experimental models for studying cancer and simplify the task of molecular characterization and genetic manipulation. Cell line-based studies have elucidated several signalling pathways involved in cancer, which have been targeted to test and develop drugs and therapeutic interventions (Table 19.1).

19.3   Genetic Manipulation

The development of methods to isolate and transfer DNA from one species to another has given rise to numerous possibilities to determine the functional and regulatory mechanisms of gene expression. Genetic engineering has been employed to genetically manipulate cells to knockout genes or overexpress specific genes. Overexpression usually involves the introduction of recombinant DNA present in a vector into the cells where it transcribes and expresses recombinant protein. This is utilized for the large-scale production of recombinant proteins of biopharmaceutical importance.

19.3.1   *General Scheme for Genetic Engineering of Animal Cells in Culture for Protein Production*

1. Requirement of mammalian cells – The common mammalian host cell lines for recombinant protein expression by genetic manipulation include the Chinese hamster ovary (CHO), mouse myeloma-derived NS0 and Sp2/0 cells, human embryonic kidney cells (HEK293) and human embryonic retinoblast-derived PER. C6 cells. Mammalian cells are preferred as they provide proper folding and necessary post-translational modification of the expressed proteins.
2. Expression vector – An expression vector contains signal sequences (promoter, ribosome binding site, terminator) that are needed to ensure gene expression in mammalian cells. Further it also carries a selective marker (e.g., the dihydrofolate reductase gene (DHFR)) for selection and amplification purposes. Transfection of mammalian cells with expression vector is carried out for either transient or stable expression of gene (s). Vectors used may be plasmid-based vectors or viral vectors (adeno-associated viral (AAV) vectors, retroviral vectors, baculovirus vectors and vaccinia virus-based vectors).
3. Transfection – Often two general methods are used for the transfection of mammalian cells. In one method the direct transfer of DNA into the cells is carried out by utilizing liposomes, chemicals such as calcium phosphate, polybrene and DEAE-dextran, physical methods such as electroporation and microinjection and other methods by virus infection. Calcium-phosphate and electroporation-mediated transfection is useful for large-scale transient gene expression. For recombinant protein expression, most often CHO cells are used for transfection with the expression vector and incubated in low-nucleotide media containing appropriate selective agent based on the marker used.
4. Expression and regulation of gene expression – Cells that have incorporated the foreign gene along with the selectable marker will survive under selection pressure. By increasing selective pressure, the expression level of protein can be increased. For example- the gene of interest is co-amplified with the DHFR gene by increasing the selective pressure with high concentrations of methotrexate

(Mtx). More high yielding clones are selected by further increasing the selection pressure and clones are cryopreserved for future use. These candidates are adapted for growth in serum-free, suspension culture and evaluated for stable product yield and growth characteristics. Expression can be transient, for a short period of time or stable with cells capable of indefinite production of proteins.

19.3.2 Applications of Genetic Manipulation of Animal Cells in Culture

19.3.2.1 Recombinant Therapeutic Protein Production

Mammalian cells are the most preferred platforms for the expression of complex glycoproteins from recombinant genes such as growth factors, cytokines, hormones, enzymes, blood products and antibodies (Table 19.2). Majority of the proteins have complex structures and require complex modifications for full biological function. Glycosylation is one of the most widely recognized and complex form of post-translational modification which involves extensive processing and trimming of the protein sequence in the Golgi apparatus and endoplasmic reticulum. Such modifications can be carried out in eukaryotic cells and hence are preferable for biopharmaceutical processes. Baby hamster kidney (BHK) and Chinese hamster ovary (CHO) cells are often used as the host cells of preference as they produce human-like glycosylation patterns (Verma et al., 2020).

19.3.2.2 Gene Therapy

The deletion, insertion or alteration of a working gene copy for curing a disease/ defect or to slow the progression of a disease is known as gene therapy. It involves the identification of the malfunctioning gene followed by isolation and generation of a correct gene construct for faultless expression. Correct delivery of the genetic material in vivo or ex vivo followed by integration of gene is pivotal for successful gene therapy. In vivo therapy involves direct introduction of the genetic material into the individual at a specific site, whereas the target cells are treated outside the patient's body in ex vivo treatment. The ex vivo technique involves gene therapy in the cultured cells, which are multiplied and ultimately introduced to the targeted tissue. Gendicine is the first product designed for gene therapy used as a medication produced by Shenzhen Sibiono Genetech, China. It is used for head and neck carcinoma treatment. The tumour suppressing gene p53 is placed in recombinant adenovirus which leads to tumour control and elimination. SBN-cel is a cell line subcloned from the human embryonic kidney (HEK) cell line 293 and is utilized for Gendicine production.

Table 19.2 List of some important approved recombinant therapeutic proteins produced in animal cell lines

Cell line	Recombinant therapeutic protein	Available as	Manufacturer	Mode of action	Disease used for
CHO	Antibody, humanized IgG1/k	Avastin	Roche	Anti- VEGF	Colorectal cancer
Sp2/0	Antibody, chimeric IgG1/k	Remicade	JNJ & Merck & Mitsubishi	TNFα inhibitor	Rheumatoid arthritis, Crohn's disease
BHK	Factor VIIa	Novo seven	Novo Nordisk	Blood clotting factor VIIa	Hemophilia A + B
CHO	Humanized IgG1; DM1	Kadcyla	Genentech	By inhibiting HER2 receptor signalling	Breast cancer
CHO	Humanized IgG4κ	Keytruda	Merck, Sharpe and Dohme	Blocks PD1 pathway	Cancer
CHO	Erythropoietin (Epoetin alfa)	Epogen	Amgen, JNJ and Kyowa	Proliferation and terminal differentiation of erythroid precursor cells	Anaemia caused by chemotherapy or chronic kidney disease
CHO	Glycosylated IgG1	Ocrevus	Genentech	Anti-CD20 antibody that depletes circulating immature and mature B cells	Multiple sclerosis
CHO	Interferon beta-1a	Rebif	Merck KGaA	Acts as interferon beta-reduction of neuron inflammation	Multiple sclerosis
CHO	Human IgG1λ	Tremfya	Janssen Biotech, Inc.	Binds to the p19 subunit of interleukin 23 (IL-23) and inhibits its interaction with the IL-23 receptor	Psoriasis
CHO	Recombinant enzyme	Brineura (Cerliponase alpha)	BioMarin Pharmaceutical	Recombinant form of human tripeptidyl peptidase (TPP-1)	Batten disease

Source: Verma et al. (2020) and Zhu et al. (2017)

19.3.3 Advanced Editing Tools

With the advent of modern gene-editing tools, mammalian host cells can be easily manipulated for the development of novel, potential and cost-effective recombinant products. Even though gene editing tools such as Transcription Activator-Like

Effector Nucleases (TALENs) and Zinc Finger Nucleases (ZFNs) are useful, the discovery of the Clustered Regularly Interspaced Short Palindromic Repeats (CRISPR)-Cas system has made gene editing more precise and easy for the study and treatment of human diseases like muscular dystrophy, haemophilia and cystic fibrosis. Recently, the dystrophic gene in Duchenne muscular dystrophy (DMD) has been corrected in patient-derived induced pluripotent stem cells by the use of CRISPR-Cas system that led to the restoration of the dystrophin protein in the cells (Li et al., 2015). CRISPR-Cas-based gene editing is being utilized for the development of animal models, which can mimic human diseases. These models can provide ample opportunities for the prediction of possible clinical trial outcomes as well as ensure the safety and efficacy of drugs against the particular disease.

19.4 Pathological Studies

Pathological Studies can be defined as the branch of science which deals with the study of causes and effects of diseases. It involves the laboratory examination of body tissue samples for diagnostic or forensic purposes. Pathology deals with four components of disease i.e., cause, development mechanisms (pathogenesis), morphologic changes like structural alterations of cells and lastly the clinical manifestations like consequences of changes. General pathology largely deals with analysis of clinical abnormalities that can be used as markers or precursors for both infectious and non-infectious disease. Cell culture can play an important role in pathological studies of a disease or infection under in vitro conditions.

One of the best examples for pathological studies is the use of cell culture for virology. Viruses are infectious agents that cause widespread infections in humans, animals and plants alike. Cell-based pathological studies include visual structural / morphologic changes of cells, change in gene and protein expression of cells and change in composition of spent medium, for example, release of specific enzymes, metabolites or proteins. Rapid detection of viruses in samples taken from patients, natural samples and further study of the properties of the isolates are of immense importance. At present, numerous approaches for virus detection are used which include PCR, CRISPR/Cas technology, NGS (next generation sequencing), immunoassays and cell-based assays. Of these, cell-based assays are the only ones that can detect viable virus particles (Dolskiy et al., 2020). Cell-based detection methods provide the added advantage of isolation and characterization of virus as well. Electron microscopy can be used for the morphological examination of viral particles isolated from cell culture. The first method utilizes the cytopathic effect (CPE) of viral infection during which dead cells after virus release and can be detected and quantified under a microscope. CPE-based detection methods suffer from certain drawbacks such as they are labour-intensive with low sensitivity and they fail to detect non-cytopathic viruses. Various cell lines are used for virus detection, such as NSK, UMNSAH/DF1, BS-C-1, Vero, MDCK, A549, HEK293 or HeLa cells. They serve as substrates for the virus culture and further aid in their differential detection based on differences in efficiency and specificity (Parker et al., 2018; Lombardo et al., 2012; Liu et al., 2019).

The second method for detection and identification of viral particles is based on the development of a reporter cell line, having specific cells that are modified to make a reporter protein when infected with a virus. The reporter must possess characteristics highly specific to the particular virus to be detected. When infected by a virus, the reporter construct is recognized as a viral genome or a chimeric protein with a specific cleavage site, which generates a quantifiable signal that can be detected. Different principles for the detection of different viruses depend on their life cycle and genome structure. Infection with virus proteins or cleavage of proteins and release of reporters by viral proteases can cause the transactivation of viral structures in the vectors of some RNA viruses. Furthermore, minigenome mimic viruses can be used for the detection of negative-strand RNA viruses with nuclear replication cycles (Kainulainen et al., 2017).

19.5 Pharmaceutical Studies

Earlier studies utilized animal models for testing drugs for toxicity, corrosion, bioavailability and activity, but recent times have seen the emergence of cell culture-based assays for drug testing and development. The reason being animal models do not effectively mimic or provide human in vivo conditions as well as unexpected results are obtained during animal trials. Cell culture-based studies have made significant contributions in this regard. Contrary to animal models, the requirement of low drug quantity and a short response time are characteristic of in vitro cell culture-based assays. Well-designed in vitro cell culture studies render a huge reduction in costs spent on animal experiments. FDA has recommended the use of human cell lines to identify metabolic pathways for drugs and have mentioned their applicability in in vitro testing in their guidelines published in the year 2004 (Şahin et al., 2017).

Cancer cell lines are used extensively for the screening and testing of drugs during drug development studies. Some commonly used cell lines for drug testing are given in Table 19.3.

19.5.1 Drug Screening in Cell Lines

Drug development begins with drug testing in cancer cell lines. Drug testing in cell culture / cell lines have both advantages and disadvantages (Table 19.4). Drug cytotoxicity is often determined in cell lines and data so obtained has found relevance in clinical value prediction. A diversity of responses to drugs are displayed by different cancer cell lines and cell line panels are useful for drug tests. The first such panel was NCI-60 which utilizes 60 cancer cell lines. It was developed to reduce animal experiments for testing of the drugs (Shoemaker, 2006). Similarly, Japanese

Table 19.3 Extensively used cell lines for drug testing and development

Cell line	Species and Disease
HeLa	Human cervix adenocarcinoma
Caco2	Human colorectal adenocarcinoma
HepG2	Human hepatocellular carcinoma
HEK 293	Human embryonic kidney 293 cells
K562	Human chronic myeloid leukaemia
A549	Human lung carcinoma
MCF7	Human breast adenocarcinoma
PC3	Human prostrate adenocarcinoma
A375	Human malignant myeloma
ND-E	Human oesophageal adenocarcinoma
CHO	Chinese hamster ovary cell line
Vero	African green monkey kidney epithelial cells

Table 19.4 Advantages and disadvantages of using cell lines for drug testing

Advantages	Disadvantages
(i) Ease of handling and manipulation.	1. HeLa cells cross-contamination
(ii) High homogeneity.	2. Loss of heterogeneity and genomic stability during testing
(iii) Similarity with initial tumour.	3. Susceptibility to contamination with bacteria and mycoplasma
(iv) Unlimited infinite source of cells.	4. Challenging maintenance of cultures for long periods
(v) High reproducibility of results.	

Foundation for Cancer Research (JFCR) developed a panel consisting of 30 tumour lines derived from the NCI-60 panel, plus nine tumour cells lines specific to the Japanese population, specifically gastric cancer cells and breast cancer cells. The panel with 39 cell lines was hence named JFCR39 (Nakatsu et al., 2005).

The release assays to analyse the release of molecules from carriers, drug diffusion tests, and toxicity testing can be used in cell lines to study the mode of action, impact on physiological processes, and therapeutic treatment of disorders. Toxicity and efficacy studies are of utmost importance as an alternative to animal testing (Michelini et al., 2010). Cell lines are mostly utilized to study toxicity of drug, effect on cell viability and to carry out permeability/bioavailability studies. Present day advances in cell culture such as co-culture with normal cells, 3D matrices for cell culture and control of levels of specific growth factors and additives by microfluidic systems have enabled better mimicking of in vivo systems and thus better screening of drugs (Kitaeva et al., 2020).

Cell viability assays utilize a number of viable cell markers to determine metabolically active cells (Riss et al., 2004; Sittampalam et al., 2016). Commonly used markers include assessing the ability to either reduce a substrate, measuring enzymaticactivityoflivecellsandmeasuringATPlevels.Theseassayscanbebroadlyclassifiedinto-

(a) *Real-Time Cell Viability Assays* – These include ATP cell viability assays whereby ATP can be measured using reagents which contain a detergent, luciferase enzyme and the substrate luciferin. Viable cells are lysed by the detergent lyses, which release ATP into the medium. Luciferase utilizes this ATP to convert luciferin into oxyluciferin to generate luminescence, which can be detected by using a luminometer. Another example of real-time cell viability assay is live cell protease activity assay. In this assay, a cell-permeable fluorogenic protease substrate (GF-AFC) is used to detect the live-cell protease activity. On entering the live cells, the substrate is cleaved by live-cell protease and generates a fluorescent signal which is directly proportional to the number of viable cells.

(b) *Tetrazolium Reduction Cell Viability Assays* – Positively charged compounds such as (3-(4,5-dimethylthiazol-2-yl)-2,5-diphenyltetrazolium bromide) MTT readily penetrate viable cells. Viable cells are able to convert MTT into a purple-coloured formazan product by dehydrogenase activity in the mitochondria. This method usually has a long incubation time of 4 hours and the formazan crystals formed are insoluble, hence a solubilizing reagent like DMSO must be added before measuring absorbance. Another category of compounds include negatively charged compounds such as MTS, XTT and WST-1 that do not penetrate the cells readily and require assistance of intermediate electron coupling reagents. These can enter cells, get reduced and then exit the cell to convert tetrazolium to the soluble formazan product (Fig. 19.1).

(c) *Resazurin Reduction Cell Viability Assay* – A dark blue-coloured cell permeable indicator dye resazurin possesses little intrinsic fluorescence. Viable cells reduce resazurin into resorufin, which is pink and fluorescent. Similar to MTT-based reduction assay after hours of incubation, the pink product is measured using a microplate fluorometer.

Cell cytotoxicity assays involve measuring of cell death. Loss of membrane integrity is a major event that occurs during cell death. This makes the cells easily permeable to chemicals or proteins or molecules. Cell death can thus be monitored by detecting efflux of particular proteins like lactate dehydrogenase or influx of chemicals like DNA binding dyes. Some common cell cytotoxicity assays are

(a) *Lactate Dehydrogenase (LDH) Release Assays* – Lactate dehydrogenase (LDH) leakage occurs from dead cells that have lost membrane integrity. LDH converts lactate to pyruvate with the concomitant production of NADH. LDH activity is measured by providing the substrates lactate and NAD+ to generate NADH. Different assay chemistries are used for measuring the NADH.

(b) *DNA Dye Cytotoxicity Assay* – DNA-binding dyes cannot enter live cells, but they enter and stain the DNA of dead cells with permeable membranes. A number of fluorescent DNA binding dyes such as propidium iodide, Hoechst 33342, SYTOX Green and CellTox Green are available for determining cell cytotoxicity.

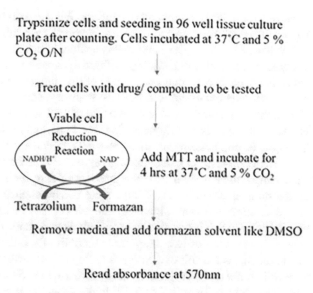

Trypsinize cells and seeding in 96 well tissue culture
plate after counting. Cells incubated at 37°C and 5 %
CO_2 O/N

Treat cells with drug/ compound to be tested

Viable cell

Reduction
Reaction
NADH·H⁺   NAD⁺   Add MTT and incubate for
4 hrs at 37°C and 5 % CO_2

Tetrazolium   Formazan

Remove media and add formazan solvent like DMSO

Read absorbance at 570nm

Fig. 19.1 MTT reduction assay

Intestinal cell permeability studies can also be carried out in cell culture models
to determine the bioavailability of the tested compounds. Cell culture models such
as the Caco2 cells have been developed and used extensively for permeability studies (Rubas et al., 1996). Caco-2 is a human colon adenocarcinoma cell line with
well-established tight junctions and capable of undergoing spontaneous enterocytic
differentiation. It is further utilized for prediction of absorption of drugs in humans
because of its similarity with permeation characteristics of human intestinal mucosa.
There is an increased use of Caco-2 cells as a screening tool in the pharmaceutical
industry at present. Lewis lung carcinoma-porcine kidney 1 (LLC-PK1) cells and
Madin–Darby canine kidney (MDCK) are other cell line models used in permeability studies.

19.6 Stem Cell Research

Animal body is made up of cells, and there are three major types of cells viz germ
cells, somatic cells and stem cells. Among these three types of cells, somatic cells
constitute the majority of the cells in an adult animal. All somatic cells in their differentiated stage carry their own copy, or copies, of the genome; the red blood cells
which are without nuclei are the only exception. Germ cells are gamete producing
cells. These cells give rise to either egg or sperm, in case of females and males,
respectively (Bongso & Richards, 2004). Whereas, undifferentiated or "blank" cells
are regarded as stem cells. This means that stem cells have the ability to differentiate
into different varieties of cells that serve different functions in various parts and

activities in the body. This ability of stem cells to differentiate into cell types beyond those of the tissues in which they normally reside is sometimes called stem cell plasticity (Biehl & Russell, 2009). The stem cells capable of not only developing into various cell types but also retaining their proliferative capacity throughout life. Therefore, stem cells are those cells that are capable of dividing indefinitely in culture and possessing the potential to differentiate into mature specialized cell types.

Depending upon the two hallmarks of stem cells, perpetual self-renewal and the ability to differentiate, they can have four possible fates. First, and most often, stem cells remain quiescent (without diving or differentiating), which ensures the maintenance of the stem cell pool. Second, they undergo symmetric self-renewal by cell division and producing daughter stem cells that are identical to their parent cells, thus increasing the number of stem cells. Third, there is asymmetric self-renewal characterized by production of two non-identical daughter cells, one identical copy of the parent and the other which is usually a more specialized somatic or progenitor cell. Fourth, there is production of two daughter cells both of which are different from the parent stem cell. As a result of stem cells taking this (fourth) fate, there is a net loss in stem cell numbers; however, there is an increase in proliferation of differentiated progenies (Biehl & Russell, 2009). With the advancements in stem cells biology, researchers have proposed some of the important attributes of "stemness" as well as the possible fates of stem cells. Some of the important attributes include (i) active Janus kinase signal transducers and activators of transcription, TGFβ and Notch signalling; (ii) sensitivity to growth factors, and integrins-mediated interaction with extracellular matrix; (iii) engagement in the cell cycle, which can be either arrested in G1 or cycling; (iv) remarkable resistance to stress with up-regulated DNA repair, protein folding, ubiquitination and detoxifier systems; (v) DNA helicases, DNA methylases and histone deacetylases regulated chromatin remodelling and modification and (vi) regulation of translation by Vasa type RNA helicases, thus regulating the Germ Cell Development and Reproductive Aging (Bongso & Richards, 2004).

There are several types of stem cells, and they are classified according to their tissue of origin viz embryonic, adult, cord blood and amniotic fluid stem cells. Embryonic stem (ES) cells are derived from 3 to 5 days old embryos. ES cells can give rise to virtually any other type of cell in the body. These pluripotent embryonic stem cells are isolated from inner cell masses (ICM) of mammalian blastocysts and cultured in a flask over a layer of "feeder" cells, which provide various factors essential for proliferation and sustaining their pluripotency. ES cell lines are developed by continuous in vitro subculture and expansion of an isolated ICM on an embryonic fibroblast feeder layer. Adult stem cells are derived from developed tissues and organs of the body. They are also known as somatic stem cells owing to the fact that they are derived from somatic cells/tissues. Unlike embryonic stem cells, adult stem cells are multipotent (not pluripotent). However, the term "adult" stem does not necessarily imply that it is originated from an adult; it can also be found in newborns and children. These cells have been extensively studied for over 50 years and have been found to play a major role in the repair and replacement of damaged tissue in the same area where they are present. A popularly studied adult stem cell is the haematopoietic stem cell (HSC), which forms the basis of successful bone marrow transplantation. Cord blood stem cells are derived from the umbilical cord. These cells are harvested and cryopreserved in cell banks for use in the future. These cells have been utilized to successfully treat children with blood cancers, such as leukaemia, and certain genetic blood abnormalities. Stem cells have also been discovered and isolated from amniotic fluid. Amniotic fluid surrounds the developing foetus inside the pregnant female's uterus. Presently, the study and application of amniotic fluid stem cells is only at its nascent stage; further research is required to understand its potential use.

With recent advancements in stem cell biology and biotechnological methods scientists are now able to transform adult stem cells into pluripotent stem cells, which are otherwise only multipotent. This transformed pluripotent adult stem cells are known as induced pluripotent stem cells (iPSCs) (Bacakova et al., 2018). iPSCs are created by genetic reprogramming of adult stem cells (known as de-differentiation) so that they behave like embryonic stem cells. Thus, iPSCs can differentiate into all types of specialized cells in the body, and potentially produce new cells for any organ or tissue.

The practice of medicine as well as other biological studies has been revolutionized by the growing knowledge of regenerative medicine and emerging biotechnologies. Developments in stem cell biology, both embryonic and adult stem cells, have transformed the prospect of tissue regeneration into an implicit reality (Sylvester & Longaker, 2004). The ability of stem cells to differentiate into various other types of cells has been exploited by scientists to devise various methods for treating and understanding diseases as well as in a variety of clinical applications. Using the perpetual potential of self-renewal, proliferation and differentiation combined with advanced targeted gene transfer technology, stem cells have been used to address numerous heritable gene defects as well as acquired diseases (Biehl & Russell, 2009). Embryonic stem cells treated with specific growth factors can be made to differentiate into specialized cells under suitable conditions. For instance, embryonic stem cells appropriately treated with growth factors, when injected into the brain can be transformed and serve as progenitors for glial cells. Likewise, new myelinated neurons or retinal epithelial cells have been created with the help of pluripotent stem cells for the treatment of animals with acute spinal cord injury or visual impairment, respectively. Similarly, diabetic animals have been treated with stem cells that produce insulin-producing cells which are responsive to blood glucose levels.

As early as the 1960s, multipotent stem cells harvested from bone marrow have been used to treat leukaemia, myeloma and lymphoma. Bone marrow-derived stem cells have been subject of great interest for the treatment of blood cancers because of their potential to differentiate and give rise to lymphocytes, megakaryocytes and erythrocytes. Also, the application of stem cells in the treatment of other diseases has been extensively explored (Kimbrel & Lanza, 2020). For example, mesenchymal stem cells that develop into bone and cartilage have been successfully used for the development of whole joints in murine models. Besides these, there have been remarkable research and application of stem cell research, some of which can be listed as below:

  (i) Correction of disorders that are associated with loss of normal cells (functions) like Alzheimer's disease, Parkinson's disease and diabetes
  (ii) Engineering, in vitro culture and replacement of damaged tissues or organs
 (iii) Study model for developmental biology
 (iv) Studying the causes of genetic defects in cells
  (v) Elucidating disease pathogenesis as well as in cancer biology in understanding mechanism of normal cells developing into cancer cells
 (vi) Discovery of new drug(s) as well as drugs safety and efficacy testing.

Besides having tremendous applicability and potential, the use of pluripotent and multipotent stem cells has their specific limitations. Pluripotent stem cells have therapeutic advantages over multipotent stem cells provided the fact that the former can become any cell type and have higher cell prolificacy (Liu et al., 2019). However, the requirement of immune-suppressive therapy entails the use of pluripotent stem cells, as these cells are not from the host and therefore trigger the host's immune response (graft-rejection). On the other hand, such rejection immune response is (theoretically) non-evident in the case of host-derived multipotent stem cells. In this context, the possibility to utilizing host's own cells has a great advantage; the immune system recognizes specific surface proteins/markers on these own multipotent stem cells and no graft-rejection immune response is developed. Contamination with animal-derived culture medium also causes serious issues leading to immunological complications. Besides these, there are various other practical and technical limitations like hurdles in manipulation and differentiation, which are also associated with stem cells culture and their therapeutic and clinical applications.

One of the major concerns in stem cell research is ethical, religious, legal and political issues. With the rapid growth and advancements in stem cell research, there is also parallel growth in various agendas and debates against it (Bacakova et al., 2018). Many people oppose the use of embryos for therapeutic purposes based on their beliefs and genuine ethical concerns. Governments in various countries banned the allocation of public money for research on human embryonic stem cell. The present rules and regulations governing study and research on stem cell also vary in different countries or societies. For clear ethical reasons, studies and experimentations involving embryonic cells, blastocyst manipulation, foetal material, foetal stem cells, ectopic grafting, etc. have generated a great deal of public interest (Bongso & Richards, 2004). Consequently, a majority of the population are still not ready to accept this cell-based therapy. In this context, it may take few more years before a scientifically-, legally- and ethically approved stem cell research application for the welfare of both humans and animals becomes a reality.

19.7 Cellular Development and Differentiation

Stem cells possess the remarkable property of self-renewal, and can simultaneously transform into more lineage-committed cells. Pluripotent stem cells (PSCs) being multipotent can give rise to all mature body cells. Human PSCs (hPSCs) were first obtained from human blastocysts and named as embryonic stem cells (ESCs) (Thompson, 1998). Further, the creation of "induced pluripotent stem cells", or iPSCs, from human fibroblasts in 2007, gave rise to the possibility of developing any cell type for regenerative medicine. In addition to regenerative medicine, the availability of human ESCs (hESCs) and iPSCs contributes to a better understanding of human development and the creation of models for the study of human disorders. Cellular differentiation especially of hESCs, which are multipotent with potential to generate any differentiated progeny, is an alluring area of study for regenerative medicine, transplantation therapy and tissue engineering. In order to reach this potential, it is imperative to control ESC differentiation and direct their development along specific pathways.

For induction of PSCs to differentiate, scientists have utilized the information available for development. Initially, differentiation comprises of the development of embryoid bodies, which undergo spontaneous differentiation as a result of signals arising from the various cell populations. For better control of the differentiation process researchers have utilized the step-wise addition of compounds such as growth factors and cytokines, or inhibitors. This controlled differentiation has given rise to a number of cell types like cells resembling cardiac muscle (Burridge et al., 2014), neural subpopulations (Shaltouki et al., 2013), or hepatocytes (Roelandt et al., 2013) and several others. Although, these cells often resemble foetal tissue rather than adult tissue in majority of cases.

Incomplete understanding of complex events that occur during development processes often make the complete differentiation of PSCs difficult. Modern day approaches to ensure lineage-specific committed differentiation of PSCs include genome editing, chemical engineering of the culture medium, recreating niche for stem cell, precursor, and mature cell growth and differentiation by using microfluidics, 3D culture, using mechanical and electrical stimulation and finally the use of vascular networks in 3D cultures for better access and availability of nutrients, oxygen and endogenous factors.

Present day advances like single-cell RNAseq data in combination with novel genome editing tools, especially the CRISPR technology, have enabled the scientists to program PSC progeny toward specific lineages. Furthermore, libraries of small molecules to identify factors that enhance differentiation have been developed by researchers. An example is differentiation of pancreatic-beta cell from PSCs, by utilizing 20 different molecules for creating insulin-responsive cells (Pagliuca et al., 2014). Recently, researchers are searching for methods such as 3D culture systems with co-culture of different cell types or PSC-derived organoid cultures to develop a "niche", in which committed cell differentiation occurs during development and ultimately giving rise to mature cells as in organs. Tailored hydrogels in 3D culture along with appropriate addition of growth factors and small molecules can copy the physicochemical properties of the in vivo environment. Apart from these, even electromechanical forces can influence the growth of developing cells similar to forces exerted by organs on their respective cells. Therefore, in numerous studies additional influence of electrical or mechanical stimulation on differentiation of cardiac and neural cells has been studied (Park et al., 2011; Eng et al., 2016).

19.8 Hybridoma Technology

A major technological breakthrough in the field of immunology is the development of monoclonal antibodies (mAb) by hybridoma technology. It is the first, most basic and successful methodology for mAb isolation. This technology has influenced numerous areas of research such as cell biology, medical microbiology, parasitology, biochemistry and physiology. Furthermore, many diagnostic and therapeutic applications of mAbs, as well as their use in the biotechnology industry, are becoming apparent. Köhler and Milstein (1975) opened up new vistas in immunology by demonstrating that somatic cell hybridization could be used to generate hybrid cells capable of growing in culture and secreting antibody molecules of predefined specificities. This noble discovery has been awarded the "Nobel" prize in 1984 for physiology and medicine. The hybridoma technology has already fulfilled the long-lasting desire of immunologists i.e., the routine production of large number of homogeneous antibodies against a wide variety of antigens.

Hybridoma technique involves the immortalization of antibody producing lymphocytes but with liming growth characteristics. The lymphocytes are fused with continuously growing tumour cell line (myeloma) cells so that hybrids produced are immortal like parent myeloma cells and continue to secrete antibodies. This technology involves selection and isolation of a hybridoma producing antibodies with predefined properties, and the mAbs so produced by a clone or family of cells derived from a single progenitor are thus homogeneous in composition (Fig. 19.2). This technology is quite robust and useful for production of mAbs for various applications such as research tools for the identification of specific epitopes of polypeptides, in vitro and in vivo identification and localization of disorders, purification of antigens and therapeutics for either prevention or treatment of diseases (Smith & Crowe Jr., 2015). Easy and efficient production of mAbs is ensured once the hybridoma clones are obtained.

The majority of antibodies in the market today are generated in cell cultures (van Dijk & van de Winkel, 2001). Preference to animal cells is given as they can carry out glycosylation and protein folding, which is essential for a protein to be effective. The CHO cell line is the most commonly used for mAb production. Apart from CHO other cell lines used include murine myelomas NSO, Sp2/0, HEK-93 and BHK. Large-scale industrial production of mAbs is possible in cell suspension culture. Anchorage free cell lines with the capacity offer an easy scale-up option. Sp2/0 and NSO cells can normally grow in suspension, whereas the CHO cells can be easily adapted for the process.

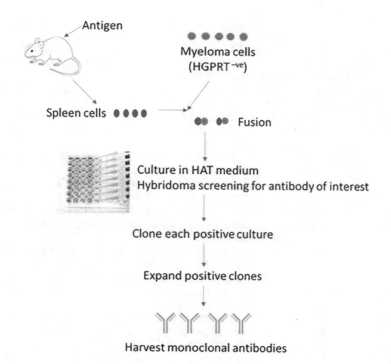

Fig. 19.2 Monoclonal antibody (mAb) production by hybridoma technology

References

Bacakova, L., Zarubova, J., Travnickova, M., Musilkova, J., Pajorova, J., Slepicka, P., et al. (2018). Stem cells: Their source, potency and use in regenerative therapies with focus on adipose-derived stem cells–a review. *Biotechnology Advances, 36*(4), 1111–1126.

Bhatt, A. N., Mathur, R., Farooque, A., et al. (2010). Cancer biomarkers – current perspectives. *The Indian Journal of Medical Research, 132*, 129–149.

Biehl, J. K., & Russell, B. (2009). Introduction to stem cell therapy. *The Journal of Cardiovascular Nursing, 24*(2), 98.

Bongso, A., & Richards, M. (2004). History and perspective of stem cell research. *Best Practice & Research. Clinical Obstetrics & Gynaecology, 18*(6), 827–842.

Burridge, P. W., Matsa, E., Shukla, P., et al. (2014). Chemically defined generation of humancar-diomyocytes. *Nature Methods, 11*(8), 855–860.

Caruso, S., Calatayud, A. L., Pilet, J., et al. (2019). Analysis of liver cancer cell lines identifies agents with likely efficacy against hepatocellular carcinoma and markers of response. *Gastroenterology, 157*, 760–776. https://doi.org/10.1053/j.gastro.2019.05.001

Chung, S., Dwabe, S., Elshimali, Y., et al. (2015). Identification of novel biomarkers for metastatic colorectal cancer using angiogenesis-antibody Array and intracellular signaling Array. 1–14. https://doi.org/10.1371/journal.pone.0134948

Dolskiy, A. A., Grishchenko, I. V., Yudkin, D. V. (2020). Cell Cultures for Virology: Usability, Advantages, and Prospects. *International Journal of Molecular Sciences, 21*(21), 7978. https://doi.org/10.3390/ijms21217978

Eng, G., Lee, B. W., Protas, L., et al. (2016). Autonomous beating rate adaptation inhuman stem cell-derived cardiomyocytes. *Nature Communications, 7*, 10312.

Geiger, T., Madden, S. F., Gallagher, W. M., et al. (2012). Proteomic portrait of human breast cancer progression identifies novel prognostic markers. *Cancer Research, 72*, 2428–2440. https://doi.org/10.1158/0008-5472.CAN-11-3711

Huss, R. (2015). Biomarkers. In A. Atala & J. G. Allickson (Eds.), *Translational Regenerative Medicine 21* (pp. 235–241). Academic.

Kainulainen, M. H., Nichol, S. T., Albariño, C. G., & Spiropoulou, C. F. (2017). Rapid determination of ebolavirus infectivity in clinical samples using a novel reporter cell line. *The Journal of Infectious Diseases, 216*, 1380–1385.

Kim, J. H., Skates, S. J., Uede, T., et al. (2002). Osteopontin as a potential diagnostic biomarker for ovarian cancer. *JAMA. 287*(13), 1671–1679. https://doi.org/10.1001/jama.287.13.1671

Kimbrel, E. A., & Lanza, R. (2020). Next-generation stem cells—Ushering in a new era of cell-based therapies. *Nature Reviews. Drug Discovery, 19*(7), 463–479.

Kitaeva, K. V., Rutland, C. S., Rizvanov, A. A., & Solovyeva, V. V. (2020). Cell culture based in vitro test systems for anticancer drug screening. *Frontiers in Bioengineering and Biotechnology, 8*, 1–9. https://doi.org/10.3389/fbioe.2020.00322

Köhler, G., & Milstein, C. (1975). Continuous cultures of fused cells secreting antibody of pre-defined specificity. *Nature, 256*, 495–497. https://doi.org/10.1038/256495a0

Kulasingam, V., & Diamandis, E. P. (2008). Tissue culture-based breast cancer biomarker discovery platform. *International Journal of Cancer, 123*, 2007–2012. https://doi.org/10.1002/ijc.23844

Li, H. L., Fujimoto, N., Sasakawa, N., et al. (2015). Precise correction of the dystrophin gene in duchenne muscular dystrophy patient induced pluripotent stem cells by TALEN and CRISPR-Cas9. *Stem Cell Reports, 4*, 143–154. https://doi.org/10.1016/j.stemcr.2014.10.013

Liu, H., Liao, H. M., Li, B., Tsai, S., Hung, G. C., & Lo, S. C. (2019). Comparative genomics, infectivity and Cytopathogenicity of American isolates of Zika virus that developed persistent infections in Human Embryonic kidney (HEK293) cells. *International Journal of Molecular Sciences, 20*, 3035.

Lombardo, T., Dotti, S., Renzi, S., & Ferrari, M. (2012). Susceptibility of different cell lines to avian and swine influenzaviruses. *Journal of Virological Methods, 185*, 82–88.

Michelini, E., Cevenini, L., Mezzanotte, L., & CoppaA, R. A. (2010). Cell based assays: Fuelling drug discovery. *Analytical Biochemistry, 397*, 1–10.

Nakatsu, N., Yoshida, Y., Yamazaki, K., Nakamura, T., Dan, S., Fukui, Y., & Yamori, T. (2005). Chemosensitivity profile of cancer cell lines and identification of genes determiningchemosensitivity by an integrated bioinformatical approach using cDNA arrays. *Molecular Cancer Therapeutics, 4*(3), 399–412.

Pagliuca, F. W., Millman, J. R., Gürtler, M., et al. (2014). Generation of functional human pancreatic β cells *in vitro. Cell, 159*(2), 428–439.

Park, S. Y., Park, J., Sim, S. H., et al. (2011). Enhanced differentiation of human neural stemcells into neurons on graphene. *Advanced Materials, 23*(36), H263–H267.

Parker, S., de Oliveira, L. C., Lefkowitz, E. J., Hendrickson, R. C., Bonjardim, C. A., WSM, W., Hartzler, H., Crump, R., & Buller, R. M. (2018). The virology of taterapox virus in vitro. *Viruses, 10*, 463.

Raju, K. L., Augustine, D., Rao, R. S., et al. (2017). Biomarkers in tumorigenesis using cancer cell lines: A systematic review. *Asian Pacific Journal of Cancer Prevention, 18*, 2329–2337. https://doi.org/10.22034/APJCP.2017.18.9.2329

Riss, T. L., Moravec, R. A., Niles, A. L., et al. (2004). Cell viability assays. *Assay Guid Man*, 1–25.

Roelandt, P., Vanhove, J., & Verfaillie, C. (2013). Directed differentiation of pluripotent stemcells to functional hepatocytes. *Methods in Molecular Biology, 997*, 141–147.

Rubas, W., Cromwell, M. E., Shahrokh, Z., Villagran, J., Nguyen, T. N., Wellton, M., Nguyen, T. H., & Mrsny, R. J. (1996). Flux measurements across Caco-2 monolayers may predict transport in human large intestinal tissue. *Journal of Pharmaceutical Sciences, 85*, 165–169.

Şahin, Ş. H. T., Mesut, B., & Özsoy, Y. (2017). Applications of cell culture studies in Pharmaceutical technology. *ACTA Pharmaceutica Sciencia, 55*(3), 63.

Shaltouki, A., Peng, J., Liu, Q., et al. (2013). Efficient generation of astrocytes from human pluripotent stem cells in defined conditions. *Stem Cells, 31*(5), 941–952.

Shoemaker, R. H. (2006). The NCI60 human tumour cell line anticancer drug screen. *Nature Reviews. Cancer, 6*(10), 813–823.

Sittampalam, G., Coussens, N., Arkin, M., et al. (2016). Assay guidance manual. *Assay Guid Man, 11*, 305–336.

Smith, S. A., & Crowe, J. E., Jr. (2015). Use of human hybridoma technology to isolate human monoclonal antibodies. *Microbiology Spectrum, 3*(1), AID-0027-2014. https://doi.org/10.1128/microbiolspec.AID-0027-2014. PMID: 26104564; PMCID: PMC8162739.

Strimbu, K., & Tavel, J. A. (2010). What are biomarkers? *Current Opinion in HIV and AIDS, 5*(6), 463–466. https://doi.org/10.1097/COH.0b013e32833ed177. PMID: 20978388; PMCID: PMC3078627.

Sylvester, K. G., & Longaker, M. T. (2004). Stem cells: Review and update. *Archives of Surgery, 139*(1), 93–99.

Thomson, J. A., Itskovitz-Eldor, J., Shapiro, S. S., et al. (1998). Embryonic stem cell lines derived from human blastocysts. *Science, 282*(5391), 1145–1147.

van Dijk, M. A., & van de Winkel, J. G. J. (2001). Human antibodies as next generation therapeutics. *Current Opinion in Chemical Biology, 5*, 368–374. https://doi.org/10.1016/S1367-5931(00)00216-7

van Staveren, W. C. G., Solís, D. Y. W., Hébrant, A., et al. (2009). Human cancer cell lines: Experimental models for cancer cells in situ? For cancer stem cells? *Biochimica et Biophysica Acta, 1795*, 92–103.

Verma A, Verma M,Singh A. (2020). Animal tissue culture principles and applications. In *Animal biotechnology* (pp. 269–293). Academic

Wang, H., Zhang, G., Wu, Z., et al. (2015). MicoRNA-451 is a novel tumor suppressor via targeting c-myc in head and neck squamous cell carcinomas. *Journal of Cancer Research and Therapeutics, 11*, 216.

Zhu, M. M., Mollet, M., Hubert, R. S., et al. (2017). *Handbook of industrial chemistry and biotechnology.* Springer.

Chapter 20
Ethical Issues in Animal Cell Culture

Divya Jindal, Vaishanavi, and Manisha Singh

20.1 Introduction

A slew of scientific breakthroughs heralded the start of a new age of scientific progress in the last century. Information technology and genetic engineering, for example, are recent examples of notable discoveries. Biotechnology has ushered in a variety of technologies during the previous few decades, including (i) bioprocessing, or cell manipulation in vitro, (ii) recombinant DNA technology and (iii) monoclonal antibodies. Biotechnology has a long and illustrious history. Since the dawn of civilization, when humans mastered the skills of "growing crops" and "breeding animals", they have also discovered how to ferment fruit juice into wine, beer and cheese, transform milk into yogurt and make spongy bread with bacteria and yeast. All of these initiatives are in the early stages of biotechnology development. Biotechnology's main goal is to develop innovative methods for the development of cells and transgenic animals. A technique for developing cells of an organism in a controlled environment is known as cell culture. Multicellular eukaryotes, pre-existing cell lines and pre-existing cell strains could all be sources for the cells. The notion of retaining live cell lines that can be separated from their primary tissue source was discovered at the starting of the nineteenth century, but animal cell culture became popular in the laboratory in the mid-1900s. Cell culture has become one of the most commonly utilized technology in life sciences studies for research with added commercial benefits. The basic culture medium development has enabled researchers to work with a diversified range of cells in controlled environments, progressing and enhancing the knowledge of cell growth and differentiation,

D. Jindal · Vaishanavi · M. Singh (✉)
Department of Biotechnology, Jaypee Institute of Information Technology (JIIT), Noida, Uttar Pradesh, India

© The Author(s), under exclusive license to Springer Nature Switzerland AG 2023
S. Mani et al., *Animal Cell Culture: Principles and Practice*, Techniques in Life Science and Biomedicine for the Non-Expert,
https://doi.org/10.1007/978-3-031-19485-6_20

identification of growth factors and mechanisms causing normal cell functioning. Bioreactors with high cell densities and culture conditions have also been investigated using new technology. Many biotechnology products (such as viral vaccinations) rely heavily on the bulk culture of animal cell lines. Although rDNA may be utilized to produce many simpler proteins in bacteria, more complex glycosylated (carbohydrate-modified) proteins must be composed in animal cells.

The goal of research in today's scenario is to figure out how culture conditions affect survival, productivity and the consistency of PTMs like glycosylation, which are important for recombinant protein biological function. Biologicals generated in animal cell cultures utilizing recombinant DNA (rDNA) technology include anticancer drugs, immunobiologicals (interleukins, lymphokines, mABs and hormones) and enzymes. Animal cell culture has been used in a wide range of studies, whether it is basic or advanced. It has served as a model system for several research initiatives working on cell-cell and cell-matrix interactions, toxicity for testing is used to investigate their effects on human cells, cell cycle mechanisms, replacement of nonfunctional genes by a functional gene in gene therapy methods, cancer cell characterization, including the effects of drugs, vaccines and pharmaceutical medicines are manufactured (Verma et al., 2020).

20.2   Ethical Concerns in Handling Animal Cell Culturing

Animal cell culture problems such as cell line misidentification, contamination with mycoplasma and phenotypic and genotypic instability are routinely overlooked by researchers. Scientific evidence is repeated by the research community quite frequently. Cells must be withdrawn or altered since they were incorrectly identified. Contamination of cell lines leads to the occult with bacteria and fungi (especially phenotypic drift caused by serial transmission (most notable mycoplasma), interactions between laboratories is common in any case, whether it is the size of the cell culture activity, big or small, and even for academic use or commercial use. There is an establishment of certain guidelines for the use of animal tissue samples in research. These recommendations and research applications emphasize the essential considerations for the proper conduct of the experiments. Cell lines can be produced in-house, purchased from a cell bank or obtained from other laboratories (if there is no other credible source) (Geraghty et al., 2014).

20.2.1   Acquisition of Cell Culture

Production of a new cell line from fresh tissue, especially when it is human, is a costly and time-consuming process. The new cell line's subsequent cost will be determined by its ability to authenticate the source that is linked with it. The cell line acquisition entails a variety of risks and it may be possible that cell lines are not exactly what they claim to be and information of a cell line with a specific feature that has been published. In the beginning, it has been characterized and contaminated. The less reliance on results, the better properties are being shown. If the laboratory that has received the cells wants to rely on data collected with a cell line in the past, it can be allowed to do so. Before permitting a cell line into broad use, the laboratory should always conduct its testing processes. Employing cell lines that are either inaccurate or anonymously infected might take a significant amount of time, money, and effort. A DNA STR profile for cell lines needs to be certified, and proof of authenticity using an approach that is suited for cell lines (Isasi et al., 2012).

20.2.2   Authentication of Cell Culture

Cell line authentication should be conducted on receiving earlier freezing of a master cell bank (MCB) or seed stock using an authorized DNA-based approach for establishing the origin of a cell line and checking for any misidentification (Muth et al., 2018; WHO, 2021). In general, the DNA of the cell line would be compared to the tissue or donor of origin. The method is currently only attainable in a small number of cell lines. However, it is preferable to create an STR profile before using a cell line, so that it can be separated from other cell lines within the same lab or unique using an international database (NCBI, 2013) or also by communicating a reputable cell bank. It is then followed to further transfers of the cell lines for further experiments. When the cell lines originate from in-bred mouse strains, STR profiles may not be able to differentiate one cell line from another cell line but can be done via a specific genetic marker for which a particular genetic modification (mutation) has been introduced (Corral-Vázquez et al., 2017; Dennert & Kumar, 2021).

20.2.3   Characterization of Cell Culture

The user should also ensure that the cell line must be suitable for their requirements. Even if a cell line is genuine, it might have lost vital traits. Long-term passaging and karyotyping are basic tests that may disclose a lot of authenticity and characterization of cell lines. Any alterations in a cell line demonstrate that an ES cell line exists in a common place. If any cells or stem cells are to be used for research, they must have a normal karyotype. Experiments using chimeras and germlines need to be conducted for the proper conduct of the experiment. CNV assays and RNA profiling are two examples of molecular assays that may be used to search for CNV. Nevertheless, on validation of the information, it becomes time-efficient and progressive work. Before starting the experiment, proper information about the cells is essential. It is also a good idea to take a picture of the cell line in the culture at different densities and do some basic characterization (For example, determining the time the cell population takes to double (Drexler & Matsuo, 1999; Naveca et al., 2021; Zhang et al., 2021).

20.2.4   Isolation of Cell Lines

Cell isolation is the process of extracting individual living cells from a tissue or a cell suspension. While certain cell types divide spontaneously, those in solid tissue require particular mechanisms to be separated into individual cells. This can be performed by using particular enzymes to degrade the proteins that connect these cells inside the extracellular matrix. After the matrix proteins are digested, the cells remain loosely linked or can be physically or chemically separated. Physical study of single isolated cells can be done using patch-clamp electrophysiology, calcium fluorescence imaging and immunocytochemistry, once the cell lines have been isolated (Hu et al., 2016).

20.3   Ethical Concerns Associated with Chimera Effect in Tissue Engineering

"Chimera" is a term used in molecular genetics to denote the combination of two DNAs from different individuals or chromosomes from the same individual, also known to be an interspecies hybrid, for example, mules (a cross between a female horse and a male horse) are examples of hybrids (Karpowicz et al., 2004). The term "chimaera" also refers to grafting tissues or cells from a post-implantation embryo injection of a different person or species, as in the case of injecting haematopoietic stem cells intraperitoneally into a sheep foetus to produce a chimeric sheep with human lymphoid and myeloid cell lineages (Srour et al., 1992).

Hans Spemann and Hilde Mangold performed initial studies that resulted in one of the earliest embryological chimeras ever generated by scientists. Embryos of an amphibian (Triturus) were transplanted into the embryos of another with varying levels of pigmentation. For cell lineage tracing investigations, Le Douarin et al. (2008) employed chimeric embryos from chicken and quailing the early stages of vertebrate development. Along with these, natural chimeras as well as man-made chimeras have also been described. For example, mothers may keep some of their children's foetal cells after birth, a condition known as foetal microchimerism. Recent technical advancements in the field of chimera research sometimes do not allow the synthesis of human organs in animals resulting in the production of human–animal chimeras. The medical demands, notably for organ transplants, are apparent, and due to the significant organ scarcity, transplantation is a viable option. Nonetheless, such a viewpoint presents significant legal and ethical issues and ethical dilemmas

20.4   Ethics Related to Development and Preservation (Cryopreservation) of Cell Cultures

Organelles, cells, tissues and other biological components are cryopreserved by freezing them at extremely low temperatures to preserve them. Theoretically and practically, living cells' reactions to ice formation are fascinating. Osmotic shock occurs with the formation of ice crystals, and membrane damage occurs during repeated freezing and thawing. Simple chilling or freezing will not keep stem cells or other viable tissues for an extended period, despite their promise for use in fundamental research and many medical applications (Karlsson & Toner, 1996). Biological and chemical activities in live cells are significantly reduced at low temperatures, a trait that may allow cells and tissues to be kept for an extended period. The majority of living cells, on the other hand, are killed by freezing because both internal and extracellular ice crystals form, generating chemical changes in cells that create mechanical limits and injury. The following are the several types of cryopreservation procedures: (1) freezing at a sluggish pace; (2) freezing at a fast pace (3) vitrification, which occurs when the cell or tissue's aqueous environment solidifies into a non-crystalline glassy phase; (4) subzero non-freezing storage and (5) preservation in a dry state.

The transition of cell components from water to ice is difficult for cells to incapacitate at low temperatures. Slow cooling causes osmotic changes with the presence of concentrated intra- and extracellular fluids, as well as mechanical interactions between ice and cells, whereas rapid cooling causes intracellular ice creation. Cell line samples are gently frozen in preservatives (often growing medium with 10% DMSO) for 1 minute. Simpler devices may suffice, but automated controlled-rate chilling equipment provides the most consistent cryopreservation, if the freezing program is tailored to the needs of the cell line. Certain cell types, such as hESC, may require "vitrification", or ultra-rapid freezing (Hunt, 2011), to trigger freezing of water molecules rather than allowing it to seep out of the cell, as occur in gradual freezing. To ensure vitality, one vial should be resuscitated immediately after a batch of cells is frozen down. After being retrieved from the bank, the cell suspension should be frozen promptly (in a water bath at 37 °C) and the cell suspension diluted progressively using a pre-warmed medium (Pegg, 2007).

20.5   Ethical Conduct in Cell Culture Handling

The necessity for constructing the facility to guarantee that excellent quality research must be generated safely and efficiently is becoming the most under-appreciated component of cell and tissue culture. Animal tissue engineering is carried out in laboratories that have been specially designed for the purpose with minimal settings. However, due to a few simple guidelines, the work should not be risked (Moysidou et al., 2021). There are various aspects to a successful tissue culture facility design. According to the advisory committee on hazardous pathogens guidelines, the laboratory should be assigned to at least two categories for the majority of cell lines ("Advisory Committee on Dangerous Pathogens (ACDP) 2004. The approved list of biological agents. Health and Safety Executive. www.hse.gov.uk "). The exact category required, however, is determined by the cell line and the nature of the intended task. The standards include recommendations for the laboratory environment, such as lighting, heating, work surfaces and flooring and the availability of handwashing facilities. Furthermore, laboratories should be conducted at air pressures that are opposite to corridors to reduce the danger of microbiological contamination.

20.5.1 Maintaining the Instability

Depending on the cell type, the artificial microenvironment wherein the cells are cultured varies widely, but it always involves appropriate glassware comprising a substrate or medium that distributes the required nutrients (essential amino acids, carbohydrates and vitamins). It is regulated by adding minerals, growth factors, gases (O_2, CO_2), pH, osmotic pressure and specific temperature in the physicochemical environment (McKeehan et al., 1990). The majority of cells require anchoring and must be grown on a solid or semi-solid substrate (anchorage-dependent culture) (Merten, 2015), whereas some can be grown floating in a medium called suspension culture. If there are extra cells after subculturing, treat them with DMSO or glycerol as a protective agent and store them at temperatures below −130 °C until needed (cryopreservation) (Segeritz & Vallier, 2017).

20.5.2 Contamination and Non-specific Identification

Microbial contamination is a significant problem in cell culture; however, several approaches can be used to reduce or eradicate contamination. The most typical sources of contamination are the operator and the laboratory environment, as well as other cells and reagents used in the lab. Contamination and aseptic procedures are the greatest defences against pathogens that pose harm to laboratory workers. If a larger issue is discovered, all contaminated cultures and related unused mediums should be destroyed, equipment should be inspected and scoured, cell culture techniques should be re-evaluated and isolation from other facilities should be enforced until the problem is resolved (Stacey, 2011).

20.5.3   Transfer of Cell Line Between Laboratories

The typical approach for shipping cells to research centres throughout the world while maintaining cell viability is cryopreserved cells frozen in dry ice or liquid nitrogen. Another option is to transport the live cells in flasks containing cell culture media. Both approaches have drawbacks, such as the requirement for a certain shipping container or the requirement for the cells to survive the transit procedure for a set period. Yang et al. (2009) devised an agarose gel-based approach for transporting live adherent cells straight from cell culture plates or dishes to room temperature. This simple technology makes it easier to send live cells over long distances while maintaining cell viability for several days (Yang et al., 2009).

20.5.4   Use of Equipment and Media

An aseptic work environment dedicated solely to cell culture activities is the most important prerequisite for a cell culture laboratory. A cell culture section within a large laboratory can suffice, while a specialized tissue culture chamber is optimal. Sterile cell culture, reagent, medium handling, incubation and storage should all be maintained inside the animal tissue culture lab (Verma et al., 2020). For producing aseptic conditions in biosafety cabinets, cell culture is the most simple and cost-effective way. To remove or reduce exposure to dangerous compounds in a cell culture laboratory, main barriers such as biosafety cabinets, enclosed containers and other engineering controls, as well as personal protection equipment, are used (Coté, 2001). PPE stands for personal protection equipment and is frequently used in conjunction with main barriers. A biosafety cabinet is a biohazard storage facility. A cell culture hood is the most crucial piece of equipment for containing bacteria and preventing contamination of your own cell culture while using various microbiological methods that produce pathogenic splashes or aerosols (*National Research Council (US) Committee on Hazardous Biological Substances in the Laboratory. Biosafety In The Laboratory: Prudent Practices for the Handling and Disposal of Infectious Materials. Appendix A, Biosafety in Microbiological and Biomedical Laboratories.*, 1989).

20.6   Dilemmas in Tissue Engineering and Tissue Banking

One of the promising fields of medical technology is Tissue Engineering (TE). In the future, it may be able to solve the problems brought on by tissue or organ damage in the therapeutic setting. It thereby helps to get healthier and longer lives. However, tissue engineering is not free of ethical challenges like other new technologies. Identifying these ethical issues early on is not only part of science's responsibility to society, but it is also in the discipline's best interests: it allows the field to thrive by avoiding it from investing time and money in paths that are unlikely to receive public support (Langer & Vacanti, 1993).

TE is still in its early stages, and despite great expectations, there are just a few therapeutic applications now available, notably for skin, cartilage and bone. Even while the sector is still mostly focused on research, clinical practice is becoming more prevalent, and clinical trials are becoming more widespread. As a result, it seems like now is a good moment to think about the moral implications of this technology. Tissue banking research is being conducted in collaboration between for-profit and non-profit groups, which has resulted in a host of ethical concerns. This trend can be attributed in part to the advancement of genetic technology, as well as the resulting surge in interest in human tissue. Private for-profit firms are establishing new tissue banks to compete with non-profit tissue banks for the recovery and processing of these issues (Bauer et al., 2004; "British Association of Tissue Banking. "General Standards for Tissue Banking." (September 29, 1999). Available at: http://www.batb.org.uk/general1.htm.,"). Furthermore, the pharmaceutical and biotech businesses frequently sponsor tissue banks in the non-profit sector, which help fund university research and rely on human biological samples collected and banked by these organizations.

Some of the current safeguards will become increasingly obsolete in protecting tissue donors (at least in the United States), while others point to a variety of cultural, organizational and professional issues that call into question the ethical appropriateness of for-profit tissue banks and academic-industry relationships. New tissue banks are emerging, many of which are for-profit firms vying with non-profit tissue banks for tissue recovery and pro-bono services discontinuation (Josefson, 2000). Furthermore, non-profit tissue banks, such as those in the pharmaceutical and biotech industries, frequently support this industry. Human biotechnology is used by businesses that donate to university research funding. These institutions have recovered and stored logical samples. Some argue that the current system of safeguards (at least in the United States) will be ineffective. Others argue that shielding tissue donors is becoming increasingly unnecessary to address a wide range of cultural, organizational and professional concerns.

20.7   Cell Culturing Associated with Intellectual Property and Legal Rights

Cell culture is the process of altering cells from an animal or plant and then cultivating them in a controlled environment. The cells can be extracted directly from the tissue and disaggregated using enzymatic or mechanical procedures before being grown, or they can be derived from an existing cell line or cell strain. All of the conditions must be maintained during cell culture. Biotechnology was a topic that initially received little attention and progressed slowly. Cell culture and manipulation, cell fusion, monoclonal antibody production, gene editing, as well as genetic engineering, and recombinant DNA (rDNA) techniques, have all ushered in a new era of biomedical innovation.

Meanwhile, researchers discovered that combining the aforementioned improvements with reproductive technologies might dramatically alter cells, embryos and animals. Concerns about intellectual property and its protection in animal biotechnology have grown as technology has advanced. Animal biotechnology includes veterinary vaccines, molecular diagnostics, transgenic or gene-edited animals for human disease detection and the use of genetically altered animals for therapies and organs.

Intellectual property (IP) is a legal term for a collection of intangible assets and regulations that protect profitable concepts, ideas and information through production and uses. IP protection laws cover patents, copyrights, trademarks, trade secrets and other related rights ("Cornish WR (1989) Intellectual property: patents, copyright, trade marks, and allied rights. 2nd ed. London: Sweet & Maxwell,"). The bulk of biotechnology techniques and processes are currently protected by law, with the vast majority of them having their origins in the United States. The majority of countries are now competing in the biotechnology business. They had no choice but to amend their national legislation to protect and promote biotechnology investment. Despite the lack of international agreement on how biotechnology should be considered, biotechnology inventors persist (Raj et al., 2015; Singh et al., 2019; Webber, 2003).

20.7.1   Laws Concerning Intellectual Property

Science is critical to the country's economic development. When science and technology are allowed to develop, the economy thrives. It is a technology that creates products and services, therefore connecting research to revenue generation directly or indirectly. The primary goal of IP protection is to allow the owner of intellectual property (patents, copyrights, trademarks and trade secrets) to prevent others from utilizing it in specific ways (Brown, 2003). The activities of practicing scientists are intricately linked to the realm of business (Poticha & Duncan, 2019). A range of intellectual property protection measures is used to protect some inventions. Before deciding on a plan, it is vital to consider both business and legal concerns (Voss et al., 2017). Many countries now have laws and codes of practice that govern the use of human and animal tissue samples in research, and these suggestions highlight the most significant legal and ethical issues that may arise.

20.8 Public and Scientific Community Opinions on Cell Culturing

Cell culture is a set of procedures used in laboratories to allow eukaryotic and pro-karyotic cells to grow in physiological settings. It was first used in the early twenti-eth century to investigate tissue growth and maturation, viral biology and vaccine development, gene function in illness and health and the production of biopharma-ceuticals using large-scale hybrid cell lines. Cultured cells can be used in as many experiments as there are cell types that can be produced in vitro. Cell culture is most typically used in clinical settings to create model systems for studying basic cell biology, replicating disease pathways and determining the toxicity of prospective therapeutic drugs. The ability to modify genes and molecular pathways is one of the benefits of employing cell culture for these applications. In addition, the homogene-ity of clonal cell populations or specific cell types, as well as well-defined culture systems, eradicates interrupting genetic or environmental factors, allowing for large data generation on reproducibility and consistency, which is impossible to ensure when trying to study whole organ systems (Segeritz & Vallier, 2017).

In the scientific community, cell culture is an important approach in cellular and molecular biology because it allows researchers to study biology, biochemistry, physiology (e.g., aging) and the metabolism of both healthy and diseased cells. Cell cultures have a great level of control over their physicochemical environment (pH, temperature, osmotic pressure, oxygen and carbon dioxide tension), as well as phys-iological parameters, which can be monitored continuously. Scientists also use stem cells to learn more about the disease and develop medications to treat it. Embryonic stem cells may be employed to regenerate more specialized tissues that have been destroyed due to disease or injury. Stem cells would most likely be replaced directly in tissues that are continually regenerated, such as blood and skin (Arango et al., 2013; "How can stem cells advance medicine?," 2007).

A pro-life ideology quickly evolved as a major driving factor behind stem cell ethical discussion and legislation, set against this apocalyptic image of science. It is safe to assume that despite a slew of other concerns regarding science's future and the ethical debate over stem cell research was pointing us in the right direction. For the past decade, cell research has been defined specifically by the controversy about the eradication of embryos. A significant minority in the United States, for example, has objected to the use of 5-day-old preimplantation human embryos. Embryos are destroyed during the harvesting process. Those who oppose embryonic stem cell research argue that it is unethical for religious or other personal reasons. All preim-plantation embryos have the same moral status as any other human being and all living beings, regardless of age (Cebo, 2019).

20.9   Conclusion

Ethical concerns in animal tissue culture play an important role and must be taken into consideration for the successful completion of studies. Animals have been utilized in the study because they are thought to mimic human biology. Animal research ethics evolved across centuries of philosophical traditions, and they are not hard rules of operation, but rather a means of expressing our moral responsibility to study animals. Our interpretation of the principles of teaching animal research ethics in organizations is presented in this chapter. We have outlined suggestions, practical recommendations and instructional tactics for instilling animal testing ethics in scientists, students and doctors clearly and concisely. These alternatives, on the other hand, can only diminish animal research dependence (via replacement and reduction) by complementing animal research.

References

Advisory Committee on Dangerous Pathogens (ACDP). (2004). The approved list of biological agents. *Health and Safety Executive*.

Arango, M. T., Quintero, R. P., Castiblanco, J., et al. (2013). *Cell culture and cell analysis*. Rosario University Press.

Bauer, K., Taub, S., & Parsi, K. (2004). Ethical issues in tissue banking for research: a brief review of existing organizational policies. *Theoretical Medicine and Bioethics, 25*(2), 113–142.

British Association of Tissue Banking. General Standards for Tissue Banking. (September 29, 1999)

Brown, W. M. (2003). Intellectual property law: a primer for scientists. *Molecular Biotechnology, 23*(3), 213–224.

Cebo, D. (2019). Public opinion about stem cell research and human cloning – The bioethics of stem cell research and therapy. *Global Journal of Community Medicine*

Cornish, W. R. (1989). *Intellectual property: Patents, copyright, trade marks, and allied rights* (2nd ed.). Sweet & Maxwell.

Corral-Vázquez, C., Aguilar-Quesada, R., Catalina, P., Lucena-Aguilar, G., Ligero, G., Miranda, B., & Carrillo-Ávila, J. A. (2017). Cell lines authentication and mycoplasma detection as minimun quality control of cell lines in biobanking. *Cell and Tissue Banking, 18*, 271–280.

Coté, R. J. (2001). Aseptic technique for cell culture. *Current Protocols in Cell Biology*.

Dennert, K., & Kumar, R. (2021). Traceability methods for cell line authentication and mycoplasma detection. *SLAS Technology, 26*, 630–636.

Drexler, H. G., & Matsuo, Y. (1999). Guidelines for the characterization and publication of human malignant hematopoietic cell lines. *Leukemia, 13*, 835–842.

Geraghty, R. J., Capes-Davis, A., Davis, J. M., Downward, J., Freshney, R. I., Knezevic, I., et al. (2014). Guidelines for the use of cell lines in biomedical research. *British Journal of Cancer, 111*, 1021–1046.

How can stem cells advance medicine? (2007). *Nature Reports Stem Cells*.

Hu, P., Zhang, W., Xin, H., & Deng, G. (2016). Single cell isolation and analysis. *Frontiers in Cell and Developmental Biology, 4*, 116.

Isasi, R., Knoppers, B. M., Andrews, P. W., Bredenoord, A., Colman, A., Hin, L. E., & Zeng, F. (2012). Disclosure and management of research findings in stem cell research and banking: Policy statement. *Regenerative Medicine, 7*, 439.

Josefson, D. (2000). Human tissue for sale: what are the costs? *The Western Journal of Medicine, 173*, 302–303.

Karlsson, J. O., & Toner, M. (1996). Long-term storage of tissues by cryopreservation: critical issues. *Biomaterials, 17*, 243–256.

Karpowicz, P., Cohen, C. B., & van der Kooy, D. (2004). It is ethical to transplant human stem cells into nonhuman embryos. *Nature Medicine, 10*, 331–335.

Langer, R., & Vacanti, J. P. (1993). Tissue engineering. *Science, 260*(5110), 920–926.

Le Douarin, N., Dieterlen-Lièvre, F., Creuzet, S., & Teillet, M. A. (2008). Quail-chick transplantations. *Methods in Cell Biology*, 19–58.

McKeehan, W. L., Barnes, D., Reid, L., Stanbridge, E., Murakami, H., & Sato, G. H. (1990). Frontiers in mammalian cell culture. *In Vitro Cellular & Developmental Biology*, 9–23.

Merten, O.- W. (2015). Advances in cell culture: Anchorage dependence. *Philosophical Transactions of the Royal Society of London* 20140040-20140040.

Moysidou, C.-M., Barberio, C., & Owens, R. M. (2021). Advances in engineering human tissue models. *Frontiers in Bioengineering and Biotechnology, 8*, 620962.

Muth, D., Corman, V. M., Roth, H., Binger, T., Dijkman, R., Gottula, L. T., et al. (2018). Attenuation of replication by a 29 nucleotide deletion in SARS-coronavirus acquired during the early stages of human-to-human transmission. *Scientific Reports, 8*, 15177.

National Research Council (US) Committee on Hazardous Biological Substances in the Laboratory. Biosafety In The Laboratory: Prudent Practices for the Handling and Disposal of Infectious Materials. Appendix A, Biosafety in Microbiological and Biomedical Laboratories. (1989). Washington (DC): Washington (DC): National Academies Press (US)

Naveca, F., Nascimentio, V., Souza, V., et al. (2021). Phylogenetic relationship of SARS-CoV-2 sequences from Amazonas with emerging Brazilian variants harboring mutations E484K and N501Y in the Spike protein.

Pegg, D. E. (2007). Principles of cryopreservation. *Methods in Molecular Biology, 368*, 39–57.

Poticha, D., & Duncan, M. W. (2019). Intellectual property-the foundation of innovation: a scientist's guide to intellectual property. *Journal of Mass Spectrometry, 54*, 288–300.

Raj, G. M., Priyadarshini, R., & Mathaiyan, J. (2015). Drug patents and intellectual property rights. *European Journal of Clinical Pharmacology, 71*, 403–409.

Segeritz, C.–. P., & Vallier, L. (2017). Cell culture: Growing cells as model systems in vitro. *Basic Science Methods for Clinical Researchers*, 151–172.

Singh, B., Mal, G., Gautam, S. K., & Mukesh, M. (2019). Intellectual property rights in animal biotechnology. *Advances in Animal Biotechnology*, 527–530.

Srour, E., Zanjani, E., Brandt, J., Leemhuis, T., Briddell, R., Heerema, N., & Hoffman, R. (1992). Sustained human hematopoiesis in sheep transplanted in utero during early gestation with fractionated adult human bone marrow cells. *Blood, 79*, 1404–1412.

Stacey, G. N. (2011). Cell culture contamination. *Methods in Molecular Biology, 731*, 79–91.

Verma, A., Verma, M., & Singh, A. (2020). Animal tissue culture principles and applications. *Animal Biotechnology*, 269–293.

Voss, T., Paranjpe, A. S., Cook, T. G., & Garrison, N. D. W. (2017). A short introduction to intellectual property rights. *Techniques in Vascular and Interventional Radiology, 20*, 116–120.

Webber, P. M. (2003). A guide to drug discovery. Protecting your inventions: the patent system. *Nature Reviews. Drug Discovery, 2*, 823–830.

WHO. (2021). COVID-19 new variants: knowledge gaps and research. WHO R&D Blueprint.

Yang, L., Li, C., Chen, L., & Li, Z. (2009). An agarose-gel based method for transporting cell lines. *Current Chemical Genomics, 3*, 50–53.

Zhang, W., Davis, B. D., Chen, S. S., Sincuir Martinez, J. M., Plummer, J. T., & Vail, E. (2021). Emergence of a novel SARS-CoV-2 variant in Southern California. *JAMA, 325*, 1324–1326.

Chapter 21
Common Troubleshooting Methods in Cell Culture Techniques

Khushi R. Mittal and Shalini Mani

21.1 Introduction

The researchers sometimes ignore the small glitches and the malfunctions in the cell cultures. Problems like bacterial, fungal or mycoplasma contamination, misidentification and unexpected cell death could simply be overlooked. However, these small glitches could cause big harm to the research work and could ruin it. Whatever kind of cell culture it is, problems could always occur. Different cell lines could be extracted from different laboratories or purchased from cell banks. From whatever source the cell lines are extracted, a contamination check should always be done. Moreover, the contamination is not only dangerous to the research but also to the person who is performing it. The instruction, training and protection are the responsibility of the owner for the worker, as said in Work Regulations under the health and safety (HSE, 1975). The Control of Substances Hazardous to Health regulations is the main component (Health and Executive S, 2005). These regulations are to make sure that the practices and procedures that are taking place in the lab are safe and justifiable. To achieve a safe product, it is important to ensure that the environment in which it is being produced is safe. Wherever the cell culture is being produced, like in a pharmaceutical company, it should be a neat and clean place, promoting good manufacturing practices (Inspection and Standards Division, 2007).

Even if full attention is given to the problems that might occur in the cell culture lab and good lab practices are followed (Davis, 2011), there is no guarantee that they would not occur. There is no other way for this, but to find approaches that could resolve these problems. However, without having the background knowledge

K. R. Mittal · S. Mani (✉)
Centre for Emerging Diseases, Department of Biotechnology, Jaypee Institute of Information Technology, Noida, India

of the problem, it is difficult to find a suitable solution for it. The reagent bottles should be marked with care to avoid confusion and save time while detecting problems. In the same way, the cell lines should be documented properly to know which cell line was used when and by whom (Geraghty et al., 2014). This may include the photographs of the cell lines while growing and the source from which it was derived.

The general approaches to resolve the problems taking place in cell culture labs may include a proper definition of the features of the problem and informing those who are in the possible chances of getting affected. Once the problem is recognized, all the possible solutions should be ruled out. The glassware and equipment, which are used during the research, should be checked for any kind of contamination and cleaned if detected. Whenever the researcher switches to the new media, he/she should store some of the old media that if any problem arises the research work does not get affected. If the problem is recognized, it should be conveyed to the supplier for which it was obtained, so that wherever the supplier has given that product, those laboratories get a head-up. When contamination is detected, all the previously used equipment and tools should be sterilized, so that they do not affect the research of others. Also, a particular equipment should be assigned for a specific purpose only, to avoid mixing and contamination. The laboratory should be designed in a way that sensitive things are kept far from contaminations. Every problem has its different unique solution, which is why the identification of the problem is the most important step to ensure safety and achieve the desired product.

21.2   Troubleshooting

21.2.1   Why Do Sometimes Cells Not Stay Viable After Thawing the Stock?

The first reason for cells not staying viable after thawing could be that maybe the cells were stored incorrectly. Taking the new stock and storing the stock in the liquid nitrogen until thawing could correct this mistake. Then another reason could be that the cells were thawed in a wrong manner, which could lead to harming of the cell. To correct this, the frozen cells should be thawed quickly, diluted slowly before plating via a pre-warmed growth medium and proper procedure should be followed. The wrong thawing medium is one more reason for cells being non-viable and it is important to check and use recommended medium only. Cells could be diluted too much and not handled properly. High-density thaw cells should be plated and should not be centrifuged, vortex or banged aggressively. Maybe the glycerol present in the freezing medium used was stored in light. This would lead to the conversion of glycerol into acrolein, which is toxic to cells, and this leaves no other option but to discard the stock and obtain new.

21.2.2   Why Do Cells Seem to Grow Slowly Sometimes?

There are several reasons for the slow-growing of cells; one could be that maybe the growth medium is not correct. Pre-warmed recommended growth medium should be used to avoid this problem. Initial inoculum increasing and making the cells adapt to new medium sequentially could also rectify the problem. Another reason could be that the cells may be passaged several times or were allowed to grow past confluency. Healthy cells with lower passage numbers are recommended and mammalian cells should be passaged during the log-phase before they reach confluence. The serum used could also be of reduced quality or the culture could be contaminated. Serum should be used from a different lot, and new stock, media and reagents should be used.

21.2.3   What Is the Troubleshooting in Cell Counting?

Cell counting is a necessary step; it is important to do quality control by counting cells before going forward. For performing various cell biology experiments, protein production or to know cell morphology, it is vital to know whether the cell is viable or not. However, as simple as this seems, it is not. The traditional counting method is error-prone, subjective and very time-consuming. Sometimes debris comes in the sample and could be counted as a cell and give false positives. Or a cell could be counted as debris and give false positives. Automated cell counters are designed to detect a specific structure that eliminates the chances of false positives and negatives. During sample preparation, sometimes it is not suspended properly and the cells settle in the bottom, giving wrong results. Proper mixing of the sample helps to avoid this problem. Cell density is also an issue that could generate errors in the readings. If the cell density is low, it is not a good representative of the stock solution, and if it is high, then it would form cell aggregates again giving wrong results.

21.2.4   What Causes Cell Clumping?

Cell clumping reduces the nutrients and hinders cell growth. There are several causes of cell clumping; one could be the overgrowth of cells. Cells should be sub-cultured before they reach 100% confluency, so that stickiness could be avoided. Mycoplasma could another reason, which could be avoided by using a kit for detecting the invisible, antibiotic-resistive bacterium. A major reason for cell clumping is the cell contents like DNA that come out in the solution after the cell lyses.

21.2.5 What Leads to a Cross-Linked or Misidentified Cell Line?

Although misidentification and cross-linking seem to be small mistakes, they could lead to bigger problems in your research works. Whenever a new, fresh and rapidly growing cell line enters a lab, it is at a high risk of contamination, as stressed in the 1960s and 1970s (Gartler, 1967; Nelson-Rees et al., 1976; Nelson-Rees et al., 1977), however repeatedly ignored. Misidentification could be a result of cross-contamination and could be caused by improper manipulations or errors in lab work, which could lead to failure of good lab practices. To avoid this, cells should be borrowed only from a reputable laboratory, otherwise not. Any kind of untested cell lines should be removed or replaced from the working area to avoid confusion or misidentification. A list of identified misidentified cell lines is also available (Committee ICLA and Others, 2012) for detecting what cell line was misread. But a lab could always add-in that lists if a new cell line is misidentified.

21.2.6 Why Do Sometimes Cells Do Not Revive After Being Cryopreserved?

This happens a lot of times that the cryopreserved cells do not revive. There are a lot of reasons to explain this, one of them is how the cells were frozen, what was the composition of the mixture, or the rate of cooling and adding. The second one could be at what temperature they were stored. It should be made sure that if the cells are stored for the long term, they should be stored at a temperature that they could survive. The way of thawing also affects the revival of cells, and that is why it is advised not to thaw violently. If the cells were not stored properly, it could also affect the survival of cells. Either the N_2 container or problems with the cryoprotectant medium could damage the cells. If the cells are stored too quickly, directly into the liquid nitrogen, it could harm the cells. Preferably the cells should be frozen in the freeze box first, with isopropanol at $-80\ °C$ and then moved to liquid nitrogen.

21.2.7 What Are the Sources of Contamination and How Should They Be Rectified?

A healthy cell culture invites contamination if it is not kept in a safe environment. There are various contaminating agents like mycoplasma, viruses, bacteria and yeasts, which could ruin your research or lab work. Mycoplasma contamination is not visible morphologically but it could also alter the function, genome and metabolism of the cell (Drexler & Uphoff, 2002). Contamination could reach to culture from any source; it could be from lab reagents, equipment, researcher's hair, nail, clothes and even breath. The first way to avoid contamination is to outline all the sources from where the contamination could reach the culture and to adopt aseptic cell culture methods (Coecke et al., 2005; Ian Freshney, 2015). Lab facilities like laminar cabinets, incubators, water baths and loose flask caps, improper handling of the culture or incorrectly sealed culture could also lead to contamination. According to FDA and various cell repositories, up to 30% of the cell cultures are found to be contaminated by mycoplasma, also known as crabgrass of cell cultures that even could not be seen visually. That is why regular tests should be done for mycoplasma using valid kits or reagents, in the stocks or cultures. Sterile conditions should be maintained and aseptic methods should be followed to avoid unnecessary contaminations. An area should be specifically dedicated to research work so that over-trafficking is not the reason for contamination.

21.2.8 What Causes Unexpected Cell Detachment?

The glassware is designed in such a way that it helps cells to stay adherent to the surface. If the cells are not able to stay adherent, then the blame goes to contamination (mycoplasma) that cannot be seen visually, improper incubator conditions, over trypsinisation of the cells, and medium without attachment factors. After taking out the cells from cryopreservation, they should be handled with care. DMSO should be separated properly from the media after thawing and centrifuging. Centrifuging should not be harsh so that there is no harm to the cells in the process. Cells should be subcultured using specific media only because cells are habitual of the environment that is provided to them initially and changing it could make the cells lose their ability to attach. The cells could be re-frozen and passaged lately, as primary cells can grow till infinite passages, and freezing these cells in good health could be beneficial to the researcher if something goes wrong in further research. Optimum concentration should be maintained of the reagents for primary cells subculturing, like a higher amount of trypsin/EDTA could damage the primary cells. For mycoplasma contamination, the culture should be segregated and tested for infection, the incubator and hood should be cleaned and if the culture is infected it should be discarded.

21.2.9 What Is the Troubleshooting in Cytotoxicity Assay?

The problems associated with cytotoxicity assay include a decrease in fluorescent signals because of pyruvate-supplemented media like Iscove's, Ham's F12 and some formulations of DMEM. This could happen because of inhibition of the product of the lactate to pyruvate conversion by catalysis of LDH enzymatic reaction. To supplement the culture medium, animal sera containing endogenous LDH activity is used, which would help to provide the background signal. Different kinds of sera have different extents of LDH activity. If a serum-free medium, a low serum medium or even a whole different type is used, it could help reduce the background signals.

21.2.10   What Triggers a Rapid pH Shift in the Medium?

The pH of the medium is really important in cell culture technology. If there is something wrong with the pH of the medium it could cause devastating effects. Improper tension of carbon dioxide could cause changes in pH. The CO_2 percentage in the incubator should be increased or decreased depending upon the sodium bicarbonate amount in the medium. Around 5–10% CO_2 should be used for 2- to 3.7-g/L concentration of sodium bicarbonate. The caps on the tissue culture flasks could be tight and could change pH. The caps could be loosened for an around one-quarter turn. Adding the wrong salts could also affect the pH of the medium. In a CO_2 environment, Earle's salt-based medium and atmospheric conditions Hank's salts-based medium could be used. Bicarbonate buffering insufficiency could also be a reason and could be resolved by adding HEPES buffer to the 10–25 mM of final concentration. Contamination by bacteria, yeast and fungi could cause pH change. Either the culture could be discarded or the culture could be decontaminated.

21.2.11   What Leads to Precipitation in the Medium and Either Causes a Change in pH or No Change in pH?

The leftovers of the residual phosphate, from washing detergent, could cause the precipitation of medium components. The glassware should be rinsed with deionized distilled water numerous times and then should be sterilized. Another reason could be frozen medium. To resolve this, the medium should be warmed to 37 °C and stirred to make sure that dissolving happens; even after this if precipitation happens then the medium should be discarded. Contamination by fungi or bacteria could also cause precipitation. Either the medium could be discarded or decontamination of the culture could be tried.

21.3   Conclusion

It is always said that prevention is better than cure and the same applies when it comes to lab works of cell culture technology. Cell culture is a field where we tackle cell lines and a minute problem could ruin the research work of a scientist. To avoid this, the cell culture labs follow several safety measures to ensure the product's safety as well as the person's safety that is experimenting. There are several processes and procedures other than this. Wherever that human cell lines are involved, the ethical issue is always raised. It is important to get ethical approval before conducting the research. Then come the issues with continuous cell lines and genetically modified cells. These two have a lot of risk to nature if leaked from the labs. A tumor produced in a lab was accidentally transferred to a healthy cell line via needle

(Gugel & Sanders, 1986) and during transplantations; cancers have also been transferred (Stephens et al., 2000). Though a person's tumor is unlikely to grow on another person, people with weak immune systems stay at high risk. Genetically modified genes could also trigger an infective agent that could damage the host cells. A viral vector could easily infect a human if leaked and could cause catastrophic effects. It is important to make sure that these kinds of genes are stored, worked with and transported with care. Several rules and regulations have been formed and passed to ensure lab safety and decrease the number of accidents in the lab as low as possible. In the future, we expect that more advanced technology could also be of great help for safeguarding our lab works.

References

Coecke, S., Balls, M., Bowe, G., et al. (2005). Guidance on good cell culture practice. A report of the second ECVAM task force on good cell culture practice. *Alternatives to Laboratory Animals, 33*, 261–287.

Committee ICLA And others. (2012). Database of cross-contaminated or misidentified cell lines.

Davis, J. M. (2011). *Animal cell culture: Essential methods*. Wiley.

Drexler, H. G., & Uphoff, C. C. (2002). Mycoplasma contamination of cell cultures: Incidence, sources, effects, detection, elimination, prevention. *Cytotechnology, 39*, 75–90.

Freshney, R. I. (2015). *Culture of animal cells: A manual of basic technique and specialized applications*. Wiley.

Gartler, S. M. (1967). Genetic markers as tracers in cell culture. *National Cancer Institute Monograph, 26*, 167–195.

Geraghty, R. J., Capes-Davis, A., Davis, J. M., et al. (2014). Guidelines for the use of cell lines in biomedical research. *British Journal of Cancer, 111*, 1021–1046.

Great Britain. Health and Safety Commission and Books HSE. (1975). Health & Safety at work Etc ACT 1974, Advice to Employers, HSE Books.

Great Britain. Medicines and Healthcare products Regulatory Agency and Great Britain. Medicines and Healthcare products Regulatory Agency. Inspection and Standards Division. (2007). *Rules and guidance for pharmaceutical manufacturers and distributors 2007*. Pharmaceutical Press.

Gugel, E. A., & Sanders, M. E. (1986). Needle-stick transmission of human colonic adenocarcinoma. *The New England Journal of Medicine, 315*, 1487.

Health and Executive S. (2005). Control of Substances Hazardous to Health [COSHH]: The Control of Substances Hazardous to Health Regulations 2002 (as Amended): Approved Code of Practice and Guidance, HSE.

Nelson-Rees, W. A., & Flandermeyer, R. R. (1976). HeLa cultures defined. *Science, 191*, 96–98.

Nelson-Rees, W. A., & Flandermeyer, R. R. (1977). Inter- and intraspecies contamination of human breast tumor cell lines HBC and BrCa5 and other cell cultures. *Science, 195*, 1343–1344.

Stephens, J. K., Everson, G. T., Elliott, C. L., et al. (2000). Fatal transfer of malignant melanoma from MULTIORGAN donor to four allograft RECIPIENTS12. *Transplantation, 70*, 232.

Index

A

Adherent cell lines, 115, 116, 169, 272
Adherent cells, 3, 4, 7, 72, 91, 99, 107, 110,
       115, 116, 126, 164, 170, 311
Analysis of culture contaminants, 236–239
Analysis of media contaminants, 44, 67,
       85, 235–248
Animal cell culture, vi, 1–8, 53–74, 120, 133,
       221, 262, 285–301, 305–315
Antibiotics, 18, 24, 45, 69, 70, 79–82, 84–86,
       93, 104, 110, 116, 124, 128–129,
       132, 178, 222, 236, 238–240, 244,
       245, 247, 319
Aseptic conditions, 65, 215, 221–222,
       246–247, 311
Aseptic techniques, 65–70, 73–74, 129, 236,
       244, 245

B

Balanced salt solutions (BSS), 81–83, 178
Bead bath, 167
Biological contamination, 65, 70–73, 235–248
Biomarker identification, vi, 285–287
Biosafety cabinets, 58, 65, 311
Biosafety measures, 53–64
Buffering system, 78, 83, 85

C

Cell characterization, 54, 138, 257, 306
Cell counting, vi, 12, 19, 29–30, 54, 103,
       131–142, 164, 169, 187, 256, 271,
       272, 319

Cell counting error, 141–142
Cell culture, v, vi, 1–8, 11–74, 77–87, 89–97,
       99–110, 115, 116, 119–125,
       127–129, 131, 133, 136, 140–142,
       155, 164, 166, 168, 173–180, 183,
       184, 186–187, 197–210, 213–233,
       235–247, 251–253, 257–259,
       261–263, 277, 278, 280, 285–301,
       305–315, 317–323
Cell detachment, 104, 120–123, 129, 321
Cell differentiation, 100–101, 105, 197, 220,
       228, 299
Cell line, 1, 14, 55, 69, 78, 90, 99, 115, 132,
       157, 164, 215, 236, 251, 267, 286,
       305, 317
Cell maintenance, 99–110, 129, 132, 148
Cell morphology, 101, 108–109, 255–256, 319
Cell population heterogeneity, 256
Cell strain, 2, 5, 95, 251, 259, 305, 312
Cell thawing, 164–167
Cellular contaminants, 81, 94, 237, 238
Cellular crosstalk, 189
Cellular density, 123
Cellular development and differentiation,
       vi, 299–300
Cellular growth kinetics
Cellular isolation, 251
Cellular synchronization, 124–126, 129
Centrifuge, 12, 14, 16, 19, 28–30, 50, 63, 118,
       169, 204, 224, 227, 272, 318
CFU counting, 134, 135
Chemical contaminants, 42, 70, 235, 238–239,
       241, 242
Chemical dissociation, 179–180

© The Editor(s) (if applicable) and The Author(s), under exclusive license to
Springer Nature Switzerland AG 2023
S. Mani et al., *Animal Cell Culture: Principles and Practice*, Techniques in Life
Science and Biomedicine for the Non-Expert,
https://doi.org/10.1007/978-3-031-19485-6

Chimera effect, 308
Co-culture, 8, 55, 183–192, 293, 299
Co-culture variables, 185–188
Complex media, 80, 82, 83, 109
Computational model, 188–189, 256
Contamination control, 128–129
Cooling rate, 46, 147–149, 152, 154,
        157–159, 165
Coulter counter, 134–136
Cross-contamination, v, 67, 73, 74, 102, 109,
        220, 235, 245, 247, 294, 320
Cross-linked cell lines, 320
Cryoinjury, 154
Cryopreservation, v, 6, 54, 147–160, 163, 164,
        169, 170, 220, 224, 227, 309,
        310, 321
Culture media, 2, 7, 15, 54, 67–70, 77, 78,
        84–87, 89–97, 126, 128, 169, 186,
        187, 225, 226, 236, 238, 241, 260,
        277, 278, 311
Culture medium, 69, 70, 78, 80–84, 90, 92–94,
        96–97, 100, 133, 164, 169,
        185–187, 204, 220, 222, 224–227,
        230, 238, 241, 242, 274, 275, 277,
        280, 281, 298, 299, 305, 321
Culture sterility, 68–70
Cytotoxicity assay, 294, 321–322

D
Dimethyl sulphoxide (DMSO), 147, 149–151,
        153, 154, 158, 163–165, 167, 168,
        170, 227, 273–275, 294, 309,
        310, 321
Dry thawing, 170
Dulbecco's Modified Eagle Medium
        (DMEM), 82–87, 94, 97, 101, 106,
        109, 321
Dye exclusion method, 29, 268–270

E
Embryo culture, 3, 179, 184, 214
Embryonic stem cells (ESCs), 80, 84, 184,
        213–219, 221, 223–226, 228, 229,
        296–299, 314
Ethic in biotechnology

F
Fetal stem cells
Flow cytometry, 102, 132, 136–139, 142,
        209, 256

G
Genetic manipulation, vi, 1, 285–291
Good lab practices, vi, 53–64, 317, 320

H
Hemocytometer, 131, 133, 134, 141, 142
Hybridoma technology, 300–301

I
Incubator, 12–15, 19, 31–33, 35, 54, 65, 74,
        77–79, 83, 85, 96, 167, 223,
        225–227, 230, 239, 241, 242, 246,
        247, 259, 321, 322
Inoculation, 7, 115–129
Intellectual property rights, 6, 312–313
In vitro cytotoxicity, 267–282
In-vitro tissue model system, 186, 189–192
Issues with cryopreservation

L
Laminar air flow, 54, 225, 227

M
Media, vi, 2, 12, 54, 66, 77, 93, 100, 115, 132,
        153, 164, 173, 186, 199, 220, 235,
        253, 272, 288, 309, 318
Mitochondrial expression
MTT/XTT assay, 267–282

O
Organ culture, 3

P
Passaging, vi, 5, 7, 91, 100, 103, 115–129,
        132, 173, 214, 215, 220, 226–227,
        229, 230, 307
Pathological studies, vi, 291–292
Pluripotent stem cells, 206, 214, 215, 218,
        219, 226, 231, 297–299
Polymers, 147, 149, 150, 152, 153,
        201–203
Primary explants, 83, 104, 174, 178–179

R
Resuscitation, 163–170
RPMI 1640, 82, 83, 85–86, 97

S

Scaffold based, 200–203, 205, 208
Scaffold free, 200, 203–205
Secondary cell culture, 5, 99
Serum free media (SFM), 3, 80, 81, 83, 84, 94, 95, 106, 110, 177, 257, 272
Spectrophotometer, 139–140, 142, 269, 273, 275–277
Stem cell culture, vi, 213–233
Stem cell research, 213, 215, 228–230, 285, 295–298, 314
Stem cells, vi, 8, 80, 99, 149, 163, 184, 197, 213, 262, 285, 291, 307
Sterilization, 12, 13, 16, 19, 21, 26, 35–52, 54, 67, 102, 122
Storage, 11–13, 16–17, 19, 25, 35, 45–47, 54, 55, 65, 66, 116, 148, 150, 155, 163, 165, 168–170, 223, 241, 242, 309, 311
Subculture, 4, 5, 7, 8, 69, 89, 92–93, 106–107, 116, 118, 123, 124, 128, 169, 173, 174, 251, 252, 258, 260, 296, 319, 321
Subculturing, 4, 5, 7, 54, 89, 92, 116–117, 119, 120, 123, 124, 128, 129, 173, 310, 321
Suspension cell lines, 115, 116, 169
Suspension cells, 4, 86, 99, 102, 103, 107, 115–129, 278

T

Three-dimensional (3D) cell culture, vi, 197–210
Tissue banking, 311–312
Tissue engineering, 155, 188, 198, 203, 206, 299, 308, 309, 311–312
Troubleshooting in cell culture, 317–323
Tryphan blue exclusion, 132, 133
Trypsin disaggregation, 176, 177
2D cell culture, 197–199, 209
Types of cell lines, 3–5, 251–263

V

Ventilation, 11–12, 17, 20, 32, 58
Vitrification, 148, 154, 159, 160, 309

W

Warming rate, 149, 158, 160, 166
Waste management, 61–64
Waste segregation, 60–61
Water bath, 12, 19, 34, 81, 164–168, 170, 224, 309, 321

Printed in the United States
by Baker & Taylor Publisher Services